Climate Change Impacts
on Freshwater Ecosystems

Climate Change Impacts on Freshwater Ecosystems

Edited by

**Martin Kernan, Richard W. Battarbee
and Brian Moss**

A John Wiley & Sons, Ltd., Publication

Library of Congress Cataloging-in-Publication Data
Climate change impacts on freshwater ecosystems / edited by Martin Kernan, Rick Battarbee and Brian Moss.
    p.   cm.
  Includes bibliographical references and index.
  ISBN 978-1-4051-7913-3 (hardback)
  1. Freshwater habitats.   2. Freshwater ecology.   3. Climatic changes–Environmental aspects.   I. Kernan, M. R.   II. Battarbee, R. W.   III. Moss, Brian, 1943–
  QH541.5.F7C65 2010
  577.6′22–dc22

                                        2010016420

A catalogue record for this book is available from the British Library.

Set in 10.5/12pt Classical Garamond by SPi Publisher Services, Pondicherry, India
Printed in Malaysia

1   2010

Contents

Preface   vii
Acknowledgements   viii
Contributors   ix

1   **Introduction**   **1**
     Brian Moss, Richard W. Battarbee and Martin Kernan

2   **Aquatic Ecosystem Variability and Climate
     Change – A Palaeoecological Perspective**   **15**
     Richard W. Battarbee

3   **Direct Impacts of Climate Change on Freshwater
     Ecosystems**   **38**
     *Ulrike Nickus, Kevin Bishop, Martin Erlandsson,
     Chris D. Evans, Martin Forsius, Hjalmar Laudon,
     David M. Livingstone, Don Monteith and Hansjörg Thies*

4   **Climate Change and the Hydrology
     and Morphology of Freshwater Ecosystems**   **65**
     *Piet F.M. Verdonschot, Daniel Hering, John Murphy,
     Sonja C. Jähnig, Neil L. Rose, Wolfram Graf, Karel Brabec
     and Leonard Sandin*

5   **Monitoring the Responses of Freshwater
     Ecosystems to Climate Change**   **84**
     *Daniel Hering, Alexandra Haidekker, Astrid Schmidt-Kloiber,
     Tom Barker, Laetitia Buisson, Wolfram Graf, Gäel Grenouillet,
     Armin Lorenz, Leonard Sandin and Sonja Stendera*

6　Interaction of Climate Change and Eutrophication　**119**
Erik Jeppesen, Brian Moss, Helen Bennion, Laurence Carvalho,
Luc DeMeester, Heidrun Feuchtmayr, Nikolai Friberg,
Mark O. Gessner, Mariet Hefting, Torben L. Lauridsen,
Lone Liboriussen, Hilmar J. Malmquist, Linda May, Mariana
Meerhoff, Jon S. Olafsson, Merel B. Soons and Jos T.A. Verhoeven

7　Interaction of Climate Change and Acid
　　Deposition　**152**
Richard F. Wright, Julian Aherne, Kevin Bishop, Peter J. Dillon,
Martin Erlandsson, Chris D. Evans, Martin Forsius,
David W. Hardekopf, Rachel C. Helliwell, Jakub Hruška,
Mike Hutchins, Øyvind Kaste, Jiří Kopáček, Pavel Krám,
Hjalmar Laudon, Filip Moldan, Michela Rogora,
Anne Merete S. Sjøeng and Heleen A. de Wit

8　Distribution of Persistent Organic Pollutants
　　and Mercury in Freshwater Ecosystems
　　Under Changing Climate Conditions　**180**
Joan O. Grimalt, Jordi Catalan, Pilar Fernandez, Benjami Piña
and John Munthe

9　Climate Change: Defining Reference Conditions
　　and Restoring Freshwater Ecosystems　**203**
Richard K. Johnson, Richard W. Battarbee, Helen Bennion,
Daniel Hering, Merel B. Soons and Jos T.A. Verhoeven

10　Modelling Catchment-Scale Responses
　　to Climate Change　**236**
Richard A. Skeffington, Andrew J. Wade, Paul G. Whitehead,
Dan Butterfield, Øyvind Kaste, Hans Estrup Andersen,
Katri Rankinen and Gaël Grenouillet

11　Tools for Better Decision Making: Bridges
　　from Science to Policy　**262**
Conor Linstead, Edward Maltby, Helle Ørsted Nielsen,
Thomas Horlitz, Phoebe Koundouri, Ekin Birol,
Kyriaki Remoundou, Ron Janssen and Philip J. Jones

12　What of the Future?　**285**
Brian Moss

Index　300

Preface

The evidence that greenhouse gas emissions, primarily from fossil fuel combustion, is and will increasingly be a principal cause of climate change has been compelling for some time. Although uncertainties remain, the threat is sufficiently real for research now to focus not only on the climate system itself but also on how changes in the climate system in future might affect the functioning of natural ecosystems.

In this book, we are concerned with how climate change might affect freshwater ecosystems. The ideas and examples presented in the book stem largely from the 'Euro-limpacs' project, a major EU-funded project on 'the impact of global change on European freshwater ecosystems'. Euro-limpacs brought together lake, river and wetland scientists from across Europe to assess not only the direct impacts of climate change on freshwaters but also its potential indirect impact through interactions with other stresses such as changes in hydromorphology, nutrient loading, acid deposition and toxic substance exposure.

A wide variety of approaches was used in the project ranging from the analysis of lake sediment and long-term instrumental records to identify past impacts of climate change, to the use of experiments, space-for-time substitution and modelling to assess what might happen in future under different climate scenarios.

The project also considered the implications of future climate change for the management of freshwater ecosystems in Europe, especially the extent to which current policies and practices designed to improve the ecological status of freshwater ecosystems need to be modified in light of projected future climate change.

This book brings together the key results from the project. Its structure follows the design of the Euro-limpacs project, first assessing the probable effects of climate change and then considering management issues.

Richard W. Battarbee

Acknowledgements

We are very grateful to Gene Likens and Curtis Richardson for encouraging us to write this book. We acknowledge European Union 6th Framework RTD programme which provided the funding for Euro-limpacs (EU Contract No. GOCE-CT-2003-505540). We would like to thank our EU project manager, Christos Fragakis, for his support throughout the 5 years of the project. We owe our considerable gratitude to the many participants involved in Euro-limpacs who provided the data and analyses underpinning much of this book. From UCL, we would like to thank Cath D'Alton for her efforts with the diagrams, and Catherine Rose and Katy Wilson for the invaluable help they provided the editorial team in putting the manuscripts together. We would also like to thank those colleagues who provided anonymous reviews for each of the chapters. We dedicate the volume to the very many scientists in Euro-limpacs who are not included as authors in the book but who contributed to the success of Euro-limpacs and whose work is drawn upon throughout the book.

Contributors

Julian Aherne
Environment & Resource Studies,
Trent University, Peterborough, ON,
Canada

Hans Estrup Andersen
NERI, Aarhus University, Silkeborg,
Denmark

Tom Barker
Institute for Sustainable Water,
Integrated Management, and
Ecosystem Research, University
of Liverpool, Liverpool, UK

Richard W. Battarbee
Environmental Change Research
Centre, Department of Geography,
University College London,
London, UK

Helen Bennion
Environmental Change Research
Centre, University College London,
London, UK

Ekin Birol
International Food Policy Research
Institute, Washington, DC, USA

Kevin Bishop
Department of Aquatic Sciences and
Assessment, Swedish University of
Agricultural Sciences, Uppsala,
Sweden

Karel Brabec
Faculty of Science, Masaryk
University, Brno, Czech Republic

Laetitia Buisson
Laboratoire Evolution et Diversité
Biologique, Université Paul Sabatier,
Toulouse, France

Dan Butterfield
Department of Geography,
University of Reading,
Reading, UK

Laurence Carvalho
Centre for Ecology and Hydrology,
Midlothian, UK

Jordi Catalan
CEAB-CSIC, Blanes, Spain

Luc DeMeester
Laboratory of Aquatic Ecology
and Evolutionary Biology, KULeuven,
Leuven, Belgium

Peter J. Dillon
Environment & Resource Studies,
Trent University, Peterborough, ON,
Canada

Martin Erlandsson
Department of Aquatic Sciences
and Assessment, Swedish University
of Agricultural Sciences, Uppsala,
Sweden

Chris D. Evans
Centre for Ecology and Hydrology,
Environment Centre Wales,
Bangor, UK

Pilar Fernandez
Department of Environmental
Chemistry, Institute of Environmental
Assessment and Water Research,
Barcelona, Spain

Heidrun Feuchtmayr
Centre for Ecology and Hydrology,
Lancaster Environment Centre,
Lancaster, UK

Martin Forsius
Ecosystem Change Unit, Natural
Environment Centre, Finnish
Environment Institute, Helsinki,
Finland

Nikolai Friberg
Department of Freshwater Ecology,
National Environmental Research
Institute, Aarhus University, Silkeborg,
Denmark

Mark O. Gessner
Department of Aquatic Ecology,
Eawag: Swiss Federal Institute
of Aquatic Science & Technology,
Dubendorf, Switzerland
and
Institute of Integrative Biology,
ETH Zurich, Zurich, Switzerland

Wolfram Graf
Department Water-Atmosphere-
Environment, Institute of
Hydrobiology and Aquatic Ecosystem
Management, BOKU–University
of Natural Resources and Applied Life
Sciences, Vienna, Austria

Gäel Grenouillet
Laboratoire Evolution et Diversité
Biologique, Université Paul Sabatier,
Toulouse, France

Joan O. Grimalt
Department of Environmental
Chemistry, Institute of Environmental
Assessment and Water Research,
Barcelona, Spain

Alexandra Haidekker
Department of Applied Zoology/
Hydrobiology, Institute of Biology,
University of Duisburg–Essen, Essen,
Germany

David W. Hardekopf
Institute of Environmental Studies,
Charles University in Prague, Prague,
Czech Republic

Mariet Hefting
Ecology and Biodiversity, Institute
of Environmental Biology, Utrecht
University, Utrecht,
The Netherlands

Rachel C. Helliwell
Macaulay Institute, Aberdeen, UK

Daniel Hering
Department of Applied Zoology/
Hydrobiology, Institute of Biology,
University of Duisburg–Essen, Essen,
Germany

Thomas Horlitz
Entera, Hannover, Germany

Jakub Hruška
Czech Geolological Survey, Prague,
Czech Republic

Mike Hutchins
Centre for Ecology and Hydrology,
Wallingford, Oxon, UK

Sonja C. Jähnig
Limnology and Conservation
Department, Natural History Museum

Senckenberg, University
of Duisburg-Essen, Gelnhausen,
Germany

Ron Janssen
Department of Spatial Analysis and
Decision Support, Institute for
Environmental Studies, VU University
Amsterdam, Amsterdam, The
Netherlands

Erik Jeppesen
Department of Freshwater Ecology,
National Environmental Research
Institute, Aarhus University, Silkeborg,
Denmark

Richard K. Johnson
Department of Aquatic Sciences
and Assessment, Swedish University
of Agricultural Sciences, Uppsala,
Sweden

Philip J. Jones
Centre for Agricultural Strategy,
School of Agriculture, Policy
and Development, The University
of Reading, Reading, UK

Øyvind Kaste
Norwegian Institute for Water
Research, Southern Branch,
Grimstad, Norway

Martin Kernan
Environmental Change Research
Centre, Department of Geography,
University College London,
London, UK

Jiří Kopáček
Biological Centre ASCR, Institute
of Hydrobiology, Ceske Budejovice,
Czech Republic

Phoebe Koundouri
Department of International
and European Economic Studies,
Athens University of Economics
and Business, Athens, Greece

Pavel Krám
Czech Geolological Survey, Prague,
Czech Republic

Hjalmar Laudon
Department of Forest Ecology
and Management, Swedish
University of Agricultural Sciences,
Umeå, Sweden

Torbden L. Lauridsen
Department of Freshwater Ecology,
National Environmental Research
Institute, Aarhus University, Silkeborg,
Denmark

Lone Liboriussen
Department of Freshwater Ecology,
National Environmental Research
Institute, Aarhus University, Silkeborg,
Denmark

Conor Linstead
Institute for Sustainable Water,
Integrated Management,
and Ecosystem Research, University
of Liverpool, Liverpool, UK

David M. Livingstone
Department of Water Resources
and Drinking Water, Eawag:
Swiss Federal Institute of Aquatic
Science and Technology, Dübendorf,
Switzerland

Armin Lorenz
Department of Applied Zoology/
Hydrobiology, Institute of Biology,
University of Duisburg–Essen, Essen,
Germany

Hilmar J. Malmquist
Natural History Museum of
Kópavogur, Kópavogur, Iceland

Edward Maltby
Institute for Sustainable Water,
Integrated Management,
and Ecosystem Research, University
of Liverpool, Liverpool, UK

Linda May
Centre for Ecology and Hydrology,
Midlothian, UK

Mariana Meerhoff
Department of Freshwater Ecology,
National Environmental Research
Institute, Aarhus University,
Silkeborg, Denmark
and
Grupo de Ecología y Rehabilitación de
Sistemas Acuáticos, Departamento de
Ecología, Facultad de Ciencias,
Universidad de la República,
Montevideo, Uruguay

Filip Moldan
Swedish Environmental Research
Institute IVL, Gothenburg,
Sweden

Don Monteith
Centre for Ecology & Hydrology,
Lancaster Environment Centre,
Lancaster, Lancashire, UK

Brian Moss
Institute for Sustainable Water,
Integrated Management and
Ecosystem Research, University of
Liverpool, Liverpool, UK

John Munthe
IVL Swedish Environmental
Research Institute, Gothenburg,
Sweden

John Murphy
FBA River Laboratory, The School of
Biological and Chemical Sciences,
Queen Mary, University of London,
Wareham, Dorset, UK

Ulrike Nickus
Institute of Meteorology and
Geophysics, University of Innsbruck,
Innsbruck, Austria

Jon S. Olafsson
Institute of Freshwater Fisheries,
Reykjavik, Iceland

Helle Ørsted Nielsen
Department of Policy Analysis,
National Environmental Research
Institute, University of Aarhus,
Roende, Denmark

Benjami Piña
Department of Environmental
Chemistry, Institute
of Environmental Assessment
and Water Research, Barcelona,
Spain

Katri Rankinen
Finnish Environment Institute,
Helsinki, Finland

Kyriaki Remoundou
Department of International
and European Economic Studies,
Athens University of Economics
and Business, Athens, Greece

Michela Rogora
Institute of Ecosystem Study, CNR,
Verbania, Italy

Neil L. Rose
Environmental Change Research
Centre, University College London,
London, UK

Leonard Sandin
Department of Aquatic Sciences
& Assessment, Swedish University
of Agricultural Sciences, Uppsala,
Sweden

Astrid Schmidt-Kloiber
Department Water-Atmosphere-
Environment, Institute
of Hydrobiology and Aquatic
Ecosystem Management,
BOKU–University of Natural
Resources and Applied Life Sciences,
Vienna, Austria

Anne Merete S. Sjøeng
Norwegian Institute for Water
Research, Oslo, Norway

Richard A. Skeffington
Department of Geography, University
of Reading, Reading, UK

Merel B. Soons
Ecology and Biodiversity Group,
Institute of Environmental Biology,
Utrecht University, Utrecht, The
Netherlands

Sonja Stendera
Department of Aquatic Sciences
& Assessment, Swedish University
of Agricultural Sciences, Uppsala,
Sweden

Hansjörg Thies
Institute of Ecology, University
of Innsbruck, Innsbruck,
Austria

Piet F.M. Verdonschot
Freshwater Ecology, Alterra, Centre
for Ecosystem Studies, Wageningen,
The Netherlands

Jos T.A. Verhoeven
Ecology and Biodiversity Group,
Institute of Environmental Biology,
Utrecht University, Utrecht, The
Netherlands

Andrew J. Wade
Department of Geography, School
of Human and Environmental
Sciences, The University of Reading,
Reading, UK

Paul G. Whitehead
Department of Geography, School
of Human and Environmental
Sciences, The University of Reading,
Reading, UK

Heleen A. de Wit
Norwegian Institute for Water
Research, Oslo, Norway

Richard F. Wright
Norwegian Institute for Water
Research, Oslo, Norway

1

Introduction

Brian Moss, Richard W. Battarbee and Martin Kernan

Changing climate and a changing planet

In June 2008, one of us chanced upon a shepherd repairing his five-ft high (he didn't deal in metres) dry limestone walls on the uplands near Asby Scar in Cumbria, north-west England. We exchanged pleasantries that inevitably, this was Britain after all, embraced the weather. It was a bright warm day. But 'Bleak in winter up here' I said. 'Not so much in the past fifteen years' he replied, 'Before that the snow lay in drifts hiding the walls, but not any more'. It was yet another anecdotal sliver of evidence to complement the mass of information assembled by the Intergovernmental Panel on Climate Change (IPCC 2007) on the reality of global warming.

That Fourth Report of the IPCC summarized changes to date (Fig. 1.1) that included an almost 1°C increase in the northern hemisphere mean air temperature, over the years since the industrial revolution accelerated the yet unabated burning of fossil fuels. It presented evidence that these processes were related and that we could have high confidence that the temperature rise was largely human-induced. Linked with it have been changes in the distribution of rainfall, with generally more falling in winter or wet seasons and less in the summer and dry seasons. There has been an increase in sea level of about 20 cm, largely due to thermal expansion of the huge mass of oceanic water, to which the melting of the mountain and polar glaciers is now making a contribution. And there has been an increase in the frequency of extreme weather events, such as cyclones, droughts and floods. In turn, there have been numerous records of changes in the phenology of species (Sparks & Carey 1995; Roy & Sparks 2000; Parmesan & Yohe 2003; Hays *et al.* 2005; Adrian *et al.* 2006) and a steady migration polewards of a variety of the more mobile species (Walther *et al.* 2002; Root *et al.* 2003).

Climate Change Impacts on Freshwater Ecosystems. First edition. Edited by M. Kernan, R. Battarbee and B. Moss. © 2010 Blackwell Publishing Ltd.

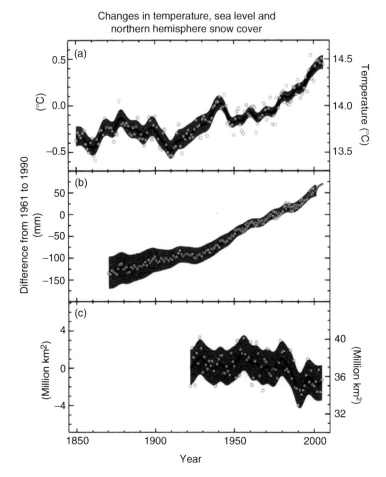

Figure 1.1   Summary of climate and sea-level change to date. (a) Global average temperature. (b) Global average sea level. (c) Northern hemisphere snow cover. (From *Climate Change 2007: The Physical Science Basis. Contribution of Working Group I to the Fourth Assessment Report of the Intergovernmental Panel on Climate Change* (eds S. Solomon, M. Manning, Z. Chen, *et al.*). Cambridge University Press, Cambridge and New York.)

Climate is a master variable, and all activity on this planet eventually depends upon it. It determines the overall structure of natural biomes, be they deserts, grasslands or deciduous or evergreen forests. It has driven the evolution of life histories, the dynamics of food webs and the development of homeostases. It fixes the circulation of the oceans, the availability of nutrients to the plankton community, the onset of rain and ripening for crops and the reflectance of radiation from the Poles. It manifests itself in the day-to-day weather, a preoccupation of everyone, not just the British. It is the greatest determinant of leisure travel, and, in its extremes, a source of extreme misery to match its delights of balmy summer days, exciting ski runs and the fresh spring rain. A major change in climate is a very considerable issue.

(a)   (b)

Figure 1.2   (a) G.E. Hutchinson and (b) James Lovelock.

Changing ideas on planetary function

Ecologists have long sought to explain the huge variation of natural systems: the tapestry of weather and soil-related detail on land and physical and chemical detail in water that fits into a grand pattern of climate zones. G.E. Hutchinson (1965) (Fig. 1.2) linked the ways that organisms evolve, as both grand and local patterns change, in his metaphor of the ecological (or environmental) theatre and the evolutionary play. His concept, in the 1960s, was very much one of the players adjusting to the nature of the theatre and then to each other. The generally accepted paradigm was that the physicochemical setting, the geology and climate, determined the biology and ecology of living organisms. Twenty years later, James Lovelock (1988) (Fig. 1.2) began an overturning of this by a spectroscopic examination of the chemistry of the atmospheres of Earth and its sister planets and a study of Earth's oceans. He calculated that the chemical state of Earth was very far from that expected by a simple chemical equilibrium of the available elements, and inferred that it was determined, and maintained, by the activities of living organisms rather than physicochemically imposed upon them for their response. Moreover, the state was regulated within the limits between which our particular biochemical system could persist. There is still controversy about the underlying mechanism of the regulation, but not about its existence. Such a change in paradigm is key to our understanding of the mutual interactions of climate and living organisms that this book is about. By altering our atmosphere, we challenge the entire biosphere system, and although we can predict some immediate physical effects, we have little idea about what the ultimate biological consequences might be.

The IPCC has made a range of predictions about how climate will change over the regions of the Earth, based on a range of assumptions about how human societies will react as the first of the changes are experienced. There is a problem, however, in these predictions. They all hold to the former model of living systems responding to imposed conditions. They are models of simple physicochemical control. They do not allow for the likelihood of positive ecological feedbacks. Temperature influences many biological processes, but not in a linear way. More usual is some sort of exponential relationship in which the process accelerates or decelerates to a point of death as temperature changes linearly. A key process in regulating the carbon dioxide content of the atmosphere is the storage of carbon as organic matter in soils and peat deposits or as calcite in the ocean sediments, derived from the scales of planktonic coccolithophorids or the matrices of corals (Lovelock 1988). If the temperature change induces more carbon dioxide or methane release, through increases of respiration using organic matter stored in soils and sediments, for example, or through inhibition of calcite formation in the walls of marine organisms, a positive feedback on further temperature increase may be induced and the greenhouse effect may be reinforced. Temperature changes predicted for the future may thus have been underestimated, and climate modellers are now attempting to rectify this.

The system that maintains the non-equilibrium, equable state of the planet is the biosphere. The biosphere has, for convenience, been divided up into atmosphere, hydrosphere and lithosphere: air, ocean and land. And the lithosphere is thought of in terms of biomes: tundra, coniferous forest, deciduous forest, tropical forest, scrub savannah, grassland and desert. In turn, these may be divided into constituent ecosystems, which Arthur Tansley (1935) defined as more or less self-contained systems of living organisms, and their biologically produced debris, in their physicochemical setting. In truth, this idea was an artefact of working in the greatly subdivided landscape of the British Isles, where several thousand years of human activity have entirely compartmented the landscape. Our upland shepherd, with his walls, in a sense influences our ecological as well as climatic thinking. For convenience we nonetheless talk of woodland, heath, saltmarsh, river and lake ecosystems. But the pristine biosphere was ultimately a continuum that adjusted mutually, gradually and in many dimensions to changing climatic and geological conditions, and in considering freshwaters in particular, the greatest understanding comes from seeing them as intimately linked with the land and atmosphere. It is sometimes convenient, however, for the process of accounting for change to see the parts rather than the whole.

A report as authoritative as that of the IPCC, the *Millennium Ecosystem Assessment*, appeared in 2005. It received much less publicity, for though weather is immediately noticeable to people everywhere, the fate of distant oceans, tundras and savannahs is not, unless you are a deep sea mariner, Inuit hunter or Masai herder. But major changes (Fig. 1.3) have happened to most natural ecosystems, and are continuing to happen to most of them, as a result of climate change and also because of many other, independent drivers that depend on the workings of global economics and the needs of a rising population. It is expected that we will have lost over half of the world's land ecosystems to agriculture or development by 2050. The urbanites may not be noticing this but the consequences will nonetheless be huge, for it is these natural ecosystems that regulate the nature of the biosphere. We have absolutely no idea how much of them can be damaged without serious consequences for human survival. All we know is

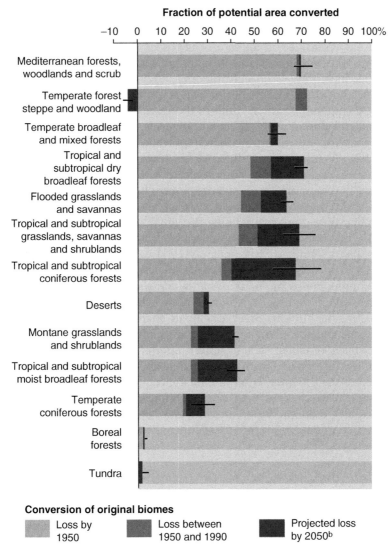

Conversion of original biomes

Loss by 1950 | Loss between 1950 and 1990 | Projected loss by 2050[b]

[a] A biome is the largest unit of ecological classification that is convenient to recognize below the entire globe, such as temperate broadleaf forests or montane grasslands. A biome is a widely used ecological categorization, and becuase considerable ecological data have been reported and modelling undertaken using this categorization, some information in this assessment can only be reported based on biomes. Whenever possible, however, the MA reports information using 10 socioecological systems, such as forest, cultivated, coastal and marine, because these correspond to the regions of responsibility of different government ministries and because they are the categories used within the Convention on Biological Diversity.

[b] According to the four MA scenarios. For 2050 projections, the average value of the projections under the four scenarios is plotted and the error bars (black lines) represent the range of values from the different scenarios.

Figure 1.3 Projected losses of major ecosystems and biomes[a]. (From *Millennium Ecosystem Assessment* 2005.)

that such systems, honed by the utterly ruthless mechanisms of natural selection to be as near fit for purpose as possible, are just as crucial to us, indeed much more fundamentally so, than the local grocer, filling station or hospital. The chemistry of the biosphere is the ultimate *sine qua non* of our existence. Damaged ecosystems, including all agricultural ones, do not store as much carbon as intact ones. James Lovelock's contribution was to point this out.

We have responded rather oddly to the increasing damage we have caused by attempting to value in classical economic terms the goods and services we draw from ecosystems, to demonstrate their importance (Costanza *et al.* 1997; Balmford *et al.* 2002). This has been influential in drawing attention to their very great apparent value and in helping communicate with economists and politicians. But perhaps we have completely missed the point. They are not items that can be used, misused, repaired, ignored or traded at will. They are outside the current economic system. What they do in maintaining the equable state of the planet for all living organisms, including us, is so fundamental as to be priceless. It would be inconceivable, as William Shakespeare (1623) well knew 400 years ago, through the wonderful speech of Portia in *The Merchant of Venice*, to value the blood as a separate component of the body. What is *sine qua non* supersedes evaluation. Yet we damage the biosphere as casually as we throw away our rubbish, and in contemplating the hitherto effects of climate change, we fail to realize that the loss of ecosystems and the changing climate are mutually linked. Indeed, we blithely cost the damage of climate change (Stern 2006) as we cost the goods and services we are losing through application of the same approach of classical economics. We have failed to see the interaction of climate, ecology and equability. Our attempts to mitigate climate change, in a desperate bid to avoid disruption of our societies, may inevitably be doomed to failure unless we begin to see the whole picture and not just the components we find most convenient to our cash economy.

Water and the freshwater biota

Though the ultimate driver of climate change effects will be temperature, the immediate executive will be the availability of freshwater. Freshwater systems stitch together the biosphere through the hydrological cycle. The stitching, however, can become undone, and the surface freshwater component is perhaps the most vulnerable part of the hydrosphere. Living organisms absolutely need liquid water. The ability of liquid water to persist is a fundamental characteristic of a planet capable of supporting life based on carbon compounds. The creation of conditions allowing its existence is the ultimate triumph of the biosphere. The Earth in chemical equilibrium would be so hot as to bear only water vapour. Moreover, human history is, at bottom, an account of the availability of water for drinking, crop growing and sanitation. It follows from the effects of climate change through floods and droughts that the next century, even the next few decades, will likely see more disruption of human activities than has been experienced in the evolution of our species.

For the freshwater systems and organisms with which this book is concerned, the detailed effects of moderate climate change could vary from being disastrous to locally positive. In the absolute scale of temperature, water has its boiling and freezing points very close to the mean surface temperature of the Earth. In its evaporation and condensation, water is the operative liquid of the earth's refrigerator. It follows that the denizens of freshwaters have had an evolutionary history in which their habitats have rather frequently frozen solid or evaporated to mud flats or rocky beds. Freshwater animals and plants are comparatively young in evolutionary terms for they have had repeatedly to recolonize newly constituted freshwaters from the land and the ocean following prolonged glaciation, volcanic disruption or periods of great aridity. They are creatures of continual disturbance (Milner 1996).

Some manifestations of this are that many aquatic insects and vascular plants retain land characteristics as adults or where they flower, respectively; the diversity of freshwaters is much lower, for example, lacking whole phyla, than that of the oceans; freshwater organisms may have particularly high rates of evolutionary change; resting spores and eggs to tide over inimical conditions are common (Pennak 1985). Marine organisms, in contrast, almost universally lack resting stages, for their medium, though changing in shape and depth, has persisted as a body of water for nearly 4 billion years. The longevity of freshwaters may sometimes be only weeks. The retention of adult flight allows movement for insects that cannot persist as resting eggs, and apart from fish, almost all the vertebrates associated with freshwaters are highly motile over land. Fish are vulnerable for few can survive drought, though they are adept at migration through river systems, using even the ocean as part of their life history in some cases. Some crustaceans, however, may respond genetically and very rapidly to thermal stress (van Doorslaer *et al*. 2007).

As climate changes, marine communities will have a continuity of habitat that will accommodate major changes in distribution, though for sedentary organisms like corals, the speed of change may cause severe difficulties. In contrast, land communities, subjected to more frequent drought and without the buffering medium of water, with its high specific heat, will be more vulnerable to extreme temperatures. But the freshwater biota might adjust most readily to climate change because of its preadaptation to disturbance. For them, however, there is a further complication. Freshwaters most immediately and most graphically reflect the many abuses an increasing human population, with its increasing demands for resources, increasing production of waste and rapidly accelerating ability to make changes through its technology, can impose. Freshwaters reflect all the activities that go on in their catchments, which means the entire land surface. Chemical and agricultural wastes, both dissolved and suspended, run into them or rain onto them. Rivers have been repeatedly used as cheap pipes to remove urban wastes. Floodplain wetlands have been embanked and drained so that their fertile soils might be cultivated. Fish communities, the main source of animal protein for many peoples, have been severely overfished. And the very ability of freshwater communities to accommodate change has led to the persistence of many introduced species that have sometimes become dominant and simplified the communities that they have invaded. Not surprisingly, the *Millennium Ecosystem Assessment* listed freshwaters as one of the most vulnerable of the ecosystems it considered (Fig. 1.4). Exactly how freshwater habitats will change,

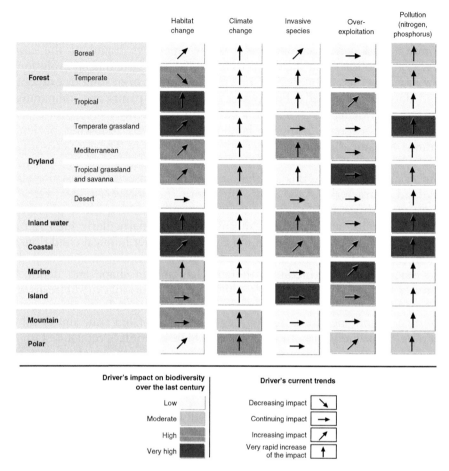

Figure 1.4 Summary of effects of major drivers on major biomes. (From *Millennium Ecosystem Assessment* 2005.)

how the adjustments of their communities will occur and what will be the detailed consequences of the changes for particular places and individual species are thus much more difficult to predict than if climate change were the only threat to them. Current attempts rest largely on expert opinion (Mooij *et al.* 2005). It is one role of this book to add to the factual basis for predictions.

Euro-limpacs, European freshwater systems and approaches to investigation

Europe provides a huge range of inland waters, from the Greenland, Icelandic and mountain glaciers to the streams and lakes of the arider parts of Spain, from the small crater lakes of the Azores to the expanses of Lakes Ladoga, Mälaren and Maggiore and from the tiny headwater streams of the hills to the large, if not

Amazonic, rivers of the Rhine and Danube. Of course, other continents contain an equal or greater variety, but Europe also offers the complication of major biological barriers to animal and plant movements in the Mediterranean, the Alps, the Baltic and the North Sea, the benefits of a long and sophisticated tradition of research in freshwater ecology, and a large concentration of freshwater scientists. Euro-limpacs, on which this book is based, has been a European-Union-funded, continent-wide research programme to further our understanding of the potential effects of climate change on freshwaters. It has contributed to our understanding of the direct physical and hydrological effects of warming in the past (Chapter 2) and present day (Chapters 3 and 4) and on the interactions with climate of nutrients (Chapter 6), acidity (Chapter 7) and toxic pollution (Chapter 7). It has looked at the implications for monitoring and restoration (Chapters 5 and 9) and the definition of reference conditions under the Water Framework Directive. Moreover, it has sought to use the results of these studies in modelling the future (Chapter 10) and in helping political organizations to make decisions on management (Chapter 11).

Euro-limpacs has been far from the last word, but it has contributed important advances, and its strength has been the wide range of approaches it has used. There is a nexus of stages in investigating any general phenomenon and climate change effects on freshwater systems are no exception. The first stage is simply in establishing their existence. There can be no doubt now that climate change is occurring and virtually no doubt that it has largely been caused by human activity. There is then a plethora of studies showing consequent effects (e.g. Carvalho & Kirika 2003; Berger *et al.* 2007), though, strictly speaking, it is rare for the consequence to be rigorously demonstrated. We are dealing with an unreplicated grand experiment with no control.

However, where changes occur in many different glaciers, rivers and lakes and where these correlate closely with changing temperatures or precipitation (Gerten & Adrian 2000; Straile 2002; Winder & Schindler 2004), there can be some confidence in the link. Such correlation, however, is made difficult because many other changes have occurred in freshwater systems over the same period as climate change, and most changes are ultimately caused by the increasing size, aspirations and technological development of human societies in the past 200 years or so.

The correlations of recent history can be placed in context by the reconstructions of the more distant past through analysis of lake and wetland sediments. The record is patchy and selective, and interpretations usually lack experimental validation, but where sediment and direct records have been compared over the past few decades, there is often a close relationship (Haworth 1980), and sophisticated statistical approaches (Birks 1998; Battarbee 2000) have been used to quantify the palaeoecological record.

For periods before the last few decades, or occasionally the last two centuries, where diary and documentary evidence exists, the sediment record is the only record and we must use it as efficiently as we can. The range of chemical and biological remains that can now be counted and calibrated against contemporary observations and sediments is very wide. It can be increasingly elaborated by the techniques of resurrection ecology where resting stages of invertebrates can be hatched and their changing characteristics and genome traced through a period of environmental change (Mergeay *et al.* 2004).

A parallel approach to palaeoecological studies is to use space-for-time investigation, where existing climate gradients provide different systems for examination. The gradient from Greenland to Greece in Europe provides a wide range of systems in which processes and food webs can be compared to predict how they might change as temperatures increase (Moss *et al.* 2004; Meerhof *et al.* 2007). There are, as with every approach, problems with this otherwise attractive endeavour. Not only does climate change along the gradient, but so do relief, geology and the intensity of human activity. Good design of observational schemes can correct for these by stratified random sampling, but one major source of variation, accidents of history, cannot. Glaciation and the nuances of biogeography impose differences that can only be judged. A formerly glaciated lake in Finland, with an Ice-Age-depleted, still recolonizing biota, may not respond to temperature increase in the same way as a long-established Mediterranean lake that may have been affected but not obliterated by the ice of the glacial period, 20,000 years ago, even if the Finnish lake eventually becomes as warm as the Mediterranean one now.

The next stage of investigation is to attempt to reproduce alleged effects through experimentation. Experiments can reveal mechanisms because the drivers of change can be controlled, and experimental designs and adequate replication allow the study of several simultaneous drivers. Experiments are thus potentially more powerful than comparative observations. They also compel the creation of mechanistic hypotheses that force the experimenter to think through the processes that are going on. But the scale of the experiment is important in ecology. Whole-system experiments (Carpenter *et al.* 2001) (clear-felled versus undisturbed sub-catchments of a forested river system, lakes subdivided by curtains and parallel-engineered river channels) are ideal but liable to pseudoreplication because the experiments are so expensive, and the subjects so individual, that generally only one system can be handled at a time. In contrast, experimental laboratory microcosms (Petchey *et al.* 1999) can be replicated extensively but lack reality. The fashion of using micro-organism communities to mimic large-scale systems (Benton *et al.* 2007) is attractive but perhaps mostly to theoreticians.

The compromise is to use subsystems of real communities: mesocosms in lakes, artificial river channels or plots in wetlands, or mesocosm tanks big enough to contain all or almost all of the structures and food-web levels of a system (McKee *et al.* 2000, 2002, 2003; Liboriussen *et al.* 2005). Usually 'almost all' is apposite, for the top predators of a fish community need much more space than is possible in replicable mesocosms, and the complete complexity of a natural system, which, in rivers, for example, might involve interactions with large land mammals (Terborgh 1988; Ripple & Beschta 2004) and tonnages of dead timber, is beyond contemplation.

Another compromise is to do the experiments on simulated systems or models using computer technology. This is, of course, the approach taken by the IPCC in modelling future climate change. *Per se* it is relatively inexpensive, but the models are reflections of the data input to them. If there are unsuspected factors involved, these cannot be included and the output of the model is a reflection of the perceptions of its perpetrators. The same is true of observational techniques and practical experiments. Through choice of variables or of initial experimental conditions, the conclusions are partly predetermined. Nonetheless, the failures of both models and experiments to replicate reality are valuable indicators of what might be missing from their designs. Such gaps are inexpensive to plug in modelling, if not in repeated large-scale experiments, and the behaviour of whole river systems, regions or the biosphere can ultimately be only the province of modelling.

The organization of Euro-limpacs reflected these advantages and uncertainties by using a range of approaches. It had to build on existing experience and facilities for the most part and could not achieve the ideal of using all the approaches on a single habitat and a single aspect of climate change, even if such singularity exists. Understanding increases nonetheless, even if tidy systems of operating are inevitably confounded by the realities of funding and personal preferences. In the end, opinion will depend on expert judgement based on all lines of evidence, for precise prediction is only possible for simple systems, and nothing in earth system science, with its underpinning of living organisms, not least the human ones, is remotely simple.

Applications and the Water Framework Directive

Euro-limpacs included substantial components concerning the application of the emerging scientific understanding. In Europe at present, water management is very much focussed on the Water Framework Directive (EC/2000/60). The Directive changes the previous approach to monitoring waters in Europe by emphasising a whole-basin approach and by requiring determination and restoration of ecological quality, as opposed simply to chemical water quality. This must be done with respect to reference systems, which are defined in the Directive as those unaltered or only negligibly altered by human activity. There are few, if any, such systems left in Europe, so great has been the impact of large population densities over several centuries, so determination of the schemes to determine ecological quality is problematic. Nonetheless, tools for determining the status of phytoplankton, aquatic plants, macroinvertebrates and fish are being developed (UKTAG 2007), often using particular indicator species or families. Climate change will inevitably upset these schemes as species become eliminated or new ones move into previous cooler habitats.

There is also the underlying issue that since climate is now strongly influenced by people, the establishment of reference pristine standards has become conceptually impossible (Moss 2007, 2008). These issues are discussed in Chapter 9. The Directive also requires restoration of aquatic systems to good ecological status, defined as only slightly different from the high ecological status of the reference standards. At this stage, the uncertainties become so great that schemes are needed to help the appraisal of the available scientific information by agencies and governments, and this issue is considered in Chapter 11.

Several reports have pointed out the economic consequences of climate change. The Stern Report (2006) concluded that climate change could be mitigated at the

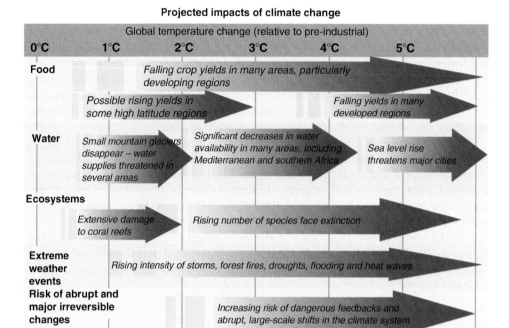

Figure 1.5 Projected effects of increasing temperatures on natural and human systems. (From Stern 2006.)

cost of a substantial but affordable sum, if there were reaction now, but much greater sums if there were delays. Governments have attempted to put in place mechanisms to generate energy by means other than burning fossil fuels, devices to encourage energy conservation and schemes to offset carbon usage by paying for trees to be planted. By and large none of these schemes has yet reduced fossil fuel consumption (Monbiot 2007) and it seems very likely that temperatures will rise later this century by several degrees. A 2°C rise may be held at a concentration of greenhouse gases equivalent to about 480 ppm carbon dioxide compared with the current value of 380 ppm carbon dioxide. It seems, however, more likely that concentrations will rise to at least 550 ppm, denoting a temperature rise of 3°C–4°C, which will bring many problems (Fig. 1.5). The possibilities of biological feedback mechanisms have not, of course, been factored into any of these targets.

A glance through any daily newspaper will reveal several pages of business and sports news that change in detail but not overall content. Pages of other news will change in scope more than business and sport and increasingly a consistent though still very small element of these will concern environmental issues. We might anticipate a time, however, when this formula will change. Sport will undoubtedly retain its hegemony, but the unfolding impacts of resource depletion, waste accumulation, ecosystem destruction, population increase and climate change must eventually displace the multi-page minutiae of stocks, shares, executive salaries and the fate of companies. A new economics will need to be in place or we may be in our final human throes. But for the moment, this book essentially takes the emerging evidence from the physical to the sociological and applies expert judgement to it to assess the interplay between freshwaters and human societies as the climate drama, now just past the Prologue, enfolds. It brings together the major concepts of Hutchinson, Lovelock and Tansley in a crucial act of the evolutionary play as the theatre itself begins to change somewhat ominously.

References

Adrian, R., Wilhelm, S. & Gerten, D. (2006) Life-history traits of lake plankton species may govern their phenological response to climate warming. *Global Change Biology*, **12**, 652–661.

Balmford, A., Bruner, A., Cooper, P., *et al.* (2002) Economic reasons for conserving wild nature. *Science*, **297**, 950–953.

Battarbee, R.W. (2000) Palaeolimnological approaches to climate change, with special regard to the biological record. *Quaternary Science Reviews*, **23**, 91–114.

Benton, T.G., Solan, M., Travis, J.M.J. & Sait, S.M. (2007) Microcosm experiments can inform global ecological problems. *Trends in Ecology and Evolution*, **22**, 516–521.

Berger, S.A., Diehl, S., Stibor, H., *et al.* (2007) Water temperature and mixing depth affect timing and magnitude of events during spring succession of the plankton. *Oecologia*, **150**, 643–654.

Birks, H.J.B. (1998) D.G. Frey & E.S. Deevey review 1 – Numerical tools in palaeolimnology – Progress, potentialities, and problems. *Journal of Palaeolimnology*, **20**, 307–332.

Carpenter, S., Cole, J.J., Hodgson, J.R., *et al.* (2001) Trophic cascades, nutrients, and lake productivity: Whole-lake experiments. *Ecological Monographs*, **71**, 163–186.

Carvalho, L. & Kirika, A. (2003) Changes in shallow lake functioning: Response to climate change and nutrient reduction. *Hydrobiologia*, **506**, 789–796.

Costanza, R., d'Arge, R., de Groot, R., *et al.* (1997) The value of the world's ecosystem services and natural capital. *Nature*, **387**, 253–260.

van Doorslaer, W., Stoks, R., Jeppesen, E. & De Meester, L. (2007) Adaptive microevolutionary responses to simulated global warming in *Simocephalus vetulus*: A mesocosm study. *Global Change Biology*, **13**, 876–886.

Gerten, D. & Adrian, R. (2000) Climate-driven changes in spring plankton dynamics and the sensitivity of shallow polymictic lakes to the North Atlantic Oscillation. *Limnology and Oceanography*, **45**, 1058–1066.

Haworth, E.Y. (1980) Comparison of continuous phytoplankton records with the diatom stratigraphy in the recent sediments of Blelham Tarn. *Limnology and Oceanography*, **25**, 1093–1103.

Hays, G.C., Richardson, A.J. & Robinson, C. (2005) Climate change and marine plankton. *Trends in Ecology and Evolution*, **20**, 337–344.

Hutchinson, G.E. (1965) *The Ecological Theatre and the Evolutionary Play*. Yale University Press, New Haven, CT.

IPCC (Intergovernmental Panel on Climate Change) (2007). Summary for policymakers l. In: *Climate Change 2007: The Physical Science Basis. Contribution of Working Group I to the Fourth Assessment Report of the Intergovernmental Panel on Climate Change* (eds S. Solomon, M. Manning, Z. Chen, *et al.*). Cambridge University Press, Cambridge and New York.

Liboriussen, L., Landkildehus, F., Meerhof, M., *et al.* (2005) Global warming: Design of a flow through shallow lake mesocosm climate experiment. *Limnology and Oceanography: Methods*, **3**, 1–9.

Lovelock, J.E. (1988) *The Ages of Gaia: A Biography of Our Living Earth*. W. W. Norton, New York and London.

McKee, D., Atkinson, D., Collings, S.E., *et al.* (2000) Heated aquatic microcosms for climate change experiments. *Freshwater Forum*, **14**, 51–58.

McKee, D., Hatton, K., Eaton, J.W., *et al.* (2002) Effects of simulated climate warming on macrophytes in freshwater microcosm communities. *Aquatic Botany*, **74**, 71–83.

McKee, D., Atkinson, D., Collings, S.E., *et al.* (2003) Response of freshwater microcosm communities to nutrients, fish and elevated temperature during winter and summer. *Limnology and Oceanography*, **48**, 707–722.

Meerhof, M., Clemente, J.M., de Mello, F.T., Iglesias, C., Pedersen, A.R. & Jeppesen, E. (2007) Can warm climate-related structure of littoral predator assemblies weaken the clear water state in shallow lakes? *Global Change Biology*, **13**, 1888–1897.

Mergeay, J., Verschuren, D., Van Kerckhoven, L. & De Meester, L. (2004) Two hundred years of a diverse Daphnia community in Lake Naivasha, Kenya: Effects of natural and human-induced environmental changes. *Freshwater Biology*, **49**, 998–1013.

Millennium Ecosystem Assessment Board (2005). *Millennium Ecosystem Assessment Synthesis Report*. United Nations Environment Programme, New York.

Milner, A.M. (1996) System recovery. In: *River Restoration* (eds G.E. Petts & P. Calow), pp. 205–226. Blackwell Science, Oxford.

Monbiot, G. (ed.) (2007) *Heat*. Penguin Books, London.

Mooij, W.M., Hulsmann, S., De Senerpont Domis, L.N., *et al.* (2005) The impact of climate change on lakes in the Netherlands: A review. *Aquatic Ecology*, **39**, 381–400.

Moss, B. (2007) Shallow lakes, the water framework directive and life. What should it all be about? *Hydrobiologia*, **584**, 381–394.

Moss, B. (2008) The Water Framework Directive: Total environment or political compromise. *Science of the Total Environment*, **400**, 32–41.

Moss, B., Stephen, D., Balayla, D.M., *et al.* (2004) Continental-scale patterns of nutrient and fish effects on shallow lakes: Synthesis of a pan-European mesocosm experiment. *Freshwater Biology*, **49**, 1633–1649.

Parmesan, C. & Yohe, G. (2003) A globally coherent fingerprint of climate change impacts across natural systems. *Nature*, **421**, 37–42.

Pennak, R.W. (1985) The fresh-water invertebrate fauna: Problems and solutions for evolutionary success. *American Zoologist*, **25**, 671–687.

Petchey, O.L., McPhearson, P.T., Casey, T.M. & Morin, P.J. (1999) Environmental warming alters food-web structure and ecosystem function. *Nature*, **402**, 69–72.

Ripple, W.J. & Beschta, R.I. (2004) Wolves, elk, willows, and tropic cascades in the upper Gallatin Range of Southwestern Montana, USA. *Forest Ecology and Management*, **200**, 161–181.

Root, T.L., Price, J.T., Hall, K.R., Schneider, S.H., Rosenzweig, C. & Pounds, J.A. (2003) Fingerprints of global warming on wild animals and plants. *Nature*, **421**, 57–60.

Roy, D.B. & Sparks, T.H. (2000) Phenology of British butterflies and climate change. *Global Change Biology*, **6**, 407–416.

Shakespeare, W. (1623) *The Merchant of Venice*. First Folio.

Sparks, T.H. & Carey, P.D. (1995) The responses of species to climate over 2 centuries – An analysis of the Marsham phenological record, 1736–1947. *Journal of Ecology*, **83**, 321–329.

Stern, N. (2006) *The Economics of Climate Change*. H.M. Treasury, London.

Straile, D. (2002) North Atlantic Oscillation synchronizes food-web interactions in central European lakes. *Proceedings of the Royal Society of London Series B – Biological Sciences*, **269**, 391–395.

Tansley, A.G. (1935) The use and abuse of vegetational terms and concepts. *Ecology*, **16**, 284–307.

Terborgh, J. (1988) The big things that run the world – A sequel to E.O. Wilson. *Conservation Biology*, **2**, 402–403.

UKTAG (United Kingdom Technical Advisory Group) (2007) *Recommendations on Surface Water Classification Schemes for the Purposes of the Water Framework Directive*. Environment Agency, Bristol.

Walther, G.-R., Post, E., Menzel, A., *et al.* (2002) Ecological responses to recent climate change. *Nature*, **416**, 389–395.

Winder, M. & Schindler, D.E. (2004) Climatic effects on the phenology of lake processes. *Global Change Biology*, **10**, 1844–1856.

2

Aquatic Ecosystem Variability and Climate Change – A Palaeoecological Perspective

Richard W. Battarbee

Introduction

Over the last decade it has become increasingly clear that there is a strong human contribution to global warming (IPCC 2007). Antarctic ice-core records (e.g. Petit *et al*. 1999; EPICA 2004) show that greenhouse gas concentrations are already higher than at any time in the last 750,000 years, temperatures in the northern hemisphere are now on average probably higher than the previous 1000 years (Mann *et al*. 1998) and climate models can only simulate temperatures accurately over the last 150 years if greenhouse gases are included as a forcing mechanism (Stott *et al*. 2001).

Evidence is also accumulating to suggest that changes in natural systems that can be unambiguously attributed to rising temperatures are also occurring. In particular, most mountain glaciers across the world are receding (Oerlemans 2005), there is evidence that the collapse of Antarctic ice shelves in the Antarctic is unprecedented in the Holocene (Domack *et al*. 2005) and ecological changes are taking place in remote Arctic lakes that appear to be outside the range of their natural variability (Douglas *et al*. 1994).

The evidence for human impact on the climate system is thought now to be so compelling that Crutzen has argued that the recent period of earth history dating from the late 18th century and associated with a significant rise in atmospheric CO_2 manifested in Antarctic ice cores (Petit *et al*. 1999) should be given a new geological name, the Anthropocene (Crutzen & Stoermer 2000). Indeed Ruddiman (2003) has argued that human activity may have affected atmospheric greenhouse gas concentrations even earlier as a result of deforestation and land-cover change associated with early agriculture in the early to middle Holocene, approximately 5–8000 years ago.

Climate Change Impacts on Freshwater Ecosystems. First edition. Edited by M. Kernan, R. Battarbee and B. Moss. © 2010 Blackwell Publishing Ltd.

Yet, despite the strength of the evidence for human-induced change, climate-change sceptics still remain, arguing that the role of natural variability is being underestimated. It can indeed be maintained that changes in aquatic ecosystems described in this book, as shown by long-term records (e.g. ice-cover loss on lakes over the last two centuries (Magnuson *et al.* 2000) or the observed increase in river and lake temperatures over recent decades (Hari *et al.* 2006)), are still within the long-term natural range of the climate system, if viewed on centennial timescales. Despite their quality, even these multi-decadal long-term data sets cover too short a time span to differentiate a recent global warming component from changes that might be caused by natural variability.

In this chapter, I examine evidence for changes in aquatic ecosystems from the palaeoecological record in an attempt to define the natural variability of aquatic systems as a baseline against which the impact of recent and projected future climate changes can be assessed. I also examine the impact of past warmer periods on surface waters as a guide to what might happen in future, although once global mean temperatures exceed +2°C, the past is unlikely to hold useful analogues as temperatures greater than about +2°C have not occurred previously for at least 100,000 years.

I describe briefly how climate has changed over the Holocene and then summarize palaeoecological evidence for the response of freshwater eco-systems, principally lakes, to climate change over different timescales, from multi-millennial to seasonal. Two principal effects of climate change are described: those that are driven by changes in temperature and those that are driven by changes in effective moisture (precipitation minus evaporation). These are considered in both high- and low-latitude settings. The chapter ends by setting evidence for greenhouse-gas-forced climate change against other causes of ecological change, specifically those associated with human activity. Note, however, that the evidence presented here for the effects of past climate change on freshwater ecosystems is entirely inferential, as inevitably is the case for all palaeoecological interpretations. Moreover, in some cases, it is difficult to avoid problems of circular reasoning as changes in the biological history of lakes revealed from sediment records are often used to reconstruct past climate changes rather than the response of lake ecosystems to climate change.

Climate over the Holocene

How far back should we go in considering past climate change and its impacts? Although most lakes in high latitudes are relatively young, formed after the recession of the last ice sheets approximately 15,000–10,000 years ago, lakes outside the glacial margin, mainly but not exclusively at lower latitudes, have survived repeated switches from glacial to interglacial conditions and contain sediment records spanning many hundreds of thousand years. A few (e.g. Lake Baikal) span the entire Pleistocene period and are of considerable interest amongst palaeoclimatologists as well as palaeoecologists as rare archives of environmental change, comparable to those found in deep oceans, and as *in situ* centres of evolution. Here I restrict the time span under consideration to the Holocene

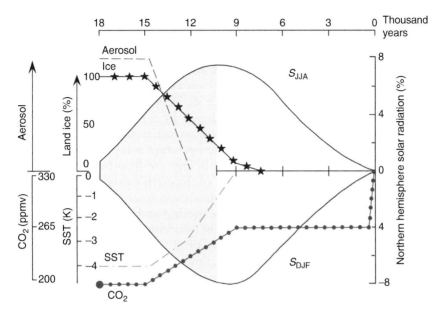

Figure 2.1 Schematic diagram of major changes since 18,000 years before present of external forcing (solar radiation (S)) and internal boundary conditions (land ice, sea-surface temperature (SST), atmospheric carbon dioxide concentration and excess glacial aerosol) used for climate simulations. (Modified from Kutzbach & Street-Perrott 1985.)

period, approximately the last 11,500 years. The onset of the Holocene is marked very clearly in Europe by a very rapid increase in air temperature and a rapid recession of land ice, clearly seen in the sediments of most lakes in Northern Europe by a switch from clays to organic lake muds. Adjustments in the 'earth system' to post-glacial warming were more gradual: global ice cover shrank, the land surface, previously depressed by the weight of ice, rose, sea level rose as a result of global ice melting, the ocean circulation pattern established a new equilibrium, new soils developed and plants and animals returned from glacial refugia to occupy land areas similar to those occupied in the previous interglacial period (cf. Roberts 1998). By approximately 8000 years ago, new 'boundary conditions' for the natural world, not significantly different from those of the present day, were established (Fig. 2.1).

Overall, the Holocene period is a warm stage, contrasting to the previous cold (or glacial) period but similar in many respects to preceding warm (or interglacial) periods, the last of which is referred to in Europe as the Eemian, approximately 130–105,000 years ago (cf. Drysdale *et al.* 2005). The Holocene, therefore, is considered to be yet another interglacial period, differing from previous ones principally because of the impact of people in practising agriculture (causing land-cover change) and in developing industrial processes (causing pollution), which together have significantly and progressively, over approximately the last 5000 years, strongly modified both terrestrial and aquatic ecosystems. It is now also entirely possible that human activity has modified or is modifying the climate system itself, even to an extent that a return to glacial conditions, as considered inevitable only a decade ago, may be prevented by the current and projected rise in greenhouse gas concentrations (Crucifix 2008).

Although the Holocene is a warm period, on average 7°C–8°C higher than the mean for the last glacial period (cf. Lowe & Walker 1997), there have been significant changes in climate within it, driven by a range of different natural forcing mechanisms. These have operated on different timescales and have affected, and will continue to affect, both temperature and precipitation patterns across the world. The principal external forcings are (i) orbital change, mainly related to changes in precessional changes of the earth's axis as it orbits the sun; (ii) solar variability, related to cyclical variations in solar activity through time; and (iii) volcanic activity, related to the scattering and absorption of incoming radiation caused by changes in volcanic dust concentrations in the stratosphere. Climate also varies in a quasi-cyclical way on a range of timescales purely in response to the internal dynamics of the climate system itself. The best known and important of these modes of internal variability is the Southern Ocean Oscillation, ENSO, although in higher latitudes of the northern hemisphere, the Arctic Oscillation (AO) and the associated North Atlantic Oscillation (NAO) can be equally or more important. Changes in these modes can also occur and can be influenced by changes in external forcing (e.g. Shindell *et al.* 2004). In addition to these natural forcings, there is now very good evidence (cf. IPCC 2007) that anthropogenically produced greenhouse gases are playing an important role in altering the climate system and are beginning to warm the planet, potentially to higher temperatures than have hitherto been caused by natural factors. However, such is the complexity and variability of the climate system that attributing changes in natural ecosystems at the present time solely to a rise in greenhouse gas concentrations is not easy (cf. Battarbee & Binney 2008).

Multi-millennial scale change

Temperate latitudes

Palaeoclimate reconstructions of sea-surface temperatures provide evidence of warmer conditions than at present prevailing in the northern hemisphere in the early Holocene (e.g. Jansen *et al.* 2008). The early warming followed by a progressive cooling to the present day agrees closely with the expected decrease in northern hemisphere insolation based on known precessional changes in the orbit of the earth around the sun. Perihelion (when the earth's orbit takes it closest to the sun) now occurs in January, whereas it occurred in July 11,000 years ago. Although proxy records vary, the cooling experienced in medium to high latitudes suggests a decrease in mean annual temperature over the Holocene of approximately 2°C (e.g. Seppä *et al.* 2005) sufficient to cause major southwards shift in the northern limit of plants and animals, including a depression of the northern timber line (Birks & Birks 2003). However, demonstrating the response of aquatic organisms to this long-term cooling is not easy, partly because human

Figure 2.2   The European pond turtle (*Emys orbicularis*). (Photograph from Biopix.dk.)

activity, in clearing forests and developing agricultural practices since the beginning of the Neolithic period, has masked the impact of climate change, and partly because changes in plant and animal records in lake sediments, especially sediments from shallow lakes, are also influenced by hydroseral processes that take place naturally.

   For example, the contraction of the northern boundary of the water chestnut (*Trapa natans*), a thermophilous floating leaved aquatic macrophyte, during the Holocene, was thought to be the result of, and provide good evidence for, Holocene cooling (cf. Alhonen 1964). Korhola and Tikkanen (1997), however, have suggested from a detailed study of shallow lake sediments in Finland that the loss of *Trapa natans* may have been caused by changes in lake habitat as lakes gradually filled in and were transformed into peatland. Similar arguments can be made for the apparent decline in the abundance of other aquatic macrophytes such as *Cladium mariscus* (Conway 1942). The one aquatic fossil record that seems unambiguous in indicating a range response to Holocene cooling in Europe is the European pond turtle (*Emys orbicularis*, Fig. 2.2). At the present day, the breeding range of this species centres on the Mediterranean, and its northern limit coincides closely with the 20°C isotherm. By contrast, its fossil record from the mid-Holocene indicates a distribution in Europe that included eastern England, Denmark and southern Sweden, where mean July temperatures today are close to 18°C (Stuart 1979), providing good evidence for the potential impact of the c. 2°C cooling over the last 5000 years suggested by climate models.

Low latitudes

In mid to low latitudes, the impact of long-term changes in insolation is registered more through changes in moisture than temperature. The changes were especially marked in regions influenced by the shifting position of the Intertropical

Figure 2.3   Exposed diatomites from a dry lake bed of mid-Holocene age in the Bodélé Depression, Northern Chad. (Photograph by J. Giles, with permission from *Nature* magazine.)

Convergence Zone (ITCZ). As the ITCZ moved to the south and the intensity of monsoon rains decreased, the extensive freshwater lakes and rivers that had formed across the Sahara, Arabia, north-west India and western China in the early Holocene progressively shrank and dried out (e.g. Gasse 2002), leaving behind residual, mainly salty, lakes and extensive dry lake beds (Fig. 2.3).

The North African region is probably the best documented, with evidence for former freshwater lakes provided by the number and extent of exposed freshwater diatomites and other lake sediments across the Sahel and Saharan regions, dating to the early and middle Holocene (Gasse 2000; Hoelzmann *et al.* 2004). These early Holocene freshwater lakes and rivers not only supported aquatic plant and animal communities but also provided a freshwater resource for indigenous herdsmen, attested by the rock paintings of giraffe, sheep and cattle in the Tibesti Mountains (e.g. Lhote 1959).

The disappearance of these freshwater systems during the middle Holocene, between approximately 6000 and 5000 years ago, is one of the most dramatic events of recent earth history. Considerable efforts have been made by palaeoclimatologists and climate modellers to document its timing and understand its causes and consequences (Hoelzmann *et al.* 2004). Of relevance to the present day debate on climate change and its effects is the evidence provided by this mid-Holocene desiccation of the rapidity with which a small change in insolation can lead to such a major response. It is very important to understand the mechanisms in the climate system responsible. In this case, climate simulations were found to match the observations only when the models were modified to include feedbacks between land cover and climate so that evapo-transperative losses from the land surface were sufficient to sustain a positive moisture balance in the atmosphere (Claussen *et al.* 1999).

The impact of this mid-Holocene drying on water quality and aquatic biodiversity in North Africa has been most strikingly exemplified by Kröpelin *et al.*'s study of Lake Yoa in the Eastern Sahara (Kröpelin *et al.* 2008), one of the very few sites in the region to retain standing water throughout the entire period. Detailed analysis of a sediment core from Lake Yoa has shown that the lake was sustained beyond the mid-Holocene desiccation by fossil groundwater recharge. Between 4200 and 3900 BP, however, there was a rapid switch from freshwater to saline conditions, with water conductivity rising to over 20,000 μS cm^{-1} leading to a decrease in lake productivity and the establishment of a salt lake macroinvertebrate community dominated by the salt-tolerant hemipteran *Anisops* and brine flies (*Ephydra*). Although it has not yet been established whether the change in climate that caused these events was gradual or rapid (cf. Holmes 2008), the history of this site illustrates clearly how climate change can force freshwater systems across critical ecological thresholds and cause comprehensive changes in ecosystem regimes, with negative consequences for human society.

Centennial to millennial scale

Whilst changes in orbital forcing can explain most of the very long-term changes in climate, palaeoenvironmental records indicate that climate during the Holocene has also varied on shorter scales. Explaining natural variability on these is less easy than for multi-millennial timescales as mechanisms are not so clearly understood. Some of the variability can be due to random fluctuations in the climate system, some to the internal behaviour of the coupled ocean-atmosphere system and some to external influences, especially to variability in solar activity. Solar activity varies on the very well-known 11-year (Schwabe) and 22-year (Hale) cycles now accurately measured by satellite radiometry (Fröhlich & Lean 1998) to the 87-year (Gleissberg), 210-year (Suess) and longer 2200-year (Hallstatt) cycles. Evidence for the longer cycles can be obtained from measurements of ^{14}C through the Holocene as solar activity, documented since 1610 from telescopic observations, has been shown to match closely the ^{14}C variability recorded by tree rings over the same time period (Stuiver & Braziunas 1993). Moreover, ^{10}Be recorded by ice cores (Beer *et al.* 1990) also varies with solar activity, and these two isotopic measures can therefore provide proxies for solar variability back through the whole Holocene period.

The problem for climate science is understanding how very small changes in solar output (c. 0.1% over an 11-year Schwabe cycle) can cause significant fluctuations in climate. However, the variability is strongly wavelength dependent with values fluctuating by more than 100% in the UV part of the spectrum (Beer & van Geel 2008). Moreover, model studies have shown that there are potential amplifying processes in the atmosphere that enable shifts in spectral solar irradiance of this magnitude to cause shifts in tropospheric circulation systems that could cause a significant response by the climate system (Haigh & Blackburn 2006).

Analysis of lake sediment records often reveals evidence of cyclical changes that occur with the same periodicity as solar cycles, but establishing definitive relationships between the two is often undermined by the relative inaccuracy and

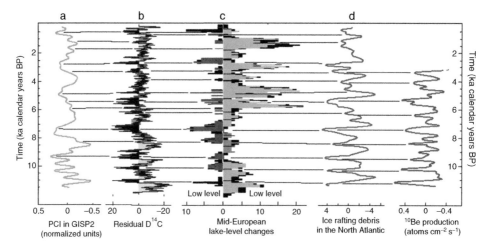

Figure 2.4 Comparison of mid-European lake-level changes (c) with (a) the Polar Circulation Index (PCI) at GISP2 (Mayewski *et al.* 1997); (b) the detrended deviation of the atmospheric $^{14}C/^{12}C$ ratio from a standard value as an indicator of solar activity; (d) the amount of ice-rafted debris found in sediment cores as a measure of the southwards drift of icebergs in the North Atlantic (after Bond *et al.* 2001) and (e) a solar modulation function derived from ^{10}Be data from the GRIP ice core (Beer & van Geel 2008). (Modified from Beer & van Geel 2008 and Magny 2004.)

imprecision of lake sediment dating (cf. Petterson *et al.* 2010). Nevertheless, and whatever the cause, there is abundant evidence to show that lake ecosystems have varied and can vary significantly on these timescales. The record includes evidence for repeated changes in moisture regimes affecting lake water level and cyclical changes in temperature affecting lake productivity as well as century-long excursions in climate, the mechanisms for which are not fully understood.

Effective moisture variability and changes in lake water level

Significant changes in lake water level over the Holocene in temperate regions of Europe and North America have been described by both Digerfeldt (1988) and Magny (1992, 2004). Digerfeldt's work in southern Sweden was based principally on the analysis of macrofossils of aquatic macrophytes in sediment cores taken along transects from lake margins into deep water. He used stratigraphic differences in the macrofossil record between cores along the transect to reconstruct changes in the position of marginal reedswamps that could then be related to lake-level change and, by extension, to climate change. Magny, on the other hand, worked on hard-water lakes in the Jura Mountains, the French pre-Alps and the Swiss Plateau. He employed essentially the same approach as Digerfeldt but used carbonate nodules instead of aquatic macrofossils as indicators of water-level change. He demonstrated (Fig. 2.4, Magny 2004) that the reconstructed water-level changes were largely in phase with other palaeo-records of climate change, including the proxy climate record from a

Greenland ice core (Mayewski *et al.* 1997), and the record of ice-rafted debris (IRD) from marine cores in the North Atlantic Ocean (Bond *et al.* 2001). As these fluctuations were also in phase with the atmospheric residual ^{14}C variation (Stuiver *et al.* 1998), a proxy for solar variability, he concluded that lake levels in this part of Europe were responding to changes in effective moisture driven by solar activity. He argued that the variability was of sufficiently high magnitude to affect not just lake levels but also the livelihood of people, whereby periods of relatively low solar irradiance might cause an increase in annual precipitation and a shortening of the growing season (Magny 2004). An association between solar activity and precipitation has also been proposed by Beer and van Geel (2008), especially with respect to an abrupt event that occurred around 850 BC that is recorded in natural archives in many regions of the world (van Geel & Berglund 2000). They proposed that a shift to a cooler and wetter climate at this time caused flooding, land abandonment and human migration in the Netherlands but, at the same time, caused population expansion in south-central Siberia where nomadic Scythians benefited from the conversion of hostile semi-desert regions to more productive steppic grasslands as a result of more humid conditions. Whilst these authors make little specific reference to changes in freshwater streams and lakes associated with this event, it can be assumed that aquatic plant and animal communities, especially those in littoral habitats, would have been significantly affected.

In low latitudes, centennial- to millennial-scale changes in lake level over the Holocene have been much more dramatic than in mid and high latitudes. Superimposed on the multi-millennial orbitally forced changes described above, relatively short-lived but extreme centennial excursions have occurred repeatedly in the Holocene in low latitudes, with palaeolimnological evidence indicating lake levels having risen and fallen sharply by tens of metres at least twice in the early-(8400–8000 BP) and mid-Holocene (4200–4000 BP) (Street-Perrott & Perrott 1991; Gasse 2000). These changes were probably related to significant but, as yet, largely unexplained weakening in the monsoon system. The impact of these climatic fluctuations was profound. The drought of 4200–4000 BP, for example, corresponded to the collapse of Old World societies from the Old Kingdom in Egypt (deMenocal 2001), the Akkadian Empire in Mesopotamia (Weiss 2000), the Indus Valley Civilization in India (Staubwasser *et al.* 2003) and the Neolithic cultures in the Central Plain of China (Wu & Liu 2004).

In Africa, the mid-Holocene multi-millennial aridification described above may also have been punctuated by significant centennial-scale variability. Although the change in insolation was gradual, the climate response may have varied strongly both temporally and regionally due to climate instability. Modelling experiments demonstrate that, during this period, relatively small random fluctuations in rainfall could trigger switches between different equilibrium states in the climate system (Claussen 2008). Proxy evidence for this is provided by Street-Perrott *et al.* (2000), who used lithological and palynological data from lake sediment records in the Manga Grasslands of the Sahel to show a stepwise process towards arid conditions, and by Jung *et al.* (2004), who used marine sediment records from the Arabian Sea to argue for a similar stepwise transition from humid to dry conditions.

Although the later Holocene in the Saharan−Sahelian region was dry with desert or semi-arid conditions prevailing and water bodies distributed and behaving substantially the same as today, there have been continuing changes in lake level on centennial timescales across the region over the last 3–5000 years (Gasse 1977; Holmes *et al*. 1997). It is difficult to document the full variability in any detail as the evidence is poor, limited by both the number and the quality of sediment records. For example, lake sediment sequences in the Saharan region in the later part of the Holocene are truncated by droughts, and previously deposited sediment is often removed by wind (cf. Kröpelin *et al*. 2008). Long core records, where they exist, therefore may contain hiatuses and become difficult to date and interpret. Nevertheless, there are a few exceptionally good, continuous records that do provide insights into the regional story. These (e.g. for Lake Edward (Russell & Johnson 2005) and Lake Naivasha (Verschuren *et al*. 2000)) suggest that centennial- and decadal-scale droughts, marked by strong fluctuations in lake level, have occurred regularly across East Africa. However, while there is evidence to suggest that some of these excursions were associated with solar forcing (cf. Verschuren & Charman 2008), there are also lake-level changes, recorded by the sediments, that are not proportional to the forcing expected from changes in solar activity, and some inferred that lake-level changes have no equivalent in the solar activity record (Verschuren & Charman 2008).

Although the mechanisms remain uncertain, the palaeo-record indicates clearly that relatively small changes in precipitation in this sensitive region can cause pronounced, rapid responses in water chemistry and aquatic biodiversity on these relatively short as well as long timescales. In particular, a small decrease in effective moisture may lead to salinization and/or complete desiccation. Such changes are not restricted to North and East Africa. Similar evidence for alternating periods of high and low lake levels is available for lakes in many sub-humid and semi-arid regions of the world. Some of the best documented are from the northern Great Plains of North America where many lakes have often switched back and forth between open and closed drainage regimes through the Holocene on multi-decadal and centennial timescales (e.g. Fritz *et al*. 2000). However, within regions, individual lakes respond to the same climate forcing differently, depending on local patterns of groundwater influence, their position in drainage networks and on differences in catchment and lake morphometries (Fritz 2008). Predicting the future response of individual lakes to future climate change, therefore, is not easy, although it is clear that a small decrease in moisture balance can quickly trigger threshold switches between fresh and saline conditions and that water levels can be rapidly lowered to the extent that even relatively deep lakes may become completely desiccated, exacerbated in populated regions by the withdrawal of freshwater surface flows and groundwater for irrigation and water supply.

Temperature

Long instrumental records together with evidence from documentary and palaeoclimate data also show that temperature has varied continually throughout the Holocene on centennial scales (Mackay *et al*. 2003). However, in contrast to low-latitude moisture variability, the amplitude of change has been relatively

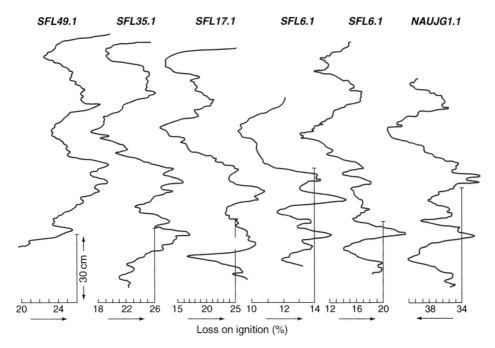

Figure 2.5 Lithostratigraphic correlation of smoothed loss on ignition curves from the uppermost (50–100 cm) sediments in six West Greenland lakes plotted against sediment core depth. Note inverted scale for NAUJG1.1. (Redrawn from Willemse & Tornqvist 1999.)

small, probably not more than 1°C–2°C, significantly less than the increases in temperature projected for the future over the next 100 years by climate models. Temperature effects on lake ecosystems over the Holocene can be obscured by independent changes in catchment and lake processes (cf. Anderson *et al.* 2008). It is difficult, consequently, to identify significant events or changes in palaeolimnological records that can be unambiguously attributed to past changes in temperature. The best evidence comes from remote sites where the confounding impact of human activity is least. In such regions, sediment records often show cyclical changes in organic matter concentration that can be interpreted as temperature-driven changes in lake productivity (e.g. Willemse & Tornqvist 1999; Battarbee *et al.* 2001). The Willemse and Tornqvist study examined the loss-on-ignition (LOI) record for six lakes in the Kangerlussuaq region of Greenland in a transect away from the margin of the West Greenland ice sheet. The results (Fig. 2.5) showed striking time-parallel variations between sites, indicating that the variability was controlled not by local factors but by a regional forcing mechanism matching the δ^{18} O–based temperature records from the GRIP and GISP ice cores nearby. The authors argued that the quasi-cyclical variability in organic matter concentration reflected a temperature-controlled variation in the length of the ice-cover period, increased light penetration and enhanced production during the longer growing season, an interpretation reinforced by the correspondence between the LOI and the abundance of *Chara* stems preserved in the sediments. It is likely that such productivity increases may be amplified by changes in external nutrient loading driven by the warming of catchment soils.

Battarbee *et al.* (2001) showed similar cyclical variations in the content of lake sediment organic matter from a study of a remote mountain lake in the Scottish Cairngorm Mountains. As the cycles had an approximately 200-year periodicity, corresponding quite closely to the periodicity of the Suess solar cycle (see above), the authors concluded, following Willemse and Tornqvist (1999), that the organic matter variability was driven by temperature-related variability in lake productivity. This conclusion was reinforced by the observation that the organic matter cyclicity was matched almost perfectly by similar cyclicity in the concentration of chironomid head capsules in the core (Battarbee *et al.* 2001). If this interpretation is correct, it is apparent, however, that the amplitude of variability in the LOI records is much higher than the amplitude of variability in air temperature, suggesting that feedbacks affecting the production and preservation of organic matter within the lake system were operating. Battarbee *et al.* (2001) speculated that small changes in air temperature might lead to more substantial changes in sediment organic matter concentration if increased algal productivity and stronger stratification caused a decrease in hypolimnetic oxygen concentrations, enhanced nutrient recycling from sediments and improved conditions for organic matter preservation.

Seasonal, inter-annual and decadal change

Most lake sediments accumulate too slowly or are too disturbed to reveal change in lake behaviour on seasonal to inter-annual timescales, although where sediment cores are very finely sampled, it is usually possible to identify decadal-scale change. The exception is where sediments are annually laminated. Such lakes are rare in most regions of the world, but where they exist, they can offer remarkably precise insights into the past.

Some of the best annually laminated sediments are found in relatively deep lakes in boreal climates that are ice covered in winter and are strongly stratified in summer. Productive lakes with low hypolimnetic oxygen concentrations are especially likely to have laminated sediments as mixing by benthic invertebrates is usually greatly reduced in anoxic environments. Extreme care is essential in coring such sites to avoid disturbance of the laminae, and specialized analytical techniques are usually needed to take advantage of the particular nature of the record. Freeze coring is often favoured followed by counting of laminae using, for example, photography and thin-section analysis (Zolitschka 2003) or image analysis (Petterson *et al.* 1999). In diatom-rich sediments, it is possible to use light microscopic examinations of thin sections (e.g. Card 2008) or tape peels (e.g. Simola 1977) to identify changes in the seasonal succession and inter-annual variability of diatom plankton that could be associated with variations in weather patterns, especially where those are known precisely from long-term instrumental records.

Annually laminated sediments favour good microfossil preservation, but it is the accuracy and precision of the chronology of such sediments that are their most important attributes because these enable stratigraphic correlations with data derived from other archives to be made with a minimum of ambiguity. In a novel study of the annually laminated sediments of Kassjön, northern Sweden, Simpson and Anderson (2009) were able to match chronologically the sediment record from the lake with the tree-ring-based temperature record from the same region in northern Sweden. They compared a 400-year period from −200 BC to AD 200, a period of known climate variability, before the introduction of agriculture into the region. The aim was to assess how much variability in the diatom record could be attributed to natural temperature variability. As both the climate proxy (tree rings) and the biological response (diatoms) were based on regularly spaced samples in time, classic time-series statistics could be used for the analysis. Although the reasons for the correlation are as yet unclear, the results show a clear correspondence between the climatic variability and the observed magnitudes of diatom assemblage shifts on a range of timescales.

Recent change

Remote regions

Given the natural variability of the climate system and the complexity of ecosystem response to climate change on different timescales, it is not easy to use contemporary observations to separate trends in aquatic ecosystem behaviour that can be attributed to anthropogenic climate change from those caused by natural climate forcing. And it is doubly difficult to identify climate impacts on aquatic systems for those systems that have been, and are, strongly influenced in other ways by human activity over recent decades and centuries. Such influences, principally from land-use change and from pollution, are far stronger than climate change. The palaeolimnological record, however, can provide a sufficiently long time frame for these different influences to be disentangled, but only the most remote regions on earth, free from pollution, provide good locations where unambiguous evidence for recent limnological change attributable to climate change can be potentially identified. Even in such regions, the confounding impact of long-distance transported pollutants, especially from nutrients, cannot always be easily discounted (cf. Wolfe *et al.* 2001).

Nevertheless, many palaeolimnologists have focussed research over recent years on lake sediment records in arctic-alpine regions above or beyond the timber line, regions also where the effects of global warming are projected to be greatest (e.g. Smol & Douglas 2007). In Europe, there has been a special interest in alpine lakes. In one large-scale study that included lakes from all the major mountain regions, the recent sediment record was compared with instrumental temperature records to assess how much of the variability in the record could be attributed to temperature change over the last 200 years (Battarbee *et al.* 2002). Homogenized instrumental temperature records constructed from meteorological station data were tuned to each mountain site using corrections for lapse-rate differences and for shading effects (Agustí-Panareda & Thompson 2002), and a full suite of physical, geochemical and microfossil analyses were conducted from dated replicate sediment cores at each site.

Despite chronological uncertainties associated with problems of matching radiometric dates from the sediment record to the calendar dates of the meteorological record, the study showed a good correlation between increasing temperature and the organic matter content in the sediment and between increasing temperature and changes in the composition of the diatom record. Both suggest that changes in algal composition and overall lake productivity have been responding to temperature increases of the last few decades and support the interpretations above describing the probable relationship between organic matter and temperature inferred for earlier periods.

One of the most striking observations at a number of mountain sites in Europe has been an increase in the abundance of planktonic diatoms, especially of the genus *Cyclotella* (Rühland *et al.* 2008). Perhaps the best example is from Lake Redo in the Spanish Pyrenees (Catalan *et al.* 2002). Sediment core analysis from this lake showed that two species of planktonic diatom, *Fragilaria nanana* and *Cyclotella pseudostelligera*, appeared in the lake in the late 19th–early 20th century and became more abundant over the last few decades. An examination of the current phytoplankton seasonality in the lake showed that these species are dominant in late summer and early autumn, appearing towards the end of the summer stratification period and reaching maximum abundance in September and October, respectively. A close correspondence between the increase in these taxa in the core record and the October air temperature (Fig. 2.6) suggests that increasing autumn water temperature has been the main reason, directly or indirectly, for the change.

Similar examples of planktonic diatom increases have been recorded from sediments of lakes in other remote high-altitude and high-latitude regions (e.g. Sovari & Korhola 1998; Koinig *et al.* 2002; Sovari *et al.* 2002; Rühland & Smol 2005; Smol *et al.* 2005; Solovieva *et al.* 2005, 2008).

Although the grounds for interpreting such changes in remote lakes as a consequence of increasing temperature are credible, the potential separate or additional effect of long-range transported nutrient pollution cannot be excluded in some cases. For example, Wolfe *et al.* (2001) showed that for two remote mountain lakes in the Rocky Mountains, the striking increase in the abundance of *Asterionella formosa* in the uppermost sediments of both lakes corresponded to an increase in the concentration of lead (Pb) in the core and to a decrease in the values of the stable isotope ^{15}N, indicating that the plankton increase was more likely to have been caused by air pollution than by global warming. Studies of present-day nutrient dynamics of remote lakes across Europe and North America have confirmed the extent to which atmospheric inorganic N deposition is an important nutrient source for such lakes (Bergström & Jansson 2006), illustrating the need for caution in attributing productivity increases entirely to the direct impact of climate change at some sites.

The problem in disentangling the separate but interacting influences of nutrient deposition and climate warming is that both can cause symptoms of eutrophication that are recorded in an identical way in lake sediments, viz. by an increase in organic matter and by a change in the relative abundance of planktonic diatoms.

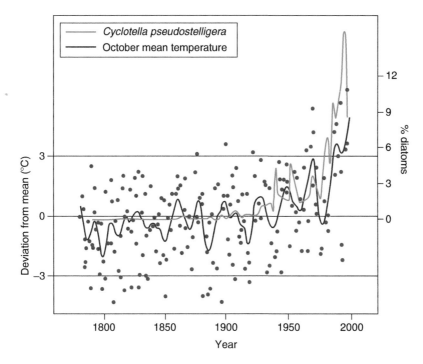

Figure 2.6 Percentage *Cyclotella pseudostelligera* vs October temperature for Lake Redo, Pyrenees. (From Catalan *et al.* 2002.)

Whilst adopting the approach of Wolfe *et al.* (2001) can help to reduce uncertainty by identifying sites with a history of N deposition, a more definitive approach is to seek analytical evidence for climate warming that is not dependent on nutrients but on other impacts of global warming, such as reduced ice cover. An example is the work of Douglas *et al.* (1994) from Ellesmere Island in the high Canadian arctic. In this case, the diatom changes in the uppermost sediments were ascribed to a reduction in the length of summer ice cover, leading to the creation of a more extensive littoral habitat with an increase in aquatic bryophytes providing a habitat for diatom epiphytes such as *Pinnularia balfouriana* (Fig. 2.7).

Separating climate change from pollution impacts at polluted sites

In populated regions of the world where most, if not all, lakes are suffering or recovering from the severe impacts of human activity, identifying the influence of climate change is more problematic, mainly because climate change is yet to become a dominant driving force and the effects of climate change are as yet masked by pollutant effects (e.g. Anderson *et al.* 1996). Attempts to address this issue, especially with respect to problems of acidification and eutrophication, have focussed both on the analysis of long-term observational data sets (e.g. Straile 2000; Jepessen *et al.* 2005; George *et al.* 2007; Ferguson *et al.* 2008) and on the palaeolimnological record.

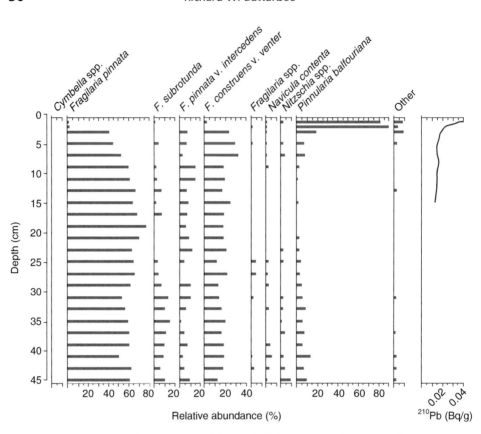

Figure 2.7   Changes in diatom composition and ²¹⁰Pb concentration in a sediment core from Elison Lake, Cape Herschel. The ²¹⁰Pb increase indicates that the uppermost 5 cm of the core covers a time span of approximately 100 years. (From Douglas *et al.* 1994. Reproduced with permission of The American Association for the Advancement of Science.)

With respect to acidification, climate change influences the acidity status of lakes primarily through the impact of changing temperature on biogeochemical processes (e.g. Koinig *et al.* 1998) and of changing precipitation patterns on catchment hydrology (Monteith *et al.* 2001). The clearest palaeolimnological evidence is provided at sites that either are remote from the effects of acid deposition or have a relatively high acid-neutralizing capacity. Sommaruga-Wögrath *et al.* (1997) showed for such a site in the Austrian Tyrol that changes in diatom-inferred pH in a core were strongly correlated with instrumental temperature records over recent decades. Larsen *et al.* (2006) described a similar response in two Norwegian lakes. Curtis *et al.* (2009) used Canonical Correspondence Analysis (CCA) of diatom records from European mountain lakes to show that air temperature, independently of acid deposition, explained a significant amount of the variation between diatom assemblages from early 19th century sediments and the present day, and Simpson and Anderson (2009) using an Additive Modelling (AM) approach demonstrated that temperature change was responsible for some of the temporal variation in diatom assemblages in a core from a lake in North-West Scotland.

These observations suggest that as sulphur deposition continues to decrease, temperature change is likely to emerge as the key driver of lake acidity at many sites; pH values at non-polluted sites may reach levels only previously experienced in the warmer early Holocene (Larsen *et al.* 2006). The results also suggest that some of the increases in alkalinity that have taken place over recent decades (e.g. Monteith & Evans 2005) might be attributed to increasing temperature as well as reducing acid deposition. However, for regions, e.g. in North-West Europe, where climate change may cause an increase in precipitation, such an effect might be offset as higher discharge and the associated deposition of sea salts in stormy conditions may depress alkalinity. So far, there is no palaeolimnological evidence for this effect, but the analysis of long-term monitoring data from the United Kingdom supports the hypothesis (D. T. Monteith, personal communication).

Identifying the role of climate change on lakes suffering from eutrophication from lake sediment records is equally difficult. All lakes with populated or agricultural catchments are influenced by nutrient pollution to some extent, and as water temperature increase tends to produce symptoms identical to those of eutrophication in terms of algal productivity, hypolimnetic oxygen stress and nutrient recycling, the palaeolimnological record is not capable of differentiating between the different stresses, at least using standard techniques. Current palaeo-ecological research is attempting to disentangle these stresses using statistical approaches beginning with sites where both long-term instrumental climatic data and nutrient concentration data are available. At Loch Leven, a large, shallow lake in lowland Scotland, the eutrophication signal in the sediment record outweighs any evidence of climate as a control on the diatom community on a decadal to centennial scale. However, at an inter-annual scale, there are several changes in species composition in the recent fossil record that may be attributed to climate variability. In particular, peaks in *Aulacoseira ambigua* and *A. granulata* and its variety *angustissima* in 2003–4, 1998–9 and 1986–7 appear to be associated with wetter, and possibly windier and cooler, summers (H. Bennion *et al.*, personal communication). At Lago Maggiore, a deep, pre-alpine lake in Northern Italy, the combination of Cladocera data from the sediment record with contemporary zooplankton and fish data has provided insights into past changes in the interaction between trophic dynamics of the lake and climate change (Manca *et al.* 2007), whereby evidence of increased instability in recent zooplankton communities follows the occurrence of extreme meteorological events and changes in fish predation, which, in turn, may be attributed to increased temperature.

Conclusions

The palaeoenvironmental record indicates that climate has varied naturally and continually on a range of different timescales influenced principally by variability in the orbit of the earth around the sun, by changes in solar irradiance and in the emission of dust from volcanic eruptions. Instrumental records also show that climate system is also naturally dynamic and has its own internal modes of variability that can lead to significant oscillations in weather patterns on inter-annual and decadal timescales. Although knowledge is still imperfect, climate scientists are beginning to establish how the timing and amplitude of climate change have varied in the past on these different timescales and in response to these different forcing mechanisms (cf. IPCC 2007).

A central objective now is to understand how such climate change has influenced natural ecosystems in the past and the extent to which such an understanding might provide insights into how climate change might influence freshwater ecosystems in the future. Palaeoecology offers a potentially powerful methodology to address this issue, although a degree of caution is needed, not only because of the fragmentary nature of the fossil record but also because changes in aquatic sediments deemed to provide evidence of the response of ecosystems to climate change are often themselves used as a proxy source of information for reconstructing past climate change. Palaeoecological interpretations and arguments can easily be open to the risk of circular reasoning. Despite these important caveats, a number of principal conclusions can be made.

First, the palaeo-evidence suggests that small changes in climate forcing, either in moisture or in temperature, can cause major changes in ecosystem response. The mid-Holocene aridification in low latitudes occurred in response to a relatively small change in insolation, but for some regions, it caused widespread desiccation and salinization, and for lakes close to the threshold between positive and negative moisture balance, the palaeo-evidence shows that only very small shifts in hydrology are needed to switch lakes between fresh and saline states. Likewise, for temperature, the large swings in the organic content of lake sediments in mid and high latitudes indicate how small temperature shifts may be magnified by in-lake feedbacks between nutrients, oxygen concentration and organic matter preservation.

Secondly, mid- and high-latitude northern hemisphere freshwaters experienced July temperatures 7–8000 years ago approximately 2°C higher than today. The cooling since then caused a range contraction in the northern limit of thermophilous taxa, but the rate of change was probably sufficiently slow for organisms and ecosystems to adapt without causing extinctions. In contrast, the future temperature increase expected from global warming may reach 2°C over the 1960–90 baseline within two decades, and the warming projected to the end of the century will be unprecedented both in rate and in magnitude. The probability that this will cause local and regional extinctions and major readjustments in the functioning of freshwater ecosystems is very high, and the extent to which these climate projections and associated ecosystem responses lie well beyond the range of natural variability is evident from the palaeo-record.

And thirdly, despite the probability of such unprecedented change in the future, it is still difficult to attribute observed changes in freshwater ecosystems over recent decades unambiguously to greenhouse-gas-forced warming. This is due to problems in separating evidence for greenhouse-gas forcing from forcing caused by natural variability and the difficulty in identifying the influence of climate on freshwater ecosystems that are already modified by pollution and other kinds of human activity. The palaeoecological record shows that cyclical changes in lake behaviour are continually taking place on decadal to centennial timescales, possibly related to variability in solar irradiance or other modes of climate variability, and that contemporary changes are not as yet significantly greater than those expected from natural variability.

The exceptions to this are lakes from high latitude and altitude where evidence for unprecedented change is becoming compelling, associated with decreased ice cover and increased productivity. For most other freshwaters situated in populated regions of the world, the lake sediment record suggests that the effects of nutrient enrichment, acid deposition, water abstraction and other human impacts, operating at local and regional scales, remain the principal stresses, masking the effects of climate change. Where climate change modifies processes in the same way as human activity (e.g. in driving nutrient dynamics, controlling alkalinity generation and modifying hydrological regimes), disentangling the role of climate change from these other stressors may continue to be difficult.

However, in Europe and other regions of the world where pollutant loadings are being reduced and lakes and streams are being progressively rehabilitated, climate change could become a more important driver of freshwater ecosystem structure and function in the future, potentially deflecting recovery trajectories from pollution away from the pre-pollution reference state and towards a new, uncertain state. The palaeoecological record in future will consequently serve not only to provide an insight into how freshwater ecosystems have varied as a result of past climate change but also to provide a measure of how different freshwater ecosystems might become in future as a result of global warming.

References

Agustí-Panareda, A. & Thompson, R. (2002) Reconstructing air temperature at eleven remote alpine and arctic lakes in Europe from 1781 to 1997 AD. *Journal of Paleolimnology*, **28**, 7–23.

Alhonen, P. (1964) Radiocarbon age of waternut (*Trapa natans* L.) in the sediments of Lake Karhejäumlrvi, SW-Finland. *Memoranda Socitatis Fauna et Flora Fennica*, **40**, 192–197.

Anderson, N.J., Odgaard, B.V., Segerstrom, U. & Renberg, I. (1996) Climate-lake interactions recorded in varved sediments from a Swedish boreal forest lake. *Global Change Biology*, **2**, 399–405.

Anderson, N.J., Brodersen, K.P., Ryves, D.B., *et al.* (2008) Climate versus in-lake processes as controls on the development of community structure in a low-arctic lake (South-West Greenland). *Ecosystems*, **11**, 307–324.

Battarbee, R.W. & Binney, H.A. (eds) (2008) *Natural Climate Variability and Global Warming: A Holocene Perspective*. Wiley-Blackwell, Chichester.

Battarbee, R.W., Cameron, N.G., Golding, P., *et al.* (2001) Evidence for Holocene climate variability from the sediments of a Scottish remote mountain lake. *Journal of Quaternary Science*, **16**, 339–346.

Battarbee, R.W., Grytnes, J.-A., Thompson, R., *et al.* (2002) Comparing palaeolimnological and instrumental evidence of climate change for remote mountain lakes over the last 200 years. *Journal of Paleolimnology*, **28**, 161–179.

Beer, J. & van Geel, B. (2008) Holocene climate change and the evidence for solar and other forcings. In: *Natural Climate Variability and Global Warming: A Holocene Perspective* (eds R.W. Battarbee & H.A. Binney), pp. 138–162. Wiley-Blackwell, Chichester.

Beer, J., Blinov, A., Bonani, G., *et al.* (1990) Use of ^{10}Be in polar ice to trace the 11-year cycle of solar activity. *Nature*, **347**, 164–166.

Bergström, A.-K. & Jansson, M. (2006) Atmospheric nitrogen deposition has caused nitrogen enrichment and eutrophication of lakes in the northern hemisphere. *Global Change Biology*, **12**, 635–643.

Birks, H.H. & Birks, H.J.B. (2003) Reconstructing Holocene climates from pollen and plant macrofossils. In: *Global change in the Holocene* (eds A. Mackay, R.W. Battarbee, H.J.B. Birks & F. Oldfield), pp. 342–357. Hodder Arnold, London.

Bond, G., Kromer, B., Beer, J., *et al.* (2001) Persistent solar influence on North Atlantic climate during the Holocene. *Science*, **294**, 2130–2136.

Card, V.M. (2008) Varve-counting by the annual pattern of diatoms accumulated in the sediment of Big Watab Lake, Minnesota, AD 1837–1990. *Boreas*, **26**, 103–112.

Catalan, J., Pla, S., Rieradevall, M., *et al.* (2002) Lake Redó ecosystem response to an increasing warming the Pyrenees during the twentieth century. *Journal of Paleolimnology*, **28**, 129–145.

Claussen, M. (2008) Holocene rapid land-cover changes – Evidence and theory. In: *Natural Climate Variability and Global Warming: A Holocene Perspective* (eds R.W. Battarbee & H.A. Binney), pp. 232–253. Wiley-Blackwell, Chichester.

Claussen, M., Kubatski, C., Brovkin, V., Ganopolski, A., Hoelzmann, P. & Pachur, H.J. (1999) Simulation of an abrupt change in Saharan vegetation at the end of the mid-Holocene. *Geophysical Research Letters*, **24**, 2037–2040.

Conway, V.M. (1942) Biological flora of the British Isles. *Cladium mariscus* (L.) R. Br. *Journal of Ecology*, **30**, 211–216.

Crucifix, M. (2008) Modelling the climate of the Holocene. In: *Natural Climate Variability and Global Warming: A Holocene Perspective* (eds R.W. Battarbee & H.A. Binney), pp. 98–122. Wiley-Blackwell, Chichester.

Crutzen, P.J. & Stoermer, E.F. (2000) The "Anthropocene". *Global Change Newsletter*, **41**, 12–13.

Curtis C.J., Juggins S., Clarke G., *et al.* (2009) Regional influence of acid deposition and climate change in European mountain lakes assessed using diatom transfer functions. *Freshwater Biology*, **54**, 2555–2572.

deMenocal, P. (2001) Cultural responses to climate change during the late Holocene. *Science*, **292**, 667–673.

Digerfeldt, G. (1988) Reconstruction and regional correlation of Holocene lake-level fluctuations in Lake Bysjön, south Sweden. *Boreas*, **17**, 165–182.

Domack, E., Duran, D., Leventer, A., *et al.* (2005) Stability of the Larsen B ice shelf on the Antarctic Peninsula during the Holocene epoch. *Nature*, **436**, 681–685.

Douglas, M.S.V., Smol, J.P. & Blake, W., Jr. (1994) Marked post-18th century environmental change in high-arctic ecosystems. *Science*, **266**, 416–419.

Drysdale, R.N., Zanchetta, G., Hellstrom, J.C., Fallick, A.E. & Zhao, J. (2005) Stalagmite evidence for the onset of the Last Interglacial in southern Europe at 129 ± 1 ka. *Geophysical Research Letters*, **32**, Article number L24708.

EPICA Community Members (2004) Eight glacial cycles from an Antarctic ice core. *Nature*, **429**, 623–628.

Ferguson, C.A., Carvalho, L., Scott, E.M., Bowman, A.W. & Kirika, A. (2008) Assessing ecological responses to environmental change using statistical models. *Journal of Applied Ecology*, **45**, 193–203.

Fritz, S.C. (2008) Deciphering climatic history from lake sediments. *Journal of Paleolimnology*, **39**, 5–16.

Fritz, S.C., Ito, E., Yu, Z., Laird, K.R. & Engstrom, D.R. (2000) Hydrologic variation in the northern Great Plains over the last two millennia. *Quaternary Research*, **53**, 175–184.

Fröhlich, C. & Lean, J. (1998) The sun's total irradiance: Cycles, trends and related climate change uncertainties since 1976. *Geophysical Research Letters*, **25**, 4377–4380.

Gasse, F. (1977) Evolution of Lake Abhé (Ethiopia and TFAI), from 70,000 b.p. *Nature*, **265**, 42–45.

Gasse, F. (2000) Hydrological changes in the African tropics since the Last Glacial Maximum. *Quaternary Science Reviews*, **19**, 189–211.

Gasse, F. (2002) Diatom-inferred salinity and carbonate oxygen isotopes in Holocene waterbodies of the western Sahara and Sahel (Africa). *Quaternary Science Reviews*, **21**, 737–767.

van Geel, B. & Berglund, B.E. (2000) A causal link between a climatic deterioration around 850 cal BC and a subsequent rise in human population density in NW-Europe? *Terra Nostra*, **7**, 126–130.

George, D.G., Hurley, M.A. & Hewitt, D.P. (2007) The impact of climate change on the physical characteristics of the larger lakes in the English Lake District. *Freshwater Biology*, **52**, 1647–1666.

Haigh, J.D. & Blackburn, M. (2006) Solar influences on dynamical coupling between the stratosphere and troposphere. *Space Science Reviews*, **125**, 331–344.

Hari, R.E., Livingstone, D.M., Siber, R., Burkhardt-Holm, P. & Güttinger, H. (2006) Consequences of climatic change for water temperature and brown trout populations in Alpine rivers and streams. *Global Change Biology*, **12**, 10–26.

Hoelzmann P., Gasse, F., Dupont, L.M., *et al.* (2004) Palaeoenvironmental changes in the arid and subarid belt (Sahara-Sahel-Arabian Peninsula) from 150 Ka Kyr to present. In: *Past Climate Variability Through Europe and Africa* (eds R.W. Battarbee, F. Gasse & C.E. Stickley), pp. 219–256. Springer, Dordrecht.

Holmes, J.A. (2008) How the Sahara became dry. *Science*, 320, 752–753.

Holmes, J.A., Street-Perrott, F.A., Allen, M.J., *et al.* (1997) Holocene palaeolimnology of Kajemarum Oasis, Northern Nigeria: An isotopic study of ostracodes, bulk carbonate and organic carbon. *Journal of the Geological Society, London*, 154, 311–319.

IPCC (Intergovernmental Panel on Climate Change) (2007) *Climate Change 2007: The Physical Science Basis. Contribution of Working Group I to the Fourth Assessment Report of the Intergovernmental Panel on Climate Change*. Cambridge University Press, Cambridge and New York.

Jansen, E., Andersson, C., Moros, M., Nisancioglu, K.H., Nyland, B.F. & Telford, R.J. (2008) In: *Natural Climate Variability and Global Warming: A Holocene Perspective* (eds R.W. Battarbee & H.A. Binney), pp. 123–137. Wiley-Blackwell, Chichester.

Jeppesen, E., Sondergaard, M., Jensen, A.P., *et al.* (2005) Lakes' response to reduced nutrient loading – Analysis of contemporary data from 35 European and North American long term studies. *Freshwater Biology*, 50, 1747–1771.

Jung, S.J.A., Davies, G.R., Ganssen, G.M. & Kroon, D. (2004) Stepwise Holocene aridification in NE Africa deduced from dust-borne radiogenic isotope records. *Earth and Planetary Science Letters*, 221, 27–37.

Koinig, K., Schmidt, R., Sommaruga-Wögrath, S., Tessadri, R. & Psenner, R. (1998) Climate change as the primary cause for pH shifts in a high alpine lake. *Water Air and Soil Pollution*, 104, 167–180.

Koinig, K.A., Kamenik, C., Schmidt, R., *et al.* (2002) Environmental changes in an alpine lake (Gossenköllesee, Austria) over the last two centuries – The influence of air temperature on biological parameters. *Journal of Paleolimnology*, 28, 147–160.

Korhola, A.A. & Tikkanen, M.J. (1997) Evidence for a more recent occurrence of water chestnut (*Trapa natans* L.) in Finland and its palaeoenvironmental implications. *Holocene*, 7, 39–44.

Kröpelin, S., Verschuren, D., Lezine, A.-M., *et al.* (2008) Climate-driven ecosystem succession in the Sahara: The past 6000 years. *Science*, 320, 765–768.

Kutzbach, J.E. & Street-Perrott, F.A. (1985) Milankovitch forcing of fluctuations in the level of tropical lakes from 18 to 0 kyr BP. *Nature*, 317, 130–134.

Larsen, J., Jones, V.J. & Eide, W. (2006) Climatically driven pH changes in two Norwegian lakes. *Journal of Paleolimnology*, 36, 175–187.

Lhote, H. (1959) *The Search for the Tassili Frescoes: The Story of the Prehistoric Rock-Paintings of the Sahara*. E. P. Dutton, New York.

Lowe, J.J. & Walker, M.J.C. (eds) (1997) *Reconstructing Quaternary Environments*. Longman, Harlow.

Mackay, A., Battarbee, R.W., Birks, H.J.B. & Oldfield, F. (eds) (2003) *Global Change in the Holocene*. Hodder Arnold, London.

Magnuson, J.J., Robertson, D.M., Benson, B.J., *et al.* (2000) Historical trends in lake and river ice cover in the Northern Hemisphere. *Science*, 289, 1743–1746.

Magny, M. (1992) Holocene lake-level fluctuations in Jura and the northern subalpine ranges, France: Regional pattern and climatic implications. *Boreas*, 21, 319–334.

Magny, M. (2004) Holocene climate variability as reflected by mid-European lake-level fluctuations and its probable impact on prehistoric human settlements. *Quaternary International*, 113, 65–79.

Manca, M., Torretta, B., Comoli, P., Amsinck, S.L. & Jeppesen, E. (2007) Major changes in the trophic dynamics in large, deep subalpine Lake Maggiore from 1943 to 2002: A high resolution comparative paleo-neolimnological study. *Freshwater Biology*, 52, 2256–2269.

Mann, M.E., Bradley, R.S. & Hughes, M.K. (1998) Global-scale temperature patterns and climate forcing over the past six centuries. *Nature*, 392, 779–787.

Mayewski, P., Meeker, L., Twickler, M., *et al.* (1997) Major features and forcing of high-latitude northern hemisphere atmospheric circulation using a 110,000-year-long glaciochemical series. *Journal of Geophysical Research*, 102, 26345–26366.

Monteith, D.T. & Evans, C.D. (2005) The United Kingdom acid waters monitoring network: A review of the first 15 years and introduction to the special issue. *Environmental Pollution*, 137, 3–13.

Monteith, D.T., Evans, C.D. & Patrick, S.T. (2001) Monitoring acid waters in the UK: 1988–1998 trends. *Water, Air and Soil Pollution*, 130, 1307–1312.

Oerlemans, J. (2005) Extracting a climate signal from 169 glacier records. *Science*, 308, 675–677.

Petit, J.R., Jouzel, J., Raynaud, D., *et al.* (1999) Climate and atmospheric history of the past 420,000 years from the Vostok ice core, Antarctica. *Nature*, 399, 429–436.

Petterson, G., Odgaard, B.V. & Renberg, I. (1999) Image analysis as a method to quantify sediment components. *Journal of Paleolimnology*, 22, 443–455.

Petterson, G., Renberg, I., Luna, S.S., Arnqvist, P. & Anderson, N.J. (2010) Climate influence on the inter-annual variability of late-Holocene minerogenic sediment supply in a boreal forest catchment. *Earth Surface Processes and Landforms*, 35, 390–398.

Roberts, N. (1998) *The Holocene: An Environmental History*. Blackwell, Oxford.

Ruddiman, W.F. (2003) The anthropogenic greenhouse era began thousands of years ago. *Climatic Change*, 61, 261–293.

Rühland, K. & Smol, J.P. (2005) Diatom shifts as evidence for recent subarctic arming in a remote tundra lake, NWT, Canada. *Palaeogeography Palaeoclimatology Palaeoecology*, 226, 1–16.

Rühland, K., Paterson, A.M. & Smol, J.P. (2008) Hemispheric-scale patterns of climate-related shifts in planktonic diatoms from North American and European lakes. *Global Change Biology*, 14, 2740–2754.

Russell, J.M. & Johnson, T.C. (2005) A high resolution geochemical record from Lake Edward, Uganda-Congo, and the timing and causes of tropical African drought during the late Holocene. *Quaternary Science Reviews*, 24, 1375–1389.

Seppä, H., Hammarlund, D. & Antonsson, K. (2005) Low-frequency and high-frequency changes in temperature and effective humidity during the Holocene in south-central Sweden: Implications for atmospheric and oceanic forcings of climate. *Climate Dynamics*, 25, 285–297.

Shindell, D.T., Schmidt, G.A., Mann, M.E. & Faluvegi, G. (2004) Dynamic winter climate response to large tropical volcanic eruptions since 1600. *Journal of Geophysical Research*, 109, D05104, doi:10.1029/2003JD004151.

Simpson, G.L. & Anderson, N.J. (2009) Deciphering the effect of climate change and separating the influence of confounding factors in sediment core records using additive models. *Limnology and Oceanography*, 54, 2529–2541.

Simola, H. (1977) Diatom succession in the formation of annually laminated sediment in Lovojarvi, a small eutrophic lake. *Annales Botanici Fennici*, 14, 143–148.

Smol, J.P. & Douglas, M.S.V. (2007) From controversy to consensus: Making the case for recent climate change using lake sediments. *Frontiers in Ecology and the Environment*, 5, 466–474.

Smol, J.P., Wolfe, A.P., Birks, H.J.B., *et al.* (2005) Climate-driven regime shifts in the biological communities of arctic lakes. *Proceedings of the National Academy of Sciences*, 102, 4397–4402.

Solovieva, N., Jones, V.J., Birks, H.J.B., *et al.* (2005) Palaeolimnological evidence for recent climate change in lakes from the northern Urals, arctic Russia. *Journal of Paleolimnology*, 33, 463–482.

Solovieva, N., Jones, V.J., Birks, J.H.B., Appleby, P. & Nazarova, L. (2008) Diatom responses to 20th century climate warming in lakes from the Northern Urals, Russia. *Palaeogeography Palaeoclimatology Palaeoecology*, 259, 96–106.

Sommaruga-Wögrath, S., Koinig, K.A., Schmidt, R., Sommaruga, R., Tessadri, R. & Psenner, R. (1997) Temperature effects on the acidity of remote alpine lakes. *Nature*, 387, 64–67.

Sovari, S. & Korhola, A. (1998) Recent diatom assemblage changes in subarctic Lake Saanajärvi, NW Finnish Lapland, and their palaeoenvironmental implications. *Journal of Paleolimnology*, 20, 205–215.

Sovari, S., Korhola, A. & Thompson, R. (2002) Lake diatom response to recent Arctic warming in Finnish Lapland. *Global Change Biology*, 8, 171–181.

Staubwasser, M., Sirocko, F., Grootes, P.M. & Segl, M. (2003) Climate change at the 4.2 ka BP termination of the Indus valley civilisation and Holocene south Asian monsoon variability. *Geophysical Research Letters*, 30, Article number 1425.

Stott, P.A., Tett, S.F.B., Jones, G.S., Ingram, W.J. & Mitchell, J.F.B. (2001) Attribution of twentieth century temperature change to natural and anthropogenic causes. *Climate Dynamics*, 17, 1–21.

Straile, D. (2000) Meteorological forcing of plankton dynamics in a large and deep continental European lake. *Oecologia*, 122, 44–50.

Street-Perrott, F.A. & Perrott, R.A. (1991) Abrupt climate fluctuations in the tropics: the influence of Atlantic Ocean circulation. *Nature*, 343, 607–611.

Street-Perrott, F.A., Holmes, J.A., Waller, M.P., *et al.* (2000) Drought and dust deposition in the West African Sahel: A 5500-year record from Kajemarum Oasis, northeastern Nigeria. *The Holocene*, 10, 293–302.

Stuart, A.J. (1979) Pleistocene occurrences of the European pond tortoise (*Emys orbicularis* L.) in Britain. *Boreas*, **8**, 359–371.

Stuiver, M. & Braziunas, T.F. (1993) Sun, ocean, climate and atmospheric $^{14}CO_2$: An evaluation of causal and spectral relationship. *The Holocene*, **3**, 289–305.

Stuiver, M., Reimer, P.J., Bard, E., *et al.* (1998) Intcal98 radiocarbon age calibration, 24,000–0 cal. BP. *Radiocarbon*, **40**, 1041–1083.

Verschuren, D. & Charman, D.J. (2008) Latitudinal linkages in late Holocene moisture-balance variation. In: *Natural Climate Variability and Global Warming: A Holocene Perspective* (eds R.W. Battarbee & H.A. Binney), pp. 189–231. Wiley-Blackwell, Chichester.

Verschuren, D., Laird, K.R. & Cumming, B. (2000) Rainfall and drought in equatorial East Africa during the past 1100 years. *Nature*, **403**, 410–414.

Weiss, H. (2000) Beyond the Younger Dryas: Collapse as adaptation to abrupt climate change in ancient West Asia and the Eastern Mediterranean. In: *Confronting Natural Disaster: Engaging the Past to Understand the Future* (eds G. Bawden & R. Reycraft), pp. 75–98. University of New Mexico Press, Albuquerque, NM.

Willemse, N.W. & Törnqvist, T.E. (1999) Holocene century-scale temperature variability from West Greenland lake records. *Geology*, **27**, 580–584.

Wolfe, A.P., Baron, J.S. & Cornett, R.J. (2001) Anthropogenic nitrogen deposition induces rapid ecological change in alpine lakes of the Colorado Front Range (USA). *Journal of Paleolimnology*, **25**, 1–7.

Wu, W.X. & Liu, T.S. (2004) Possible role of the "Holocene Event 3" on the collapse of Neolithic Cultures around the Central Plain of China. *Quaternary International*, **117**, 153–161.

Zolitschka, B. (2003) Dating based on freshwater- and marine-laminated sediments. In: *Global Change in the Holocene* (eds A. Mackay, R.W. Battarbee, H.J.B. Birks, & F. Oldfield), pp. 92–106. Hodder Arnold, London.

3

Direct Impacts of Climate Change on Freshwater Ecosystems

Ulrike Nickus, Kevin Bishop, Martin Erlandsson, Chris D. Evans, Martin Forsius, Hjalmar Laudon, David M. Livingstone, Don Monteith and Hansjörg Thies

Introduction

Before the 1990s, most environmental scientists considered climate – in spite of its variability – to exert a relatively constant influence on freshwater ecosystems. In recent years, however, it has become very clear that climate change exerts additional stress to surface waters and that it interacts with other drivers such as hydromorphological change (Chapter 4), eutrophication (Chapter 6), acidification (Chapter 7) and toxic substance contamination (Chapter 8). The main impacts of climate change on freshwater ecosystems result from changes in air temperature, precipitation and wind regimes. Freshwater systems respond by changes in their physical characteristics including stratification and mixing regimes of lake water columns, catchment hydrology or changes in ice-cover which, in turn, may induce chemical changes in habitats, e.g. alterations to oxygen concentration, nutrient cycling and, possibly, water colour. Biological responses include changes in the phenology and species distribution of most organism groups. Links between changes in climate and freshwater ecological responses have already been reported and are predicted to continue under a projected future climate. However, the modelling of this behaviour is complicated by likely non-linearities; responses may be punctuated, with abrupt shifts occurring as thresholds are crossed.

This chapter focuses on the direct impacts of climate change that are independent of other human-induced drivers such as land-use change, nutrient enrichment, acid deposition and the input of toxic substances. We first briefly describe how climate has changed during the past few decades, globally and in Europe, and show, using selected Global and Regional Climate Models, how climate is expected to change in future under different emission scenarios. We then outline some of the principal physical and chemical responses of freshwater ecosystems to climate change that have been revealed by recent research.

Climate Change Impacts on Freshwater Ecosystems. First edition. Edited by M. Kernan, R. Battarbee and B. Moss. © 2010 Blackwell Publishing Ltd.

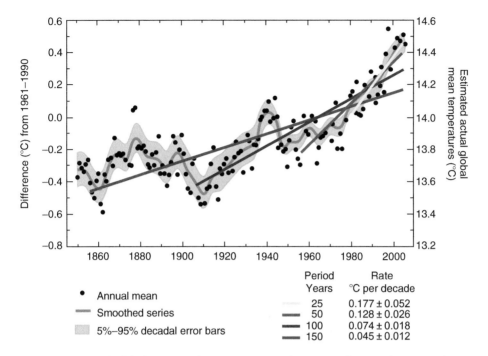

Figure 3.1   Annual global mean surface temperature since 1850 showing linear trends over the past 25 yr (yellow), 50 yr (orange), 100 yr (blue) and 150 yr (red). (From *Climate Change 2007: The Physical Science Basis. Contribution of Working Group I to the Fourth Assessment Report of the Intergovernmental Panel on Climate Change* (eds S. Solomon, D. Qin, M. Manning, *et al.*). Cambridge University Press, Cambridge and New York.)

Climate change

According to the IPCC Fourth Assessment Report (2007), global surface air temperatures of the last few years (1995–2006, with the exception of 1996) are among the 12 highest on record since 1850. Global average air temperature increased by more than 0.7°C between 1906 and 2005, and decadal warming has almost doubled over the past 50 years with an average value of 0.13°C per decade (Fig. 3.1). Warming has been greater over land than over the oceans. Since 1979, air temperature over land has risen, on average, by 0.27°C per decade (versus 0.13°C per decade over the oceans). The greatest warming has occurred at higher northern latitudes, with mean arctic winter and spring surface air temperatures having increased by approximately twice the global average, with strong decadal fluctuations during the past 100 years. A similar trend has been found for what Auer *et al.* (2007) call the Greater Alpine Region (the area between 4–19°E and 43–49°N), which has warmed twice as much as the global or northern hemispheric mean since the late 19th century at all elevations.

   Precipitation, though spatially and temporally highly variable, has also changed significantly in many parts of the globe during the past 100 years with respect to its amount, intensity, frequency and type. In Europe, average precipitation has

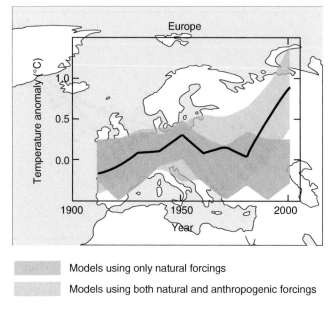

Models using only natural forcings

Models using both natural and anthropogenic forcings

Figure 3.2   Temperature change relative to the 1901–50 mean for the period 1906–2005. The black line indicates observed values, the coloured bands give the modelled data covered by 90% of the recent model simulations. (Modified from *Climate Change 2007: The Physical Science Basis. Contribution of Working Group I to the Fourth Assessment Report of the Intergovernmental Panel on Climate Change* (eds S. Solomon, D. Qin, M. Manning, *et al.*). Cambridge University Press, Cambridge and New York.)

increased in northern parts, while precipitation has decreased in the Mediterranean (IPCC 2007). These tendencies may be associated with changes in the North Atlantic Oscillation (NAO), a north–south dipole in sea-level pressure across the Atlantic (e.g. Hurrell *et al.* 2003), which has its strongest signature in winter. The prevalence of more positive winter NAO values from the 1970s to the 1990s reflects the enhanced westerly air flow across the North Atlantic, moving warm moist air over much of Europe in winter and resulting in wet conditions in Northern Europe and dry conditions in the south. However, topography may generate fine spatial scales in climate, and observed changes in air temperature or precipitation may thus differ from the average picture.

The IPCC Fourth Assessment Report (2007) notes that the increase in air temperature since the middle of the last century is very likely (i.e. >90% probability) to be due to the increase in anthropogenic greenhouse gas concentrations. The observed changes can be accurately simulated only if climate models include greenhouse gas forcing (Fig. 3.2).

What can we expect for a future climate? General Circulation Models (GCMs) give a temperature increase of about 0.2°C per decade over the next two decades for a range of emission scenarios (IPCC 2007). Further warming will be caused by the continued emission of greenhouse gases at or above current rates. The projected changes in the climate system during this century are very likely to be larger than those observed during the 20th century. The best estimate of the global average surface warming, expressed as the temperature change from 1980–99 to 2090–99, is projected to range from 1.8°C for the low emissions scenario, B1, to 4.0°C for the high scenario, A1F1. Even for the case of a constant radiative forcing, if greenhouse gases and aerosols were kept constant at year 2000 levels, models give a temperature increase of a further 0.6°C by the end of the century.

For Europe, warming is projected to be greater than the global mean. Regional climate model simulations run under the PRUDENCE project (http://prudence. dmi.dk) indicate that warming is likely (i.e. >66% probability) to be greatest in winter in Northern Europe and in summer in the Mediterranean area. Similarly, the increase in minimum winter temperatures is likely to be higher than average in Northern Europe, while maximum summer temperatures are likely to rise more than average in Southern and Central Europe (Räisänen *et al.* 2004; Christensen *et al.* 2007a, b) (Fig. 3.3a).

Geographical patterns of changes in annual precipitation are projected to be similar to those already observed over the last few decades, with a high probability of an increase across most of Northern Europe and a decrease in the Mediterranean. In Central Europe, precipitation is likely to increase in winter and decrease in summer (Fig. 3.3b). However, changes in precipitation may vary considerably at local scales, particularly in areas of complex topography, such as the Alps, where there are strong orographic effects, and there is still considerable uncertainty in the projection of future precipitation.

Climate change is often perceived through the occurrence of extreme events, although extreme weather events occur, and will occur, in most regions even with an unchanged climate. Extremes are infrequent events at the low and high ends of a probability distribution of any variable. Assuming the probability of occurrence of values is shaped like a Gaussian (or bell shaped) curve, a small shift in the average or the centre of the distribution also moves the extremes accordingly. The higher frequency of hot days in a warmer climate, for instance, may be accompanied by a decreasing number of cold days or frosts (Fig. 3.4a). A changing probability of extremes may be caused not only by a shift in the mean value of a variable, but also by a change in its variance, and most likely by the interaction of changes in both mean and variance (Fig. 3.4b and c). The IPCC Fourth Assessment Report (2007) states that although more models, ensembles and statistical techniques have been used in the simulation and projection of extreme events, some assessments still rely on simple reasoning about how climate warming might change extremes and others rely on the qualitative similarity between observed and simulated changes.

Heat waves, i.e. episodes of several consecutive high temperature days with maxima exceeding the 90th percentile of the daily distribution in the 1961–90 IPCC reference period, are projected to show an increase in frequency, intensity and duration (IPCC 2007). Global and regional climate models project a likely increase in the inter-annual variability of summer mean temperature, in particular, in Central Europe (e.g. Schär *et al.* 2004; Vidale *et al.* 2007). The heat wave in Europe in summer 2003 may be taken as an example of what might become more common in a future warmer climate. For instance, the average summer temperature

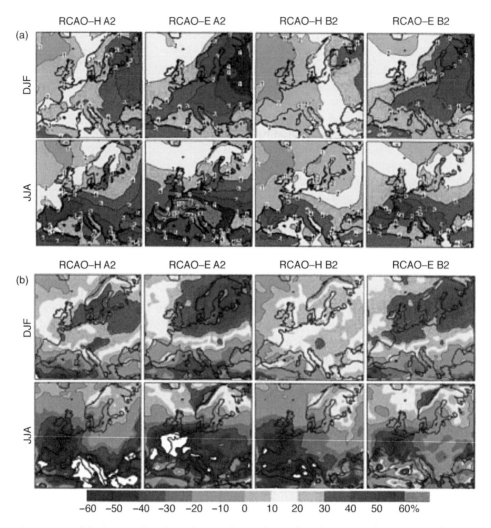

Figure 3.3 (a) Changes in winter (December–February) and summer (June–August) mean surface air temperature in the Rossby Centre coupled regional climate model (RCAO) simulations (scenario minus control period). Contours and shading are at intervals of 1°C. (b) Changes in winter and summer mean precipitation (percent differences from the control period). (RCAO-H indicates RCAO driven by the HadAM3 Global Climate Model (Hadley Centre, the United Kingdom), RCAO-E indicates RCAO driven by the ECHAM/OPYC3 Global Climate Model (Max Planck Institute, Germany); A2 and B2 give the emission scenarios the simulations are based on.) (Modified from Räisänen *et al.* 2003.)

in Switzerland (June to August 2003) exceeded the long-term mean by more than 5 standard deviations (Fig. 3.5). Schär *et al.* (2004) showed that this extremely unlikely event (in statistical terms) is explained well by GCM and Regional Climate Model (RCM) projections for the end of this century, which predict a regime with increases in both mean temperature and temperature variability (Fig. 3.4c).

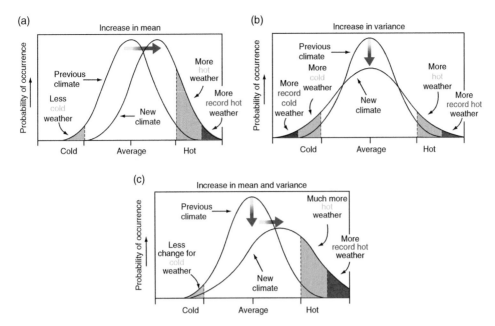

Figure 3.4   Schematic diagram showing the effect on extreme temperatures of (a) an increase in mean temperature, (b) an increase in the variance and (c) an increase in both mean temperature and variance for a normal distribution of temperature. (Modified from *Climate Change 2001: The Scientific Basis. Contribution of Working Group I to the Third Assessment Report of the Intergovernmental Panel on Climate Change* (eds J.T. Houghton, Y. Ding, D.J. Griggs, *et al.*). Cambridge University Press, Cambridge and New York.)

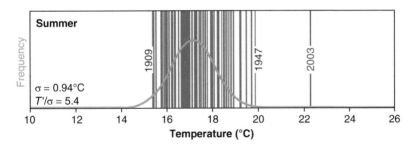

Figure 3.5   Distribution of Swiss summer temperatures (June–August) for 1864–2003. (From *Climate Change 2007: The Physical Science Basis. Contribution of Working Group I to the Fourth Assessment Report of the Intergovernmental Panel on Climate Change* (eds S. Solomon, D. Qin, M. Manning, *et al.*). Cambridge University Press, Cambridge and New York after Schär *et al.* 2004.)

In winter, extreme daily precipitation events are very likely to increase in both magnitude and frequency in the northern parts of Europe, with a larger relative change in frequency than in magnitude (e.g. Räisänen 2005). In summer, too, extreme daily precipitation is projected to increase despite decreasing mean precipitation, but the expected magnitude of the change strongly depends on the climate model employed.

Future climate scenarios and Euro-limpacs

Global circulation models (GCMs), as mathematical representations of the climate system, based on well-established physical principles and on observations of the atmosphere, ocean, cryosphere and land surface, provide credible quantitative estimates of future climate change, particularly at larger scales (e.g. Räisänen 2007; IPCC 2007). A comparison of observed and simulated present-day climate generally shows good agreement for many basic variables and thus provides considerable confidence in the ability of climate models to deliver reliable future climate projections, although individual models can differ in their simulations. Climate projections from model runs depend on which assumptions for future greenhouse gas emissions are used. The standard approach is to use the SRES (Special Report on Emission Scenarios) emission scenarios A1, A2, B1 and B2 based on storylines of 'how the world will develop until the end of this century'. They comprise distinct potential future scenarios of greenhouse gas emissions, population growth and economic development (Nakićenović *et al.* 2000) (Fig. 3.6).

Physical impacts

Long-term data from surface waters already show changes associated with climate warming. Rising air temperatures are reflected in increasing surface temperatures in lakes and streams, in higher thermal lake stability and in a longer ice-free season in lakes, with a later freezing in autumn or winter and an earlier melt in spring or summer. Increasing hypolimnetic temperatures in lakes may lead to a higher risk of deep-water anoxia. Changing wind patterns may alter the input of mixing energy to lakes, and hence affect their overall heat balance and internal heat distribution. Changes in wind and air temperature will be reflected in changes in the physical behaviour of lakes, which may go hand in hand with a modification of the chemical and biological characteristics of surface waters. Changing precipitation patterns, like changes in the total amount, seasonality or intensity, may alter hydrological cycles including river runoff regimes. Wetlands, in particular, may be affected by changes in flooding. A change in the amplitude, frequency, duration or timing of floods may affect biogeochemical processes, plant nutrient dynamics and plant communities.

Regional climate variability is often related to recurrent patterns of atmospheric circulation such as the North Atlantic Oscillation (NAO), the Northern Annular Mode or the El Niño-Southern Oscillation. For Europe, the NAO, as pointed out above, is the most prominent pattern of atmospheric variability. It corresponds to changes in the westerly winds, and the NAO index is a measure of the strength of the meridional sea-level pressure gradient between the Icelandic Low and the Azores High. Potentially, the NAO can have an impact on temperature and precipitation over large areas of Western and Northern Europe, and freshwater ecosystems have been shown to be sensitive to changes in the NAO.

The Emission Scenario of the Special Report on Emission Scenarios (SRES)

A1. The A1 storyline and scenario family describes a future world of very rapid economic growth, global population that peaks in mid-century and declines thereafter and the rapid introduction of new and more efficient technologies. Major underlying themes are convergence among regions, capacity building and increased cultural and social interactions, with a substantial reduction in regional differences in per capita income. The A1 scenario family develops into three groups that describe alternative directions of technological change in the energy system. The three A1 groups are distinguished by their technological emphasis: fossil intensive (A1FI), non-fossil energy sources (A1T), or a balance across all sources (A1B) (where balanced is defined as not relying too heavily on one particular energy source, on the assumption that similar improvement rates apply to all energy supply and end-use technologies).

A2. The A2 storyline and scenario family describes a very heterogeneous world. The underlying theme is self-reliance and preservation of local identities. Fertility patterns across regions converge very slowly, which results in continuously increasing population. Economic development is primarily regionally oriented and per capita economic growth and technological change more fragmented and slower than other storylines.

B1. The B1 storyline and scenario family describes a convergent world with the same global population that peaks in mid-century and declines thereafter, as in the A1 storyline, but with rapid change in economic structures toward a service and information economy, with reductions in material intensity and the introduction of clean and resource-efficient technologies. The emphasis is on global solutions to economic, social and environmental sustainability, including improved equity, but without additional climate initiatives.

B2. The B2 storyline and scenario family describes a world in which the emphasis is on local solutions to economic, social and environmental sustainability. It is a world with continuously increasing global population, at a rate lower than A2, intermediate levels of economic development, and less rapid and more diverse technological change than in the A1 and B1 storylines. While the scenario is also oriented towards environmental protection and social equity, it focuses on local and regional levels.

An illustrative scenario was chosen for each of the six scenario groups A1B, A1FI, A1T, A2, B1 and B2. All should be considered equally sound.

The SRES scenarios do not include additional climate initiatives, which means that no scenarios are included that explicitly assume implementation of the United Nations Framework Convention on Climate Change or the emission targets of the Kyoto protocol.

Figure 3.6 Description of the SRES emission scenarios (*Climate Change 2007: The Physical Science Basis. Contribution of Working Group I to the Fourth Assessment Report of the Intergovernmental Panel on Climate Change* (eds S. Solomon, D. Qin, M. Manning, *et al.*). Cambridge University Press, Cambridge and New York.)

There is now much evidence for these observed trends and potential future changes to freshwaters as a result of climate change. Here, we present a series of examples and case studies based on the analysis of long-term data series, field experiments and physical modelling.

Thermal regimes of lakes and streams

The thermal regime of water bodies is mainly determined by the local weather. The net heat exchange across the air–water interface is given by the sum of energy fluxes related to radiation, latent and sensible heat (Edinger *et al.* 1968; Imboden & Wüest 1995). A shift in climate variables such as air temperature, radiation, cloud cover, wind or humidity will influence these heat fluxes and thus alter the heat balance of lakes and rivers. Model studies predict that lake temperatures, especially in the epilimnion, will increase with increasing air temperature, so that temperature profiles, thermal stability and mixing patterns are expected to change as a result of climate change (e.g. Hondzo & Stefan 1993; Stefan *et al.* 1998).

Analyses of long-term data series demonstrate that such a change has already occurred in recent decades. One of the first studies on increasing water temperatures was on boreal soft water lakes in the Experimental Lakes Area of north-western Ontario (Canada) by Schindler *et al.* (1990). The authors showed that these lakes experienced an increase in water temperature of c. 2°C between 1969 and 1988 and that water renewal rates decreased as a result of higher-than-normal evaporation and lower-than-average precipitation. At Lake Tahoe in the south-west USA, volume-weighted lake temperature increased by about 0.15°C per decade between 1970 and 2002 with a concomitant increase in lake thermal stability (Coats *et al.* 2006). At Lake Baikal, the world's largest lake, with a maximum depth of 1600 m, surface waters have warmed at a rate of 0.2°C per decade over the past 60 years (Hampton *et al.* 2008). Lake Baikal was expected to be rather resistant to climate change due to its enormous volume, but even here, increasing water temperatures and a longer ice-free season are having major implications for nutrient cycling and food-web structure. In Lake Constance, a warm monomictic lake in Central Europe, the mean annual water temperature has increased by 0.17°C per decade since the 1960s (Straile *et al.* 2003). This warming is strongly related to increasing winter air temperatures and has affected the duration and extent of winter lake mixing, the heat content of the lake and the vertical distribution of oxygen and nutrients. Reduced winter cooling favours the persistence of small temperature gradients and may result in an incomplete mixing of the lake.

Fluctuations in lake surface water temperatures are transported downwards by vertical mixing, and can reach the deep waters when the thermal stratification is weak. In particular, the hypolimnetic temperatures of deep lakes, which are determined by winter meteorological conditions and the amount of heat reaching deep-water layers before the onset of thermal stratification may act as a 'climate memory'. Increasing air temperatures may thus lead to a progressive rise in deep-water temperatures, as found, for instance, by Ambrosetti & Barbanti (1999) for lakes in Northern Italy.

Dokulil *et al.* (2006) reported a coherent warming in the hypolimnia of 12 deep lakes across Europe. Annual mean hypolimnetic temperatures increased by about 0.1°C–0.2°C per decade during the past 20–50 years, despite differences between lakes and years. Hypolimnetic temperatures in most lakes tended to reflect fluctuations in the North Atlantic Oscillation (NAO) with the winter-to-spring NAO index explaining 20%–60% of the inter-annual variability in deep-water temperatures. In particular, winter-to-spring periods with a high positive NAO index were associated with high hypolimnetic water temperatures, when warmer-than-average surface water was transported downwards during spring overturn. However, the strength and persistence of the climate signal with time and depth are determined by the lakes' geographical location, landscape topography, mixing conditions and lake morphometry.

The thermal regime of Lake Zurich

Lake Zurich, a 136-m deep peri-alpine lake in Switzerland, has one of the longest temperature data series in Europe with water temperature profiles measured at approximately monthly intervals since the 1940s (Livingstone 1993). This lake

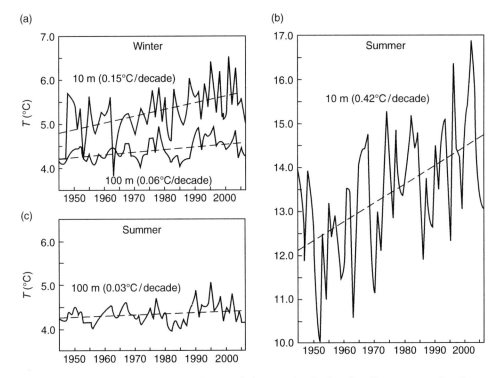

Figure 3.7 Warming trends in Lake Zurich (Switzerland) showing linear regression lines and their gradients: (a) at 10 and 100 m depth in winter (December, January, February), (b) at 10 m depth in summer (June, July, August) and (c) at 100 m depth in summer.

provides a good example of the response of deep temperate lakes to long-term changes in air temperature. The lake has undergone long-term warming at all depths. However, the temperature increase has been more rapid in the surface layers than in the deep-water layers, resulting in increased thermal stability in all seasons and in an extended period of summer stratification by about 2–3 weeks from the 1950s to the 1990s (Livingstone 2003).

In winter (December–February), highly significant increasing trends in water temperature have occurred at all depths (Mann-Kendall test (MK), $p < 0.001$), but the highest long-term winter warming rates (~0.15°C per decade) are found in the uppermost part of the water column (0–10 m) (Fig. 3.7a). From 10 to 80 m, long-term warming rates decrease with increasing depth, but from 80 m to the lake bottom, no further change occurs; warming rates remain approximately constant at about 0.06°C per decade. In summer, the highest rate of long-term warming (0.42°C per decade; MK, $p < 0.01$) has occurred at 10-m depth (Fig. 3.7b), while at the lake surface, the temperature increase since the 1950s has been comparatively low (~0.07°C per decade; MK, $p < 0.01$). Surface water temperatures approximately reflect the behaviour of the regional summer air temperature, and reveal a period of rapid cooling from 1945 to 1970 (-1.0°C per decade) followed by strong warming from 1970 to 2006 (0.5°C per decade).

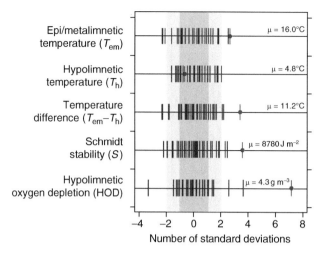

Figure 3.8   Standardized values of water temperature, Schmidt stability (the quantity of work required to mix a lake to a uniform vertical density with no addition or substraction of heat) and hypolimnetic oxygen depletion in Lake Zurich (Switzerland). Data are expressed as standard deviations from the long-term mean (μ) of 1956–2002. The summer 2003 values are shown in red. (Modified from Jankowski *et al.* 2006. Reproduced with permission of the American Society of Limnology and Oceanography.)

In the hypolimnion (i.e. below 20 m), summer temperatures show a low long-term warming trend of 0.03°C–0.06°C per decade, which is, however, not statistically significant even at the $p < 0.1$ level (Fig. 3.7c).

As in most deep, temperate-zone lakes, temperatures in the uppermost 10–20 m of Lake Zurich are influenced directly by weather in all seasons, whilst temperatures in the deeper water respond most strongly to weather in winter and spring. Physically, Lake Zurich appears to be quite sensitive to climate variability since it can behave either as a dimictic, monomictic or oligomictic lake, mixing twice or once per year or not mixing at all, respectively. Over the past few decades, the frequency of dimixis (after ice melt and in autumn) has decreased, while the frequency of years with incomplete mixing has increased as winters have become warmer (Livingstone 1997; Peeters *et al.* 2002).

In Central Europe, the summer of 2003 was exceptionally hot, with air temperatures similar to those predicted for an average summer during the late 21st century (Schär *et al.* 2004). During that summer, Lake Zurich experienced the highest epilimnetic temperatures ever recorded, exceeding the long-term mean (1856–2002) by almost three standard deviations (Fig. 3.8). By contrast, hypolimnetic temperatures were slightly lower than average. The resulting high thermal stability of the water column resulted in hypolimnetic oxygen depletion exceeding its long-term mean by more than seven standard deviations. The potential ecological consequences of this, such as the release of phosphorus from the sediments, possibly ultimately resulting in an increase in the intensity of algal blooms, may thus counteract management and restoration efforts undertaken in the past to mitigate anthropogenic eutrophication.

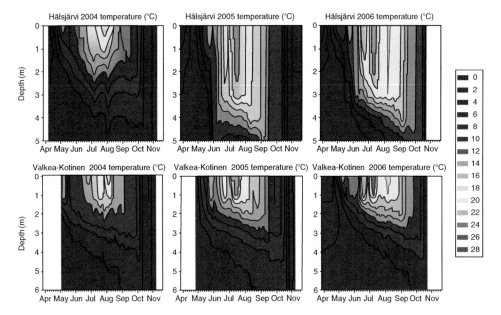

Figure 3.9   Interpolated seasonal development of temperature in the experimental Lake Halsjärvi (top) and the reference Lake Valkea-Kotinen (bottom) in the reference year (2004) and in the two experimental years 2005 and 2006 (M. Forsius unpublished data, 2009).

The whole-lake mixing experiment THERMOS

Increasing average wind speed is expected to raise the input of mixing energy to lakes. This scenario, i.e. increasing geostrophic winds due to a northward shift in cyclone activity and a higher air surface temperature by the end of this century, is projected for the north of Europe.

A whole-lake manipulation experiment in the small humic Lake Hälsjärvi in Southern Finland simulated the increasing input of mixing energy and its impact on the stratification cycle and heat balance of the lake (M. Forsius unpublished data, 2009). A submerged propeller caused the thermocline to deepen by about 2 m during two summer periods in 2005 and 2006 compared with the reference years and the nearby reference lake Valkea-Kotinen (Fig. 3.9).

The observed deepening of the thermocline in Lake Hälsjärvi corresponds to modelled impacts on the lake thermal regime under the A2 SRES emission scenario. Simulations with the MyLake model (Multi-year Lake simulated model; Saloranta & Andersen 2007) simulated an increase in the mean heat content of about 9.5 MJ m^{-3}, which is equivalent to an increase in the average water temperature by 2.3°C in summer and early autumn compared with the reference period 1961–90 (Saloranta *et al.* 2009). The increased mean heat content in the manipulation experiment was about 11 MJ m^{-3} (equivalent to a 2.6°C increase in the water mass temperature) in the same period, indicating that the lake manipulation experiment well represented the mean simulated future increase in heat content in the summer and autumn. A consequence of the lowered thermocline in Lake Hälsjärvi was a change in the oxygen stratification during the summers of 2005 and 2006. While the minimum oxygenated layer was only 1 m deep in 2004, mixing increased its depth to 3.5 m in the experimental years 2005 and 2006. The manipulation experiment also affected nutrient levels in Lake Hälsjärvi, causing a statistically significant decline in total nitrogen and ammonia (M. Forsius unpublished data, 2009). Changes in lake temperature, oxygen distribution and nutrient levels in the lake also affected the phytoplankton (L. Arvola unpublished data, 2009). The biomass of diatoms and non-flagellated green algae increased as well as the rate of change of the phytoplankton community, while the zooplankton and fish communities remained steady.

Another whole-lake manipulation experiment conducted in the deeper, oligotrophic clear-water lake Breisjon in Norway increased the thermocline depth (from 6 to 20 m) and the mean temperature at the time of maximum heat content (from 10.7°C to 17.4°C) and delayed freezing by about 20 days (Lydersen *et al.* 2008). During the experimental manipulation, only minor changes in water chemistry, nutrients and water transparency occurred, but planktonic chlorophytes and diatoms decreased, mixotrophic dinoflagellates increased and periphyton biomass increased. The manipulation did not affect zooplankton biodiversity and did not affect fish populations (i.e. size, condition factor or population density of perch and brown trout).

Water temperature in Swiss rivers and streams

Rivers and streams have also warmed during the past few decades, and stream water temperatures are projected to increase further in a future warmer climate (e.g. Stefan & Sinokrot 1993; Webb 1996). Many studies have used linear regression models of stream temperature versus air temperature to explain the variance in water temperature (cf. Mohensi & Stefan 1999 and references therein). However, in addition to meteorological variables, the energy budget and thermal capacity of rivers and streams may be heavily affected by human activities such as the increased input of heated cooling waters from power plants.

Within Euro-limpacs, water temperatures in rivers and streams across Switzerland, covering an altitudinal range of more than 4000 m, were studied by Hari *et al.* (2006). The study showed parallel warming at all altitudes, reflecting the changes in regional air temperature during the past 40 years (Fig. 3.10). Regional coherence (i.e. spatial correlation between time series within a region) was high for rivers and streams on the Swiss Plateau and in the foothills of the Alps, but decreased with increasing altitude. In catchments containing glaciers or hydroelectric power stations, the warming was strongly reduced, presumably owing to the influence of inflowing melt water from glaciers and deep-water from reservoirs.

The effects of warming on populations of cold-water fish, such as brown trout, are expected to be deleterious at the warmer boundaries of their habitat and positive at the cooler boundaries. On the Swiss Plateau, brown trout populations inhabit the upper limit of their range of temperature tolerance. There, a potential upward migration of fish due to increasing water temperatures is often impeded by natural and artificial physical barriers, and a climatically driven upward habitat shift increases the likelihood of a population decrease. Hari *et al.* (2006)

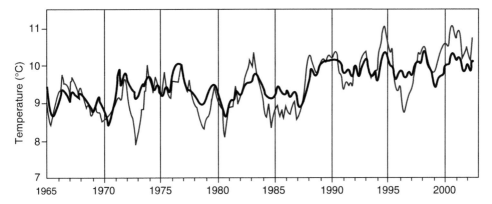

Figure 3.10   Mean water temperature of rivers and streams measured at 25 stations in Switzerland (black line) and mean air temperature at Basle and Zurich (thin blue curve). The data illustrated are annual running means. (Modified from Hari *et al.* 2006.)

found that such a climate-related decrease in brown trout population had already occurred in alpine rivers and streams in Switzerland, and was accelerated by an increasing incidence of temperature-dependent Proliferative Kidney Disease.

Ice-cover

The seasonal ice-cover of lakes and rivers plays an important role in freshwater systems, and changes in the thickness and duration of the ice layer are of ecological importance and have consequences for human activities. According to IPCC (2007), freeze-up is defined conceptually as the time at which a continuous and immobile ice-cover forms, while break-up is generally the time when open water becomes extensive in a lake or when the ice-cover starts to move downstream in a river. Major variables affecting duration and thickness of lake and river ice are air temperature, wind, snow depth, heat content of the water body and rate and temperature of potential inflows. Dates of freeze-up and ice break-up have proved to be good indicators of climate variability at local to regional scales, and as a response to large-scale atmospheric forcing (e.g. Walsh 1995; Livingstone 1999, 2000; Yoo & D'Odorico 2002; Blenckner *et al.* 2004). Thus, as climate changes, and air temperatures, particularly in the winter, tend to increase, these shifts should be reflected in ice-cover. Moreover, several authors have used the correlation of ice-cover dates and air temperature to translate shifts in freeze-up and break-up into estimated changes in air temperature (e.g. Palecki & Barry 1986; Robertson *et al.* 1992; Assel & Robertson 1995; Magnuson *et al.* 2000). A typical value for lakes at mid-latitudes is a 4–5-day shift in mean freeze-up or break-up dates for each degree Celsius change in mean autumn or spring temperatures. Relationships tend to be stronger for freezing dates and in colder climates (Walsh 1995).

Generally, time series of ice phenology data from lakes and rivers in Eurasia and North America provide evidence of later freezing and earlier break-up of the seasonal ice-cover. For instance, long-term ice-cover records of 26 selected lakes and rivers in the northern hemisphere revealed that from 1884 to 1995, the average date of freezing and break-up, both advanced by about 6 days per 100 years (Magnuson *et al.* 2000). This prolongation of the ice-free season translates to a 1.2°C increase in air temperature per 100 years. Inter-annual variability in the ice phenology data from the selected rivers and lakes has increased since the 1950s. At Lake Baikal, the ice-free season has lengthened by about 16 days per 100 years, with a more consistent trend towards later freezing despite variability from decade to decade. The trend towards earlier break-up occurred mainly before 1920 (Livingstone 1999; Magnuson *et al.* 2000; Todd & Mackay 2003) and reflects the close relationship of melting to the April mean air temperature, which has shown no trend since the 1920s, in contrast to other months of the winter. In Canada, most lakes have experienced earlier break-up and a longer ice-free season during the past decades (e.g. Futter 2003; Duguay *et al.* 2006). The trends in lake ice-cover provide further evidence of a spring warming that has been observed over North America since the second half of the 20th century. Ice phenology data from Lake Mendota (Wisconsin, the United States), the lake with the longest uninterrupted ice record in North America, dating back to 1855, showed three climatic periods with relatively stable ice conditions between 1890 and 1979 but decreasing trends in ice-cover duration before and after this period due to increasing winter/early spring air temperatures (Robertson *et al.* 1992).

Long series of lake ice data are available from Fennoscandia. Korhonen (2006) analysed freezing and break-up records of almost 90 lakes in Finland dating back to the early 19th century and ice thickness records of about 30 lakes dating back to the 1910s. There were significant changes with earlier ice break-up, except for the very north of Finland, and later freezing, resulting in shorter ice duration. Palecki & Barry (1986) conducted a statistical study of the relationship between ice phenology data and air temperature for Finnish lakes, and found that the same change in freeze-up date indicated a larger shift in autumn air temperature in the north of Finland than in Southern Finland. Ice phenology dates revealed a strong dependence on latitude, and the maritime influence of the Baltic Sea caused a northward deflection of average freezing and melting date isolines near the coast. A similar dependence of ice phenology on latitude was described by Blenckner *et al.* (2004) for 50 lakes in Fennoscandia. The prevalence of zonal or meridional winds in autumn and spring, expressed by regional circulation indices, was used to explain the temporal and regional variability of freeze-up and break-up dates.

An analysis of 54 Swedish lakes (Weyhenmeyer *et al.* 2005) confirmed the existence of trends in the timing of melting during the IPCC reference period (1961–90), but also showed that these trends were dependent on latitude. Trends towards earlier melting were substantially greater in warmer, southern Sweden than in the colder, northern regions. The non-linear dependence of break-up dates on latitude can be described in terms of an arc cosine function of lake-specific annual mean air temperature for 196 lakes across Sweden (1961–2002) (Weyhenmeyer *et al.* 2004). Thus, an increase in air temperature is expected to have greater impacts on the melting date in warmer southern Sweden compared with the colder north with mean annual air temperatures of −2°C to 2°C. The authors showed that the average temperature increase of 0.8°C from 1991 to 2002 across Sweden (compared with the reference period 1961–90) caused ice break-up to occur about 17 days earlier in the south of Sweden, i.e. south of 60°N, but only 4 days earlier in the northern part of the country.

An altitudinal gradient of ice phenology was studied in the Tatra Mountains of Poland and Slovakia (Šporka *et al.* 2006). Mini-thermistors with data loggers measured surface water temperatures in 19 morphologically different lakes, which covered an elevation range of almost 600 m (1580–2157 m a.s.l.). For practical reasons, freezing date was defined as the calendar date on which the measured lake surface water temperature decreased to 0°C, and the date on which it increased above 0°C again, represented melting date. Freeze-up dates spanned a period of 52 days, but exhibited no detectable dependence on altitude. Break-up dates, occurring between beginning of May and end of June, however, depended strongly on altitude. The average gradient in the timing of melting was about 9 days per 100 m and explained more than 60% of the variability. Ice-cover duration varied from 136 to 232 days and exhibited a significant linear dependence on altitude, at a rate of about 10 days per 100 m. In contrast to ice break-up, which depends strongly on altitude (as a proxy for air temperature), freezing appears to be governed not only by air temperature, but to a considerable extent by local factors like lake morphometry, exposure to radiation and wind, or inflows, which influence lake surface water temperature in autumn and early winter.

Chemical impacts

Climate not only has an impact on the physical characteristics of surface waters, but also is a master variable for ecologically important chemical processes. Here, we discuss two examples of how climate may directly or indirectly affect surface water chemistry, first with respect to changes in concentrations of dissolved organic carbon (DOC) and secondly related to increases in the release of major ions and heavy metals from active rock glaciers in high mountain regions.

Dissolved organic carbon in surface waters

DOC is an important constituent of many natural waters. It is generated by the partial decomposition of organic matter and may be stored in soils for varying lengths of time before transport to surface waters. The humic substances generated by organic matter decomposition impart a characteristic brown colour to the water due to the absorption of visible light by these compounds. DOC thus influences light penetration into surface waters, as well as their acidity, nutrient availability, metal transport and toxicity. During the past two decades, rising DOC concentrations have been observed across much of the British Isles (Fig. 3.11), large areas of Fennoscandia, parts of Central Europe, and northeastern North America (e.g. Freeman *et al.* 2001; Evans *et al.* 2005, 2006; Vuorenmaa *et al.* 2006; Monteith *et al.* 2007). When first observed, these increases were widely interpreted as evidence of climate-change impacts on terrestrial carbon stores due to rising temperatures and the increasing frequency and severity of summer droughts (e.g. Freeman *et al.* 2001; Hejzlar *et al.* 2003; Worrall *et al.* 2004). Increasing precipitation could also lead to increasing DOC concentrations, first

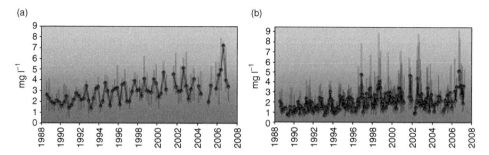

Figure 3.11 Median DOC concentration (mg l^{-1}) of surface waters in the UK Acid Water Monitoring Network, 1988–2007. (a) Lakes. (b) Streams. Bars show the 25th and 75th percentile concentrations at each sampling interval.

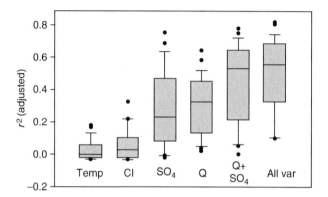

Figure 3.12 The variance in OM explained by the different potential drivers. Box-plot (10th, 25th, 50th, 75th, and 90th percentiles marked) of the r^2-values for COD, modelled with linear regression for 28 rivers, using temperature, [Cl$^-$], [SO$_4$$^{2-}$] and flow as explanatory variables. Combinations of these parameters were modelled using multiple linear regression. (Modified from Erlandsson *et al.* 2008.)

by increasing the proportion of DOC-rich water derived from the upper organic horizons of mineral soils and secondly by reducing water residence time, and hence DOC removal, in lakes (Hongve *et al.* 2004). Rising levels of atmospheric CO_2, influencing plant growth and litter quality were also proposed to explain increased rates of DOC production (Freeman *et al.* 2004).

Erlandsson *et al.* (2008) showed that for the 35-year period (1970–2004), most of the inter-annual variability of concentrations of dissolved organic matter (OM) in 28 large Scandinavian river catchments were explained by discharge and concentrations of sulphate, with discharge being the more important driver. Despite the heterogeneity of the catchments with regards to climate, size and land use, there was a high degree of synchroneity in chemical oxygen demand (COD), a common proxy for OM, across the entire region. Multiple regression models with discharge and sulphate concentration explained up to 78% of the annual variability in COD, while two other candidate drivers, air temperature and chloride concentration in river water, added little explanatory value (Fig. 3.12).

During the period 1990–2004, the observed increases in DOC concentrations in Europe and eastern North America have clearly been driven mainly by reductions in atmospheric sulphur deposition, resulting from international legislation to regulate pollutant emissions since the 1980s (Evans *et al.* 2006; Vuorenmaa *et al.* 2006; Monteith *et al.* 2007). As sulphur deposition has fallen, reductions in acidity and ionic strength have allowed more DOC to remain in solution in soil water, and therefore to be leached to surface waters.

European sulphur emissions have fallen dramatically (by around three quarters) since their 1980 peak (Vestreng *et al.* 2007), and while some further reductions are likely, they clearly will not be on the scale of those observed in the past. On the other hand, climatic changes are expected to continue, or even accelerate, into the future. Some field studies have suggested that a large proportion of DOC is derived from recently fixed carbon and any response by net primary production to changes in temperature or soil moisture may therefore be expected to influence the size of the organic matter pool available for decomposition and, hence, DOC production. Further, laboratory studies have confirmed that DOC production *per se* increases with rising temperature as a result of microbial activity, particularly in peats subject to drying (e.g. Evans *et al.* 2005)

Experimental data suggest that rates of soil DOC production are increased under higher temperatures and in response to a shift from anaerobic to aerobic conditions in saturated soils (Clark *et al.* 2009). A study by Tipping *et al.* (1999) also indicated that climatic warming will increase the production of potentially soluble organic matter. On the other hand, more severe drying during droughts may have the opposite effect on DOC leaching; field manipulation experiments on podsolic heathland soils in Wales (see below) showed decreasing microbial activity and DOC concentrations in response to experimental drought, with increased DOC observed following soil re-wetting (Toberman *et al.* 2008). Overall, Evans *et al.* (2006) concluded for UK surface waters that warming (of around 0.6°C) since the 1980s could account for a small proportion of the observed increase in DOC, but that reductions in acid deposition were the dominant driver of change to date. Statistical analysis of long-term data records from Storgama, Birkenes and Langtjern (Norway) indicate that climate variables explain a significant part of the seasonal variation in DOC concentrations (de Wit *et al.* 2008), while the long-term increase in TOC is related to reduced acid deposition. Most of the seasonal variation is apparently related to temperature and precipitation.

In coastal areas, climate may also impact DOC through decadal-scale variations in the deposition of sea-salt ions (Evans *et al.* 2001). High levels of sea-salt deposition, which are linked to high wind speeds and hence to positive phases of the North Atlantic Oscillation, may affect DOC solubility through a mechanism analogous to the effect of sulphur deposition, by causing transient periods of acidification and increased ionic strength, suppressing DOC release. This mechanism also appears to have contributed to observed DOC increases in some areas (Monteith *et al.* 2007).

In summary, these findings suggest that, as sulphate deposition returns towards background (i.e. pre-industrial) levels, climatic factors such as discharge and temperature may become the major drivers of variability in dissolved organic matter.

(a)   (b)

Figure 3.13 Experimental design – (a) deep frost plot and (b) shallow frost plot. (Photographs by Peder Blomkvist.)

The effect of soil frost and snow cover on DOC in surface waters – a manipulation experiment in northern Sweden

Snow cover in northern regions provides a major fraction of the annual water budget, and it plays a fundamental role in regulating the winter biogeochemistry of soils in northern forests (Groffman *et al.* 2001). Snow cover limits or even prevents the development of soil frost. Changes in the timing, extent and duration of the snow cover, as projected under a future warmer climate, may result in an increased number of freeze–thaw events, or longer snow-free periods during winter (Stieglitz *et al.* 2003; Mellander *et al.* 2007). As DOC in some streams is strongly controlled by soil solution chemistry in the riparian zone, changes here could alter both quantity and bioavailability of DOC exported to the adjacent streams during the spring flood – up to half of the annual runoff and DOC flux may occur during snow melt in small streams and rivers in northern Sweden (Ågren *et al.* 2007). Moreover, many boreal surface waters experience a pH decline of one to two pH units during snow melt, driven primarily by a transient increase in DOC export from the terrestrial systems (Buffam *et al.* 2007).

The multi-year field manipulation experiment at the Svartberget Research Station in northern Sweden, initiated in 2002, was designed to study the effect of changed winter temperature on DOC in riparian soils. Soil frost was manipulated by insulation to prevent below-ground freezing (shallow frost plot) and by delaying snow accumulation by 3 months to increase soil frost (deep frost plot) and was compared with natural conditions (control plot) (Fig. 3.13).

Delaying winter snow cover by 3 months also delayed soil warming well into July. As a consequence, DOC concentration in the previously frozen soil layers was clearly increased, compared with the unfrozen reference plots, with a soil-frost-induced doubling of DOC in the uppermost soil horizons during late spring and early summer (Fig. 3.14). Soil frost also seemed to have affected DOC characteristics, measured as changes in the overall shape of the absorption spectrum, which were interpreted as suggesting that the bioavailability of organic matter increased after freezing (Berggren *et al.* 2007).

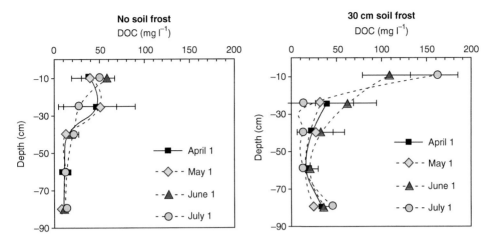

Figure 3.14 Concentration profiles of DOC in the soil during winter, spring and early summer (April–July) at five soil depths in plots without and with deep soil frost.

The quality and quantity of DOC in the soil frost experiment during the winter seemed to be controlled by lysis of cell structures and by limiting the soil microbial activity. While freeze-out processes are believed mainly to control the concentration of DOC in soil solution below the expanding ice during freezing, lysis of cells may release highly bioavailable organic compounds of low molecular weight and low C/N ratio (Stepanauskas *et al.* 2000). Temperatures below freezing are unfavourable for heterotrophic microbial activity. This may result in undecomposed organic material of high substrate quality that subsequently will be decomposed during unfrozen conditions. As a consequence, the soil frost experiment has shown that the thermal conditions in the soil ecosystem influence the soil organic matter decomposition rate and CO_2 production (Öquist & Laudon 2008).

The effect of summer droughts on soil solution DOC – the Clocaenog experiment (the United Kingdom)

In many areas of Europe, climate change is projected to lead to increased frequency and severity of summer droughts. At Clocaenog, a heathland site in North Wales, the United Kingdom, repeated summer droughts have been experimentally induced each year since 1999, initially as part of the CLIMOOR (Climate Driven Changes in the Functioning of Heath and Moorland Ecosystems) and VULCAN (Vulnerability assessment of shrubland ecosystems in Europe under climatic changes) projects. A retractable transparent roof system was used to reduce summer rainfall by around 60%, for a set of replicated 20 m² plots (Beier *et al.* 2004).

Droughts were found to reduce rates of soil respiration, as biological activity became moisture limited. Measurements of soil solution DOC (Fig. 3.15) suggest that DOC production is similarly affected; in the control plots, DOC concentrations consistently increase in summer, but in the drought plots concentrations can fall

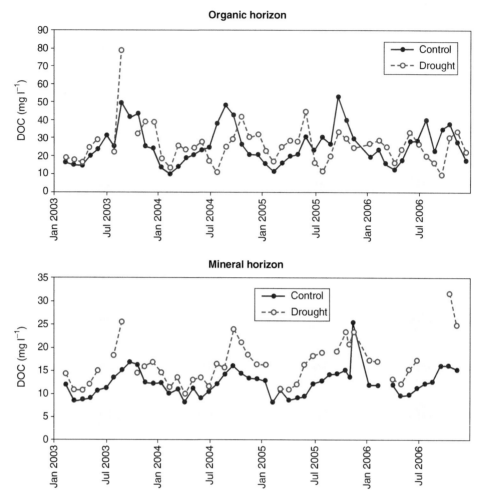

Figure 3.15 DOC concentrations over 4 years of experimentally induced summer drought on a heathland at Clocaenog, North Wales, the United Kingdom. Shaded areas indicate experimental drought periods.

to much lower levels. In some years (e.g. in 2004), it appears that this drought-induced reduction was followed by increased concentrations on re-wetting, although this was not observed in all years. Similar patterns of reduced DOC concentrations during droughts, and raised concentrations on re-wetting, have been noted in several previous studies of peatland soil solutions and surface waters (e.g. Watts *et al.* 2001; Clark *et al.* 2005). Other work undertaken at the Clocaenog site suggests that reduced DOC losses in summer are linked to decreased activity of the phenol oxidase enzyme, which has an important role in litter decomposition and DOC production (Toberman *et al.* 2008). In the mineral soil solutions at Clocaenog, on the other hand, changes to the seasonal cycle are less evident; instead, there is some indication that repeated droughts are leading to progressive, year-round DOC increases (Fig. 3.15).

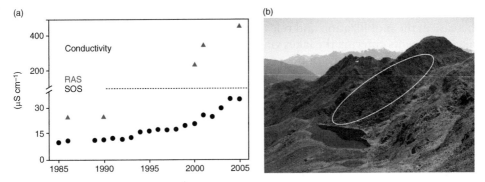

Figure 3.16 (a) Conductivity in two alpine lakes in the period 1985–2005, Rasass See (RAS: triangles) and Schwarzsee ob Sölden (SOS: circles). Horizontal dotted line indicates a break in the vertical *y*-axis scale. (Modified from Thies *et al.* 2007, Copyright 2007 American Chemical Society.) (b) Rasass See and major parts of its catchment. The ellipsis indicates the position of the active rock glacier. (Photograph by V. Mair.)

One possible explanation for this could be that sustained reductions in soil moisture as a result of treatment may have altered the soil structure (Sowerby *et al.* 2008), reducing the capacity of the mineral soil to retain DOC leached from the organic horizon. Overall, results of the Clocaenog experiment demonstrate that summer drought has the potential to significantly alter the timing, and/or the amount, of DOC leaching to surface waters, with potentially important consequences for freshwater ecological status.

Solutes in high mountain lakes

Remote high mountain lakes are excellent sensors of environmental and climate change for entire mountain regions. Over the past two decades, a substantial rise in solute concentration at two remote high mountain lakes in catchments of metamorphic bedrock (gneiss, micaschists) in the European Alps has been observed (Thies *et al.* 2007). At Rasass See (2682 m, Italy), a high altitude lake south of the main alpine divide, electrical conductivity has increased by a factor of 18 during the last two decades (Fig. 3.16) and the concentrations of the most abundant ions, magnesium, sulphate and calcium, by 68-fold, 26- and 18-fold, respectively.

At Schwarzsee ob Sölden (2796 m, Austria), a high mountain lake north of the main alpine divide, the solute increase was less pronounced. Electrical conductivity has increased by a factor of 3 during the same period (Fig. 3.16) and the concentrations of magnesium, calcium, and sulphate have increased six-fold compared with values in 1985.

These high solute values cannot be explained by weathering of the metamorphic bedrock as has been postulated earlier for corresponding high mountain lake waters in the Alps (Sommaruga-Wögrath *et al.* 1997). Neither do current levels of atmospheric deposition nor any recent direct anthropogenic impact account for the solute increase. This is particularly relevant for the nickel concentrations of 243 µg l^{-1} at Rasass See, which are more than 20 times above the drinking water limit. The high concentrations can only be explained by an increase in the mobilization and release of solutes from active rock glaciers in the lake catchments entering the lakes via melt water, related to the observed increase in average air temperature in the region over recent decades (Auer *et al.* 2007). The findings are supported by studies on active rock glaciers and meltwater streams in the Austrian Alps by Krainer & Mostler (2002, 2006). Increasing conductivity values of a similar magnitude have also been recorded in a high elevation stream draining from a rock glacier in the US Rocky Mountains (Williams *et al.* 2006).

An important question is why have Rasass See and Schwarzsee reacted so differently in respect to the solute increase, although their catchments are situated only 45 km apart and are characterized by the same geology. The probable explanation is that the catchments differ in the size of the active rock glaciers in relation to the size and volume of the lakes. At Rasass See, rock glaciers occupy c. 20% of the catchment equivalent to c. 200% of the lake surface, whereas at Schwarzee ob Sölden, rock glaciers occupy only c. 5% of the catchment, which is equivalent to c. 30% of the lake surface, and the volume of Rasass See is four to five times smaller than Schwarzee. In addition, Rasass See is situated at an altitude 100 m lower than Schwarzee. Although these factors probably explain the differences between the lakes, the specific sources and pathways of solutes and heavy metals released from the melting rock glaciers into adjacent surface waters are still unknown.

Conclusions

Changing climate is already having an impact on the physical, chemical and biological characteristics of freshwater ecosystems, both directly through changes in air temperature and precipitation and indirectly through interaction with other stressors. In future, non-climatic impacts should be reduced if pollutant loadings decrease and surface-water ecosystems are progressively restored. But global warming is very likely to continue, even if greenhouse gases and aerosols are kept constant at year 2000 levels, giving rise to a minimum projected average further increase in air temperature by 0.6°C by the end of this century (IPCC 2007). Changes in the characteristics of freshwater ecosystems as illustrated here are likely to continue and will become much more pronounced as greenhouse gas emissions rise and ecosystems cross critical thresholds that cause abrupt non-linear system shifts to occur.

References

Ågren, A., Buffam, I., Jansson, M. & Laudon, H. (2007) Importance of seasonality and small streams for the landscape regulation of DOC export. *Journal of Geophysical Research-Biogeosciences*, **112**, G03003, doi:10.1029/2006 JG000381.

Ambrosetti, W. & Barbanti, L. (1999) Deep water warming in lakes: An indicator of climatic change. *Journal of Limnology*, **58**, 1–9.

Assel, R. & Robertson, D.M. (1995) Changes in winter air temperature near Lake Michigan, 1851–1993, as determined from regional lake-ice records. *Limnology and Oceanography*, **40**, 165–176.

Auer, I., Böhm, R., Jurkovic, A., *et al*. (2007) HISTALP-historical instrumental climatological surface time series of the Greater Alpine Region. *International Journal of Climatology*, **27**, 17–46.

Beier, C., Emmett, B., Gundersen, P., *et al*. (2004) Novel approaches to study climate change effects of terrestrial ecosystems in the field – Drought and passive night time warming. *Ecosystems*, **7**, 583–597.

Berggren, M., Laudon, H. & Jansson, M. (2007) Landscape regulation of bacterial growth efficiency in boreal freshwaters. *Global Biogeochemical Cycles*, **21**, GB4002, doi:4010.1029/2006GB002844.

Blenckner, T., Järvinen, M. & Weyhenmeyer, G.A. (2004) Atmospheric circulation and its impact on ice phenology in Scandinavia. *Boreal Environment Research*, **9**, 371–380.

Buffam, I., Laudon, H., Temnerud, J., Mörth, C.M. & Bishop, K. (2007) Landscape-scale variability of acidity and dissolved organic carbon during spring flood in a boreal stream network. *Journal of Geophysical Research-Biogeosciences*, **112**, G01022, doi:10.1029/2006JG000218.

Christensen, J.H., Hewitson, B., Busuioc, A., *et al*. (2007a) Regional climate projections. In: *Climate Change 2007: The Physical Science Basis. Contribution of Working Group I to the Fourth Assessment Report of the Intergovernmental Panel on Climate Change* (eds S. Solomon, D. Qin, M. Manning, *et al*.). Cambridge University Press, Cambridge and New York.

Christensen, J.H., Carter, T., Rummukainen, M. & Amanatidis, G. (2007b) Evaluating the performance and utility of regional climate models: The PRUDENCE project. *Climatic Change*, **81**, 1–6.

Clark, J.M., Chapman, P.J., Adamson, J.K. & Lane, S.N. (2005) Influence of drought-induced acidification on the mobility of dissolved organic carbon in peat soils. *Global Change Biology*, **11**, 791–809.

Clark, J.M., Ashley, D., Wagner, M., *et al*. (2009) Increased temperature sensitivity of net DOC production from ombrotrophic peat due to water table draw-down. *Global Change Biology*, **15**, 794–807.

Coats, R., Perez-Losada, J., Schladow, G., Richards, R. & Goldman, C. (2006) The warming of Lake Tahoe. *Climate Change*, **76**, 121–148.

Dokulil, M., Jagsch, A., George, G.D., *et al*. (2006) Twenty years of spatial coherent deepwater warming in lakes across Europe related to the North Atlantic Oscillation. *Limnology and Oceanography*, **51**, 2787–2793.

Duguay, C.R., Prowse T.D., Bonsal, B.R., Brown, R.D., Lacroix, M.P. & Ménard, P. (2006) Recent trends in Canadian ice cover. *Hydrological Processes*, **20**, 781–801.

Edinger, J.E., Duttweiler, D.W. & Geyer, J.C. (1968) The response of water temperatures to meteorological conditions. *Water Resources Research*, **4**, 1137–1143.

Erlandsson, M., Buffam, I., Fölster, J., *et al*. (2008) Thirty-five years of synchrony in the organic matter concentration of Swedish rivers explained by variation in flow and sulphate. *Global Change Biology*, **14**, 1191–1198.

Evans, C.D., Monteith, D.T. & Harriman, R. (2001) Long-term variability in the deposition of marine ions at west coast sites in the UK Acid Waters Monitoring Network: Impacts on surface water chemistry and significance for trend determination. *The Science of the Total Environment*, **265**, 115–129.

Evans, C.D., Monteith, D.T. & Cooper, D.M. (2005) Long-term increases in surface water dissolved organic carbon: Observations, possible causes and environmental impacts. *Environmental Pollution*, **137**, 55–71.

Evans, C.D., Chapman, P.J., Clark J.M., Monteith, D.T., & Cresser, M.S. (2006) Alternative explanations for rising dissolved organic carbon export from organic soils. *Global Change Biology*, **12**, 2044–2053.

Freeman, C., Evans, C.D. & Monteith, D.T. (2001) Export of organic carbon from peat soils. *Nature*, **412**, 785.

Freeman, C., Fenner, N., Ostle, N.J., *et al*. (2004) Export of dissolved organic carbon from peatlands under elevated carbon dioxide levels. *Nature*, **430**, 195–198.

Futter, M. (2003) Patterns and trends in southern Ontario lake ice phenology. *Environmental Monitoring and Assessment*, **88**, 431–444.

Groffman, P.M., Driscoll, C.T., Fahey, T.J., Hardy, J.P., Fitzhugh, R.D., & Tierney, G.L. (2001) Colder soils in a warmer world: A snow manipulation study in a northern hardwood forest ecosystem. *Biogeochemistry*, **56**, 135–150.

Hampton, S.E., Izmest'Eva, L.R., Moore, M.V., Katz, S.L., Dennis, B. & Silow, E.A. (2008) Sixty years of environmental change in the world's largest freshwater lake – Lake Baikal, Siberia. *Global Change Biology*, **14**, 1947–1958.

Hari, R.E., Livingstone, D.M., Siber, R., Burkhardt-Holm, P. & Güttinger H. (2006) Consequences of climatic change for water temperature and brown trout populations in Alpine rivers and streams. *Global Change Biology*, **12**, 10–26.

Hejzlar, J., Dubrovsky, M., Buchtele, J. & Ruzicka, M. (2003) The apparent and potential effects of climate change on the inferred concentration of dissolved organic matter in a temperate stream (the Malse River, South Bohemia). *The Science of the Total Environment*, **310**, 142–152.

Hondzo, M. & Stefan, H. (1993) Regional water temperature characteristics of lakes subjected to climate change. *Climatic Change*, **24**, 187–211.

Hongve, D., Riise, G. & Kristiansen, J.F. (2004) Increased colour and organic acid concentrations in Norwegian forest lakes and drinking water – A result of increased precipitation? *Aquatic Science*, **66**, 231–238.

Hurrell, J.W., Kushnir, Y., Ottersen, G. & Visbeck, M. (2003) An overview of the North Atlantic Oscillation. In: *The North Atlantic Oscillation; Climate Significance and Environmental Impacts*, Vol. 134 (Geophysical Monographs Series) (eds J.W. Hurrel, Y. Kushnir, G. Ottersen & M. Visbeck), pp. 1–35. American Geophysical Union, Washington, DC.

Imboden, D.M. & Wüest, A. (1995) Mixing mechanisms in lakes. In: *Physics and Chemistry of Lakes* (eds A. Lerman, D.M. Imboden & J.R. Gat), pp. 83–138. Springer Verlag, Dordrecht.

IPCC (Intergovernmental Panel on Climate Change) (2001) *Climate Change 2001: The Scientific Basis. Contribution of Working Group I to the Third Assessment Report of the Intergovernmental Panel on Climate Change* (eds J.T. Houghton, Y. Ding, D.J. Griggs, *et al.*). Cambridge University Press, Cambridge and New York.

IPCC (Intergovernmental Panel on Climate Change) (2007) *Climate Change 2007: The Physical Science Basis. Contribution of Working Group I to the Fourth Assessment Report of the Intergovernmental Panel on Climate Change* (eds S. Solomon, D. Qin, M. Manning, *et al.*). Cambridge University Press, Cambridge and New York.

Jankowski, T., Livingstone, D.M., Forster, R., Bührer, H. & Niederhauser, P. (2006) Consequences of the 2003 European heat wave for lakes: Implications for a warmer world. *Limnology and Oceanography*, **51**, 815–819.

Korhonen, J. (2006) Long-term changes in lake ice cover in Finland. *Nordic Hydrology*, **37**, 347–363.

Krainer, K. & Mostler, W. (2002) Hydrology of active rock glaciers: Examples from the Austrian Alps. *Arctic, Antarctic and Alpine Research*, **34** (2), 142–149.

Krainer, K. & Mostler, W. (2006) Flow velocities of active rock glaciers in the Austrian Alps. *Geografiska Annaler*, **88A** (4), 267–280.

Livingstone, D.M. (1993) Temporal structure in the deep-water temperature of four Swiss lakes: A short-term climatic change indicator? *Verhandlungen Internationale Vereinigung für theoretische und angewandte Limnologie*, **25**, 75–81.

Livingstone, D.M. (1997) An example of the simultaneous occurrence of climate-driven "sawtooth" deep-water warming/cooling episodes in several Swiss lakes. *Verhandlungen Internationale Vereinigung für theoretische und angewandte Limnologie*, **26**, 822–826.

Livingstone, D.M. (1999) Break-up dates of alpine lakes as proxy data for local and regional mean surface air temperatures. *Climatic Change*, **37**, 407–439.

Livingstone, D.M. (2000) Large-scale climatic forcing detected in historical observations of lake ice break-up. *Verhandlungen Internationale Vereinigung für theoretische und angewandte Limnologie*, **27**, 2775–2783.

Livingstone, D.M. (2003) Impact of secular climate change on the thermal structure of a large temperate central European lake. *Climatic Change*, **57**, 205–225.

Lydersen, E., Aanes, K.J., Andersen, S., *et al.* (2008) Ecosystem effects of thermal manipulation of a whole lake, Lake Breisjøen, southern Norway (THERMOS project). *Hydrology and Earth System Sciences*, **12**, 509–522.

Magnuson, J.J., Robertson, D.M., Benson, B.J., *et al.* (2000) Historical trends in lake and river ice cover in the Northern Hemisphere. *Science*, **289**, 1743–1746.

Mellander, P.E., Ottosson, M. & Laudon, H. (2007) Climate change impact on snow and soil temperature in boreal Scots pine stands. *Climatic Change*, **85**, 179–193.

Mohensi, O. & Stefan, H.G. (1999) Stream temperature/air temperature relationship: A physical interpretation. *Journal of Hydrology*, **218**, 128–141.

Monteith, D.T., Stoddard, J.L., Evans, C.D., *et al.* (2007) Dissolved organic carbon trends resulting from changes in atmospheric deposition. *Nature*, **450**, 537–540.

Nakićenović, N., Alcamo, J., Davis, G., *et al.* (2000) *Emission scenarios. A Special Report of Working Group III of the Intergovernmental Panel on Climate Change*. Cambridge University Press, Cambridge and New York.

Öquist, M. & Laudon, H. (2008) Winter soil-frost conditions in boreal forests control growing season soil CO_2 concentration and its atmospheric exchange. *Global Change Biology*, **14**, 2839–2847.

Palecki, M.A. & Barry, R.G. (1986) Freeze-up and break-up of lakes as an index of temperature changes during the transition seasons: A case study for Finland. *Journal of Climate and Applied Climatology*, **25**, 893–902.

Peeters, F., Livingstone, D.M., Goudsmit, G.H., Kipfer, R. & Forster, R. (2002) Modeling 50 years of historical temperature profiles in a large central European lake. *Limnology and Oceanography*, **47**, 186–197.

Räisänen, J. (2005) Impact of increasing CO_2 on monthly-to-annual precipitation extremes: Analysis of the CMIP2 experiments. *Climate Dynamics*, **24**, 309–323.

Räisänen, J. (2007) How reliable are climate models? Review article. *Tellus*, **59A**, 2–29.

Räisänen, J., Hansson, U., Ullersteig, A., *et al.* (2003) GCM driven simulations of recent and future climate with the Rossby Centre coupled atmosphere – Baltic Sea regional climate model RCAO. *SMHI Reports Meteorology and Climatology*, **101**, 61.

Räisänen, J., Hansson, U., Ullersteig, A., *et al.* (2004) European climate in the late twenty-first century: Regional simulations with two driving global models and two forcing scenarios. *Climate Dynamics*, **22**, 13–31.

Robertson, D.M., Ragotzkie, R.A. & Magnuson, J.J. (1992) Lake ice records to detect historical and future climatic changes. *Climatic Change*, **21**, 407–427.

Saloranta, T.M. & Andersen, T. (2007) MyLake – A multi-year lake simulation model code suitable for uncertainty and sensitivity analysis simulations. *Ecological Modelling*, **207**, 45–60.

Saloranta, T.M., Forsius, M., Järvinen, M. & Arvola, L. (2009) Impacts of projected climate change on thermodynamics of a shallow and deep lake in Finland: Model simulations and Bayesian uncertainty analysis. *Hydrology Research*, **40**, 234–248.

Schär, C., Vidale, P.L., Lüthi, D., *et al.* (2004) The role of increasing temperature variability in European summer heat waves. *Nature*, **427**, 332–336.

Schindler, D.W., Beaty K.G., Fee, E.J., *et al.* (1990) Effects of climatic warming on lakes of the central boreal forest. *Science*, **250**, 967–970.

Sommaruga-Wögrath, S., Koinig, K.A., Schmidt, R., Sommaruga, R., Tessadri, R. & Psenner, R. (1997) Temperature effects on the acidity of remote alpine lakes. *Nature*, **387**, 64–67.

Sowerby, A., Emmett, A., Tietama, A. & Beier, C. (2008) Contrasting effects of repeated summer drought on soil carbon efflux in hydric and mesic heathland soils. *Global Change Biology*, **14**, 2388–2404.

Šporka, F., Livingstone, D.M., Stuchlík, E., Turek, J. & Galas, J. (2006) Water temperatures and ice cover in the lakes of the Tatra Mountains. *Biologia*, **61** (Suppl. 18), S77–S90.

Stefan, H.G & Sinokrot, B.A. (1993) Projected global climate change impact on water temperatures in five North Central U.S. streams. *Climatic Change*, **24**, 353–381.

Stefan, H.G., Fang, X. & Hondzo, M. (1998) Simulated climate change effects on year-round water temperatures in temperate zone lakes. *Climatic Change*, **40**, 547–576.

Stepanauskas, R., Laudon, H. & Jørgensen, N. (2000) High DON bioavailability in boreal rivers during spring flood. *Limnology and Oceanography*, **45**, 1298–1307.

Stieglitz, M., Dery, S.J., Romanovsky, V.E. & Osterkamp, T.E. (2003) The role of snow cover in the warming of arctic permafrost. *Geophysical Research Letters*, **30**, 541–544.

Straile, D., Jöhnk, K. & Rossknecht, H. (2003) Complex effects of winter warming on the physicochemical characteristics of a deep lake. *Limnology and Oceanography*, **48**, 1432–1438.

Thies, H., Nickus, U., Mair, V., *et al.* (2007) Unexpected response of high alpine lake waters to climate warming. *Environmental Science & Technology*, **41**, 7424–7429.

Tipping, E., Woof, E., Rigg, A., *et al.* (1999) Climatic influences on the leaching of dissolved organic matter from upland UK moorland soils, investigated by a field manipulation experiment. *Environment International*, **25**, 83–95.

Toberman, H., Evans, C.D., Freeman, C., *et al.* (2008) Summer drought effects upon soil and litter extracellular phenol oxidase activity and soluble carbon release in an upland *Calluna* heathland. *Soil Biology & Biochemistry*, **40**, 1519–1532.

Todd, M.C. & Mackay, A.W. (2003) Large-scale climatic controls on Lake Baikal ice cover. *Journal of Climate*, **16**, 3186–3199.

Vestreng, V., Myhre, G., Fagerli, H., Reis, S. & Tarrasón, L. (2007) Twenty-five years of continuous sulphur dioxide emission reduction in Europe. *Atmospheric Chemistry and Physics*, 7, 3663–3681.

Vidale, P.L., Lüthi, D., Wegmann, R. & Schär, C. (2007) European summer climate variability in a heterogeneous multi-model ensemble. *Climatic Change*, 81, 209–232.

Vuorenmaa, J., Forsius, M. & Mannio, J. (2006) Increasing trends of total organic carbon concentrations in small forest lakes in Finland from 1987 to 2003. *The Science of the Total Environment*, 365, 47–65.

Walsh, J.E. (1995) Long-term observations for monitoring of the cryosphere. *Climatic Change*, 31, 369–394.

Watts, C.D., Naden, P.S., Machell, J. & Banks, J. (2001) Long term variation in water colour from Yorkshire catchments. *The Science of the Total Environment*, 278, 57–72.

Webb, B.W. (1996) Trends in stream and river temperature. *Hydrological Processes*, 10, 205–226.

Weyhenmeyer, G.A., Meili, M. & Livingstone, D.M. (2004) Nonlinear temperature response of lake ice breakup. *Geophysical Research Letters*, 31 (7), L07203, doi:10.1029/2004GL019530.

Weyhenmeyer, G.A., Meili, M. & Livingstone, D.M. (2005) Systematic differences in the trend towards earlier ice-out on Swedish lakes along a latitudinal temperature gradient. *Verhandlungen Internationale Vereinigung für theoretische und angewandte Limnologie*, 29, 257–260.

Williams, M.W., Knauf, M., Caine, M., Liu, F. & Verplanck, P.L. (2006) Geochemistry and source waters of rock glacier outflow, Colorado Front Range. *Permafrost and Periglacial Processes*, 17, 13–33.

de Wit, H.A., Hindar, A. & Hole, L. (2008) Winter climate affects long-term trends in stream water nitrate in acid-sensitive catchments in southern Norway. *Hydrology and Earth System Sciences*, 12, 393–403.

Worrall, F., Harriman, R., Evans, C.D., *et al.* (2004). Trends in dissolved organic carbon in UK rivers and lakes. *Biogeochemistry*, 70, 369–402.

Yoo, J.C. & D'Odorico, P. (2002) Trends and fluctuations in the dates of ice break-up of lakes and rivers in Northern Europe: The effect of the North Atlantic oscillation. *Journal of Hydrology*, 268, 100–112.

4

Climate Change and the Hydrology and Morphology of Freshwater Ecosystems

Piet F.M. Verdonschot, Daniel Hering, John Murphy, Sonja C. Jähnig, Neil L. Rose, Wolfram Graf, Karel Brabec and Leonard Sandin

Introduction

This chapter focuses on existing knowledge and new data, as well as hypotheses, on how climate–hydromorphological interactions might alter freshwater ecosystems in future. We include projected changes in climate and land use, and their effects at the catchment, reach and habitat scales. The emphasis is on streams and rivers, although some aspects of lake hydromorphology are also considered. We focus on streams and rivers mainly because more knowledge exists on the small-scale effects of climate change on stream and river hydrology and morphology and ultimately on their biological effects, compared with lakes. Finally, we discuss how climate change might affect attempts to restore stream and river ecosystems.

   Climate affects freshwater ecosystems indirectly through societal and economic systems, such as land management, as well as directly by temperature and precipitation. In many cases, climate change is an additional stressor adding to the impacts of human activity. Freshwater biodiversity, for example, is at present affected by the over-exploitation of natural resources, water pollution, flow modification, habitat degradation and by invasive alien species (Dudgeon *et al.* 2006). In future, however, the effects of climate change are expected to become more prominent, especially if the magnitude and rate of climate change are at the higher end of the projected range (cf. Solomon *et al.* 2007; European Environmental Agency 2008), which, on current evidence, seems likely. The principal changes expected in Europe are described in Chapter 3. For streams and rivers, changes in precipitation and discharge regimes are as important as changes in temperature. Already there is evidence for an increase in annual river discharge in eastern Europe and a decrease in southern Europe, and annual discharge is also projected to decline strongly in southern and south-eastern Europe in future

Climate Change Impacts on Freshwater Ecosystems. First edition. Edited by M. Kernan, R. Battarbee and B. Moss. © 2010 Blackwell Publishing Ltd.

but to increase in almost all parts of northern and north-eastern Europe. Changes in seasonality are also expected, with most climate models showing a general precipitation decrease in summer months and an increase in autumn and winter months. Furthermore, extreme daily (especially summer storm) precipitation events may become more frequent (Räisänen *et al.* 2003), discharge may vary more, with more extremes (higher spates and longer droughts), and annual discharge regime may become less predictable (Arnell 1999).

A key concern is how this will influence the morphology, hydrology, habitats and species diversity in rivers and lakes. In large parts of Europe, channel straightening, weir and dam construction, disconnection of the river from its flood plain and alteration of riparian vegetation have already severely reduced the ecological quality of streams and rivers (Kristensen & Hansen 1994; Armitage & Pardo 1995; Hansen *et al.* 1998). In lakes, water-level fluctuations affect littoral habitats through changes in light, sediment and wave regimes (Wantzen *et al.* 2008), and the modification of shorelines affects macrophytes (Radomski & Goeman 2001; Elias & Meyer 2003), macroinvertebrate communities, littoral fish communities and overall biodiversity (Jennings *et al.* 2003; Scheuerell & Schindler 2004).

Predicted changes in land use

Land-use change

Stream hydrology is also influenced by changes in land use (e.g. Poff *et al.* 1997) through (i) changes in infiltration, evaporation, run-off and discharge regime (Knox 1987); (ii) changes in riparian vegetation and channel morphology; and (iii) changes in the input of sediment (Reid & Page 2003; Wissmar & Craig 2004), organic matter (Allan *et al.* 1997), nutrients, pesticides and other pollutants. Most studies concerned with the influence of land-use change on rivers have been conducted at a large spatial (e.g. Feld 2004; Townsend *et al.* 2004), but local land-use change is also important, for example, through the removal of riparian vegetation, which may increase erosion and sediment inwash. In addition, land use can intensify or buffer the effects of climate change. If, for example, land is drained, an increase in precipitation will result in a greater run-off and, consequently, more peak hydrographs. In contrast, forests can retain rainwater and reduce the impact of increased precipitation on run-off.

European land use is expected to change significantly over coming decades (Busch 2006) with the amount of agricultural land decreasing and the amount of forest and urban land increasing (Schröter *et al.* 2005; Verburg *et al.* 2008), though the loss of agricultural land elsewhere, the rising world population and its need to be fed and a demand for biofuels may reverse these expectations. Such land-use changes have had and will continue to have a strong influence on stream and river hydrology and morphology (Sandin 2009). A rare example of a site with a long-term record of changes in river morphology embraces the tributaries of the River Vecht in the Netherlands. From the end of the 19th century until the

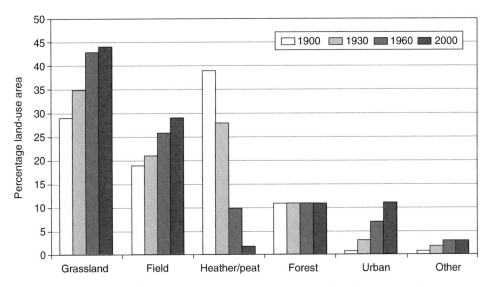

Figure 4.1   Land-use changes in the river Vecht catchment (the Netherlands) in a period when climate change has been relatively minor.

present day, there has been only a minor change in climate. In contrast, there have been major changes in land use and river morphology (canalization) taking place simultaneously in the periods ±1895–1905, 1925–35 and 1955–65. The area of heather and moorland peat decreased dramatically as agricultural, urban and other human uses increased (Fig. 4.1).

The streams in the Vecht catchment show consequent degradation in structure over the last 100 years. The total stream length was shortened by about 20%; 40% of the connected backwaters were lost; the number of oxbows increased in the 1930s due to straightening of the major streams but has decreased since then, with only 38% of the oxbows remaining. Over the last 100 years, an increase in both temperature and precipitation has been observed, but average annual discharge has not changed significantly over the last 30 years. In general, most streams followed meandering channels in 1900; between 1930s and 1960s, some were still meandering but currently most are straight.

Effects of climate change at the catchment scale

Direct climate impacts on stream hydrology

The effects that changes in climate may have on stream hydrology are illustrated by the expected changes in discharge for the River Lambourn in southern England (Fig. 4.2). The Lambourn catchment (265 km²) has a limited river network owing to its underlying porous chalk geology and is dominated by arable land and improved pasture. Based on RCAO HadAM3H model outputs under the B2 scenario (see Chapter 3) and a catchment-scale rainfall-stream flow model

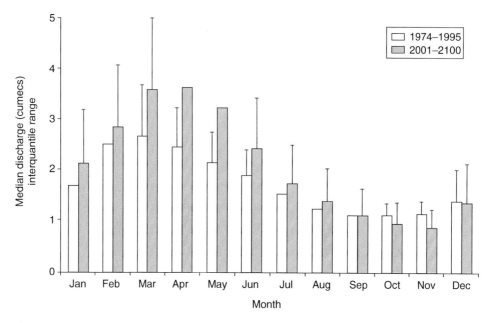

Figure 4.2   Current and projected discharge of the river Lambourn, southern England.

(Littlewood 2002), the mean monthly discharge for 2071–2100 was predicted. The current and modelled discharge time series were compared for magnitude, frequency and timing of extreme discharge events.

A more variable discharge regime for the river Lambourn by the end of this century is predicted. The biggest difference between now and then is the size of the winter peaks. During 1974–95, 45% of annual peaks exceeded $3 \, m^3 \, s^{-1}$, while for 2071–2100, 83% of peaks are predicted to exceed this level. There were two prolonged drought events in 1974–95, caused by a lack of sufficient winter recharge in 1976 and again in 1991 and 1992. The modelled discharge data for 2071–2100 predict that equivalent winter droughts are less likely to occur. Even the poorest winter recharge levels for the projected period are comparable with current normal recharge levels. The median and range of summer and autumn low flows are not predicted to change to such a severe extent as in winter and spring. However, midsummer discharges may generally be slightly increased and the winter recharge may be delayed by about 1 month, shifting from November to December. These predicted changes to discharge magnitude, frequency and timing are likely to affect the physical structure of the River Lambourn and its riparian zone, as well as the flora and fauna it supports.

Climate change will thus alter the dominant pattern of precipitation, which in turn changes run-off and discharge regimes, including spates and droughts in streams and rivers. Spates and droughts are important drivers that contribute to stream patchiness. Climate change, therefore, presents a challenge to the structure and function of current stream ecosystems (Meyer *et al.* 1999; Wright *et al.* 2004). During spates, habitats may be destroyed; during low flows, they will be silted, and during base-flow conditions, habitats will be generated again (Lake 2000).

This fluctuation in habitat conditions will become greater in future, and stream communities will fluctuate more (Milne 1991). The predictability of resources will decrease, and species will have to adapt or become locally extinct.

Several studies have focused on the biological implications of lower resource predictability (Palmer *et al.* 1995; Palmer & Poff 1997; Townsend *et al.* 1997), and many show how ecosystems are directly influenced by flow (Death & Winterbourn 1995; Gasith & Resh 1999; Poff 2002; Bond & Downes 2003; Fritz & Dodds 2004). High flows and spates scour accumulated sediment and debris, redistribute streambed material and organic matter in the channel, change channel morphology and form new erosion (runs and riffles) and deposition (point and mid-channel bars, pools, sand accumulations) zones. High flows and spates may also disturb in-channel and encroaching riparian vegetation, homogenize water chemistry among the stream channel and adjacent water bodies and increase shear stress on organisms. In contrast, low flows and droughts bring siltation of fine mineral and organic material, decrease oxygen concentrations and increase those of some nutrients and minerals. They promote mineralization of organic material in the stream bottom and drying of the banks; they reduce bank stability, expose large parts, or all, of the stream bed and stress many organisms. Change in flow variability and timing can also be important, resulting in unpredictable erosion–deposition processes with frequent shifts in channel morphology and habitat availability and loss of synchronization of flow stages with stages in organisms' life cycles, such as egg deposition, growth and pupation.

Direct climate change impacts on lake hydrology

Climate change will impact lake hydrology mainly through effects on residence time and water level as well as through receptors and sources of stream flow. Short residence times mean that pollutants such as excess nutrients from point sources are flushed out of the lake ecosystem, whereas with decreasing precipitation and longer residence times, they will accumulate, with likely changes in phytoplankton communities (Schindler *et al.* 1990, 1996; Hillbricht-Ilkowska 2002) and in food-web composition and structure. In lakes with long residence times, internal processes may become more important. For example, phytoplankton production may increase with higher temperatures due to increased nutrient availability, and eutrophication problems may thereby become more severe (Mooij *et al.* 2005). A decline in water level due to decreased precipitation may cause changes in the nutrient status and acidity of lakes with low buffering capacities (Carvalho & Moss 1999).

Water-level change can also directly affect phytoplankton development in a lake. For example, a strong influence of the North Atlantic Oscillation on the lake water level was observed in Lake Vortsjarv (Estonia) (Nõges *et al.* 2007) where water-level changes influenced phytoplankton composition and biomass independently of the nutrient loading. In addition, less severe winters cause a reduction in winter ice cover, which can lead to lower lake water levels and lake system changes through the following summer months (Croley 1990). Longer ice-free periods potentially lengthen the growing season for algae and aquatic macrophytes. Higher temperatures may raise the rate of mineralization of organic matter in catchment soils, releasing carbon, phosphorus and nitrogen, and particulate phosphorus input may also be raised from increased erosion of catchment soils. Increased nutrient loading, coupled with water temperature increases, could thereby increase autochthonous productivity and greater autochthonous biogenic contributions to the sediment.

Effects on lake hydrology may become more apparent when considering seasonal patterns. For example, higher winter air temperatures will alter the balance of precipitation input so that there will be more winter rain and less snow, an effect most marked in upland lakes. This will lead to lower spring 'high-flow' peaks as a result of reduced snow melt and, as a consequence, less flushing in spring (Jenkins & Boulton 2007). Furthermore, reduced lake ice cover and increased frequency of winter storms will lead to reduced winter stratification and hence changes in in-lake biogeochemistry. In contrast, increased summer air and water temperatures will increase potential evaporation, which, coupled with lower summer rainfall, leads to lower flows, which will affect both lake levels and downstream river communities. Increasing episodic events such as storms, at whatever season they occur, will lead to greater flushing and run-off. More frequent or more severe episodes will affect lake and outflow chemistry such that, for example, acid episodes may be more frequent. Greater winter run-off will also lead to a higher suspended sediment load, which will lead to increases in lake sediment accumulation rate (SAR).

Sediment accumulation rate is one of the most important physical variables affecting the functioning of lake ecosystems. It affects lake morphology, physical and chemical stratification, the characteristics of lake habitats and hence the distribution of aquatic flora and fauna particularly in littoral areas. In shallower lakes, rapid accumulation can raise the sediment surface into the zone of resuspension, thus affecting light penetration and plant growth, and ultimately may lead to complete terrestrialization. In many lakes, an increase in SAR has been observed over the last 100 years (e.g. Rose *et al.* 2010). Reasons for this include changes in land use and land management, causing accelerated catchment soil erosion, and increased biogenic sedimentation from eutrophication. Climate change can influence both of these processes and is likely to play a more important role in the future.

More rainfall and increased frequency of extreme events (both summer droughts and winter storms) could increase catchment soil erosion, resulting in more allochthonous material reaching lake sediments. Susceptibility to erosion will depend on soil type, vegetation and land use, but upland peatland soils are already known to be eroding quickly, and this could be further exacerbated by continuing climate change. Climate change may also have indirect impacts on erosion rates through changes to catchment vegetation and land use, while susceptibility to erosion may also be affected by changes in soil temperature and moisture regimes, pH, base saturation and microbial activity (Helliwell *et al.* 2007). Increases in both autochthonous and allochthonous inputs will tend to increase SARs in all lakes, but shallow, eutrophic lakes are most likely to be altered by accelerated in-filling processes. As lakes fill in, there will be changes in sediment distribution as mean basin slopes decrease and sediment accumulation zones widen (Blais & Kallf 1995). This may lead to more rapid encroachment of reedswamp and faster filling in.

The effects of climate change on hydraulics and morphology at the reach scale

The pattern and variation of current velocity within a stream reach has a major influence on longitudinal and transverse channel morphology, species diversity, food-web structure and ecological processes (Jowett & Duncan 1990; Poff *et al.* 1997). It is often assumed that monthly or daily means, which are often readily available, are sufficient to characterize flow regime (e.g. Clausen & Biggs 2000; Olden & Poff 2003). However, even single events can cause substantial changes in the physical habitat and can affect ecological functioning (Schlosser 1995; Arndt *et al.* 2002). Gore *et al.* (2001) listed five major hydraulic conditions that most affect the distribution and ecological success of stream biota: suspended load, bed-load movement, water-column effects, such as turbulence and velocity profile, and substratum interactions (near-bed hydraulics). However, stream organisms are generally adapted to a wide variability in stream discharge (Allan 1995; Petts *et al.* 2000) and can accommodate large changes.

A study carried out in the River Lambourn assessed how hydraulic changes at the reach scale affected benthic macroinvertebrate community composition. The results suggested considerable resilience in stream macroinvertebrate communities to future changes in discharge regimes. Changes in the cover of the five main mesohabitats (*Berula, Ranunculus* and *Callitriche* plant stands, gravel and silt (Fig. 4.3)) were recorded over 9 years. The macroinvertebrate community associated with each mesohabitat was also sampled (Wright *et al.* 2003). *Berula* and *Ranunculus* stands were more abundant in early summers preceded by high-discharge winters, while gravel habitats tended to be more prominent after winters with lower discharge. The plant and mineral mesohabitats proved to be distinct from each other, with the caddis fly families, Glossosomatidae and Goeridae, being relatively more abundant in gravel, the bivalves, Sphaeriidae, favouring silt patches and the mayfly families, Ephemerellidae and Caenidae, preferring the plants. However, taxa did not strongly associate with just one particular mesohabitat; they tended to be equally abundant on two or more mesohabitats. Perhaps because of the lack of strong obligate relationships between taxa and specific mesohabitats, the early summer density of most taxa was not correlated with previous winter discharge levels. But distribution across a range of available habitats means that the macroinvertebrate community is buffered against yearly variations in discharge regime, as long as there are no extreme spates resulting in strong currents. These can cause catastrophic downstream drift due to increased shear stress (Layzer *et al.* 1989), especially for small animals (De Jalon *et al.* 1994).

The increase in discharge extremes (both spates and droughts) and the decrease in discharge predictability expected with changing climate may increase the relative importance of abiotic parameters, both hydraulic conditions and physical habitat features, in shaping future stream communities (Poff & Ward 1989). Spates have substantial effects. Some organisms may move or be moved to habitats with less disturbance (e.g. stream margins, the hyporheic zone (Boulton *et al.* 1998) or other patches protected from high flow) and, thus, will seek refuges (Townsend & Hildrew 1994; Fritz & Dodds 2004). Others may be able to remain in or on

Figure 4.3   Habitats in the river Lambourn, southern England.

relatively stable habitats where they are sheltered from shear stress. Many may lack adaptations and be washed downstream (Blanch *et al.* 2000; Imbert *et al.* 2005).

When disturbances become highly unpredictable, species might display risk-spreading strategies, for example, asynchronous hatching of eggs (Lytle & Poff 2004). Behavioural adaptations enable organisms to respond directly to individual floods and droughts and unstable substrates by moving to refuges, following use of cues that they can detect (Lytle & Poff 2004). Droughts and low flows elicit other strategies. There may be physiological resistance (Smith & Pearson 1985), adapted life cycles, allowing early maturation and flight of adults, or resting stages (Ladle & Bass 1981; Williams 1987; Bunn 1988). There may also be increased mobility with organisms seeking out remaining higher-flow habitats (Harrison 1966; Olsson & Söderström 1978; Delucchi 1989). Climate change may also result in death through sediment injury or burial, increased drift downstream, physiological problems for respiration, as particles settle on gill filaments, and through low oxygen concentrations in the water due to the deposition of fine organic sediments (Stazner & Holm 1982; Strommer & Smock 1989; Hart & Finelli 1999; Holomuzki & Biggs 2003; Gordon *et al.* 2004). There may also be food shortages as food sources (organic matter, periphyton) are buried by settling particles or scoured by the current. There can be an increase of the sediment concentration in suspension in the water column (affecting filter feeders positively) or a decrease in potential prey (loss of individuals due to drift, burial) for predators.

Current, streambed composition and sediment transport are strongly linked (Gordon *et al.* 2004). Simultaneously, an increase in discharge often results in an increase in bed-sediment movement and more resuspension of bottom sediments or soil eroded from banks (Wood & Armitage 1997). Erosion, transport and sedimentation thus result in local changes within the streambed. Short-term equilibria typically persist between disturbances, but spates, especially with anthropogenic disturbances (canalization, agriculture in the watershed), can lead to changes in stream slope and mean velocity, sediment particle size distribution, sediment discharge and the planform of the stream (e.g. cut-off meanders). Climate-change-induced extreme spates will reshape the gross form of the channel. Maintenance of that form and its smaller-scale features, such as substrate patterns, may be better related to more frequent, less-extreme discharge events. Streambed and bank stability determine the influence of high discharges on the channel form. For example, a high density of plant roots in the bank reduces erosion. The discharge which just fills the stream to its banks, termed the bankfull discharge, is often assumed to control the form of streams (Allan 1995; Gordon *et al.* 2004).

In a stream reach, different current velocities occur along meanders, around obstacles like debris and in pool-riffle sequences. Spates erode, establish or maintain these structures. Stream organisms respond differently to changes in current extremes, substrate composition and stability. The patchy distribution of aquatic macrophytes is related to variation in disturbance frequency and intensity brought about by changes in discharge and by colonization success and growth rates (Sand-Jensen & Madsen 1992). Climate change will cause changes, probably increases, because of lower summer flows, in macrophyte abundances (Rørslett *et al.* 1989; Walker *et al.* 1994). A more variable discharge pattern may reduce macroinvertebrate species richness and standing crop within the streambed (Layzer *et al.* 1989; Munn & Brusven 1991) but increase plant species diversity. Cobb *et al.* (1992) show that a further increase in current velocity variability will lower macroinvertebrate diversity due to substrate instability.

Fish are often adapted to a certain level of hydrological variability (Poff & Allan 1995), but changes in variability may have negative effects on fish populations (Ficke *et al.* 2007). With a change from predictable to unpredictable hydraulic conditions (Poff & Allan 1995), the fish community will be prone to move from 'environmental specialists' to a community with only generalist species, that is, those that are able to exploit a wide variety of resources and changing environmental conditions (Fausch & Bestgen 1997). Also, if climate change results in more lake-like conditions in summer with low to zero velocities along with macrophyte growth, then pest species, such as mosquitoes, could become highly abundant (De Moor 1986). Under such conditions, more tolerant, ubiquitous fish species may become dominant (Pusey *et al.* 1993).

Effects of climate change on hydraulic conditions and channel morphology at the habitat scale

At the habitat scale, subtle variations in current, near-bed velocities, bottom roughness, grain sizes and distribution together influence the distribution and abundance of particular species of plants and animals. The primary need of stream macroinvertebrates is related to current, as it provides oxygen and food particles, shapes habitats and transports wastes. A cross-section profile of current velocity in a straight stream reach has a more or less parabolic shape. The velocity is highest in the middle of the profile close to the surface and decreases towards the banks and bottom due to friction. In a straight gutter, this pattern continues over long distances giving low habitat variability. Natural streams are not straight gutters; they are irregularly shaped channels that induce complex current velocity profiles. Turbulent flow patterns characterize natural streams, and a continuously changing discharge causes the turbulence patterns to change continuously. These processes, varying in space and time, determine substrate composition and stability, creating a heterogeneous environment at the habitat scale. In a review of seven case studies concerned with links among land use, hydromorphology and river biota at different spatial scales (Sandin 2009) found that the strongest relationship was between in-stream variables (Pedersen & Friberg 2009) or hydraulic variables (Syrovátka *et al.* 2009) and biota rather than catchment characteristics (see also Wiberg-Larsen *et al.* 2000; Mérigoux & Dolédec 2004).

The relationship between the benthic macroinvertebrates and the immediate hydrology was investigated in the river Becva in the Czech Republic. Following strong disturbance in winter, the spring distributions of animals were much more similar in different parts of the river than those in autumn, when quiet summer flow had allowed a differentiation of habitats. In July 1997, a high spate occurred and changed large parts of the regulated channel structure. In five re-formed stretches, the interactions between habitat patterns and benthic macroinvertebrates were monitored in spring and autumn of 2004–05 (Fig. 4.4). During autumn, at low and stable flow regimes, a habitat gradient develops from the side arms to the main channel (lower right area of Fig. 4.4). In contrast, the spring samples are more uniform and clustered, except for the samples from the side arms, following the disturbance of the winter spates (upper left part of Fig. 4.4).

Stream and river species are adapted to unidirectional flow, modified by turbulence. Morphological and behavioural adaptations of the organisms can be classified into those involving position of the organisms, such as locomotion, attachment and body shape (e.g. flattening, concealment, flexibility), those associated with feeding or nutrient intake (e.g. functional feeding groups, growth forms), those associated with physiology (e.g. respiration, temperature) and those pertaining to reproduction. The habitat and such characteristics are related, and there is considerable information available on the differences between organism assemblages living on substrata, for example, on rocks, in shifting sands, on logs and in leaf packs.

With fewer, but more extreme rainfall events, together with changes in land use, siltation will be an even more serious problem in stream and river ecosystems. Siltation degrades habitat diversity with generally negative consequences for stream communities. This was exemplified in the River Waldaist in Upper Austria, where the importance of habitat availability and quality for different life stages of selected trichopterans in relation to siltation was studied (Fig. 4.5). Trichopterans usually have five larval instars and one pupal instar (developmental stage), with most instars depending on a different

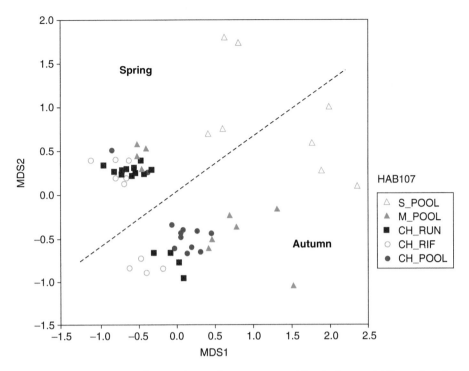

Figure 4.4 Comparison of macroinvertebrate communities (using multidimensional scaling) among habitat types in stretches of the river Becva. (S_POOL, side arm; M_POOL, main-channel margin; CH_RUN, CH_RIF and CH_POOL, main-channel run, riffle and pool, respectively.)

habitat type. Furthermore, organisms need places for hiding, resting, mating, copulation and egg deposition, increasing the importance of in-stream habitat variability. The survey included an unsilted reference and a degraded, silted site. The glossosomatid *Agapetus ochripes* scrapes algae and other micro-organisms from stable substrates and is found in large numbers at the reference site but is scarce at the degraded site. Pupae are only found at the reference site, suggesting that conditions at the degraded site prevent larvae of this species from becoming adults. *Lype phaeopa* is a scraper that only occurs on woody debris and is found at eight times higher densities at the reference site in comparison with the degraded site. The carnivorous species *Brachycentrus montanus* needs stable substrates on which to fix its case, to be able to stretch its spiny legs as a filtering apparatus into the current. Three different instars could be found at the silted site in quite high densities, but again there were no pupae, whereas at the reference site, all instars and a high number of pupae were recorded. Another example is *Lasiocephala basalis* (*Lepidostoma basale*), which mainly lives on woody debris, feeding on bacteria and fungi. Density is comparable between sites, but again pupation takes place exclusively at the reference site (Fig. 4.5).

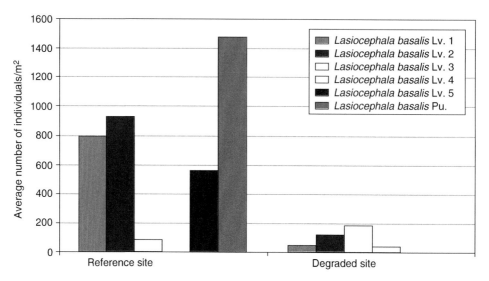

Figure 4.5   The occurrence of the life stages (Lv 1–5 = larval stages 1–5, Pu = pupal stage) of the trichopteran *Lasiocephala basalis* in a reference and a degraded site of the river Waldaist, Austria.

The effects of climate change on stream and river restoration success

After a long period of modification of streams and rivers, as well as their catchments, to the needs of agriculture, industry and households, there has been an increasing awareness of the large negative impact of these alterations. In the Netherlands, only about 4% of the streams still have a natural morphology and a (more or less) natural hydrology. In Denmark, only 2% are more or less natural (Brookes 1987), and in Germany, this is 2%–5%. Environmental awareness and concern for the loss of stream and floodplain habitats and associated biodiversity have stimulated a major programme of stream rehabilitation and restoration, especially physical stream restoration in Europe. For example, in the Netherlands, 70 projects were carried out in 1991, 170 in 1993 and over 200 in 1998 with a total cost by 2006 estimated at 1.3 billion euro (Verdonschot & Nijboer 2002).

There are many ways of physically restoring streams such as reforestation of the floodplain, re-meandering and the removal of dams and bank structures. Newer approaches include the addition of coarse woody debris (Gippel & Stewardson 1996; Gerhard & Reich 2000), the removal of sediment deposits in floodplains (cf. Kern 1994) and various methods to combat the deep cutting of streams.

For effective stream restoration, the complex links among physical parameters, habitat diversity and biodiversity need to be understood. When a stream has been physically restored, success, measured mainly by an increase in biodiversity, depends on the extent of re-colonization by the original (indicator) species. This is the 'field of dreams' hypothesis, which states 'if we build it, they will come'.

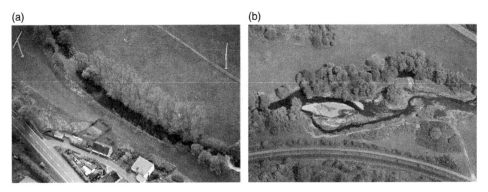

Figure 4.6   (a) Single-channel and (b) restored multiple-channel sections along the river Lahn. (Photographs by S. Jähnig and A. Lorenz.)

However, this is by no means a guaranteed outcome (Palmer *et al.* 1997). Whether the original species are able to re-colonize the restored stream depends not only on the quality of the restored habitat but also on a number of factors such as the dispersal capacities of the species and the presence or absence of migration barriers between the source populations and the restored areas (Hughes 2007).

Establishment of an invasive or non-native species may also hinder re-colonization, and biodiversity may, in general, be threatened by invasive species replacing native ones. Colonization may be less successful than hoped for, as illustrated by a study from the German Central Highlands, where the benthic macroinvertebrate biodiversity of multiple channels re-created in seven rivers was compared with that of seven engineered straight channels (Fig. 4.6).

The hydromorphological diversity of the multiple-channel sections approached the intended reference condition. Habitat diversity increased, and increased sediment dynamics were observed. However, substrata in single- and multiple-channel sections still had similar macroinvertebrate community composition, and the alpha diversity of substratum-specific communities did not change. Different substrata, however, did host distinct macroinvertebrate communities, so beta diversity was greater in the restored channels as they tended to have a greater range of substrata. A comparison of representative communities in single- and multiple-channel sections showed very high Bray–Curtis similarities (69%–77%). Mean similarity analyses (using MEANSIM) revealed that the macroinvertebrate community composition within a channel type was less similar (single-channel within-group similarity: 0.61, multiple-channel within-group similarity: 0.63) than the composition between channel types (between-group similarity: 0.66).

Differences between paired stream sections could mainly be attributed to single taxa that occurred solely in either the single- or multiple-channel sections. These exclusive taxa were mostly found on organic substrates such as living parts of terrestrial plants, large wood, coarse particulate organic matter and mud. Whether this occurrence is related to the specific substrata or to chance is still to be resolved. The overall high similarity of macroinvertebrate communities from single- or multiple-channel sections could be due to prevailing large-scale catchment pressures, the small scale of the restoration or a lack of potential re-colonizers. In general, river restoration schemes have proved similarly disappointing for these reasons.

One of the most important keys to restoration is understanding the complex links between physical, chemical and biological components, ultimately at the catchment scale. Stream restoration is usually focussed on individual stream reaches, largely ignoring the importance of connectivity, such as channel movement for terrestrial–aquatic linkages (e.g. stream–floodplain exchanges) and biotic dispersal (Verdonschot & Nijboer 2002). Catchment-scale processes largely determine the structure and function of streams and their floodplains. Thus, the scale of restoration needs to be adjusted to the scale of the dominant shaping processes. The key themes in stream ecology deal with the four dimensions (Ward 1989) of hierarchy of physical organization (Frissell *et al.* 1986), adaptation, response of species and human disturbance. A catchment approach includes all these themes. Climate change, and thus hydromorphological change, affects the stream at the highest hierarchical levels, implying that under changing climate conditions, either the measures to meet restoration targets must be increased or the targets must be less ambitious.

Conclusions

Climate change will alter the hydromorphological conditions of lakes, streams and rivers in Europe. The magnitude of change induced by climate is still relatively small in comparison with the impact of anthropogenic land use, but in future, climate change may cause significant change in hydrology and, at the same time, impose land-use changes in catchments. Changes in stream hydrology are best characterized in terms of dynamics, particularly associated with increases in drought as well as in spate frequency. In lakes, hydrological changes are expressed in terms of more dynamic fluctuations as well as overall changes in water level and their impact on eutrophication (cf. Chapter 6). In rivers, climate change may also increase flow variability resulting in higher scouring and siltation rates, and where rivers flow into lakes, sediment loads may increase, leading to accelerated SARs. To some extent, biota in streams are adapted to changes in habitat-scale conditions and will tolerate such dynamics. However, different life stages require different environmental conditions, and a disturbed timing, for example, of snow melt and the connected high-flow conditions, can result in a loss of taxa. More dynamic conditions and a loss of native taxa widen the opportunity for non-native species to enter ecosystems. Globalized transport systems and hydrological links between large catchments further enhance this process.

Restoration of streams and rivers is promoted by the EU Water Framework Directive and other legislations (e.g. EU Habitats Directive, EU Floods Directive, Natura 2000). Restoration success, however, has been lower then expected. This failure is mainly due to the scale of the restoration being too small in many cases, a lack of an ecosystem approach and a lack of understanding of key biological processes governing dispersal and colonization rates and the role of barriers. Future climate change is likely to reduce further the chances of restoration success.

References

Allan, J.D. (1995) *Stream Ecology: Structure and Function of Running Waters.* Kluwer Academic Publishers, Dordrecht.

Allan, D.J., Erickson, D.L. & Fay, J. (1997) The influence of catchment land use on stream integrity across multiple spatial scales. *Freshwater Biology*, **37**, 149–161.

Armitage, P.D. & Pardo, I. (1995) Impact assessment of regulation at the reach level using macroinvertebrate information from mesohabitats. *Regulated Rivers: Research & Management*, **10**, 147–158.

Arndt, S.K.A., Cunjak, R.A. & Benfey, T.J. (2002) Effect of summer floods and spatial-temporal scale on growth and feeding of juvenile Atlantic salmon in two New Brunswick streams. *Transactions of the American Fisheries Society*, **131** (4), 607–622.

Arnell, N.W. (1999) The effect of climate on hydrological regimes in Europe: A continental perspective. *Global Environmental Change*, **9**, 5–23.

Blais, J.M. & Kalff, J. (1995) The influence of lake morphometry on sediment focusing. *Limnology and Oceanography*, **40**, 582–588.

Blanch, S.J., Walker, K.F. & Ganf, G.G. (2000) Water regimes and littoral plants in four weir pools of the River Murray, Australia. *Regulated Rivers: Research & Management*, **16**, 445–456.

Bond, N.R. & Downes, B.J. (2003) The independent and interactive effects of fine sediment and flow on benthic invertebrate communities characteristic of small upland streams. *Freshwater Biology*, **48**, 455–465.

Boulton, A.J., Findlay, S., Marmonier Stanley, P.E.H. & Valett, H.M. (1998) The functional significance of the hyporheic zone in streams and rivers. *Annual Review of Ecology and Systematics*, **29**, 59–81.

Brookes, A. (1987) Restoring the sinuosity of artificially straightened stream channels. *Environmental Geology and Water Science*, **10**, 3341.

Bunn, E.S. (1988) Life histories of some benthic invertebrates from streams of the northern Jarrah Forest, Western Australia. *Australian Journal of Marine and Freshwater Research*, **39** (6), 785–804.

Busch, G. (2006) Future European agricultural landscapes – What can we learn from existing quantitative land use scenario studies? *Agriculture Ecosystems and Environment*, **114**, 121–140.

Carvalho, L. & Moss, B. (1999) Climate sensitivity of Oak Mere: A low altitude acid lake. *Freshwater Biology*, **42**, 585–591.

Clausen, B. & Biggs, B.J.F. (2000). Flow variables for ecological studies in temperate streams: Groupings based on covariance. *Journal of Hydrology*, **237**, 184–197.

Cobb, D.G., Galloway, T.D. & Flannagan, J.F. (1992) Effects of discharge and substrate stability on density and species composition of stream insects. *Canadian Journal of Fisheries and Aquatic Sciences*, **49**, 1788–1795.

Croley, T.E., II (1990) Laurentian Great Lakes double CO_2 climate change hydrological Impacts. *Climatic Change*, **17**, 27–47.

De Jalon, D.G., Sanchez, P. & Camargo, J.A. (1994) Downstream effects of a new hydropower impoundment on macrophyte, macroinvertebrate and fish communities. *Regulated Rivers: Research & Management*, **9**, 253–261.

De Moor, F.C. (1986). Invertebrates of the Lower Vaal River, with emphasis on the Simuliidae. In: *The Ecology of River Systems* (eds B.R. Davies & K.F. Walker), pp. 135–142. Dr W. Junk Publishers, Dordrecht.

Death, R.G. & Winterbourn, M.J. (1995) Diversity patterns in stream benthic invertebrate communities: The influence of habitat stability. *Ecology*, **76** (5), 1446–1460.

Delucchi, M.C. (1989) Movement patterns of invertebrates in temporary and permanent streams. *Oecologia*, **78** (2), 199–207.

Dudgeon, D., Arthington, A.H., Gessner, M.O., *et al.* (2006) Freshwater biodiversity: Importance, threats, status and conservation challenges. *Biological Reviews*, **81**, 163–182.

Elias, J.E. & Meyer, M.W. (2003) Comparisons of undeveloped and developed shorelands, Northern Wisconsin, and recommendations for restoration. *Wetlands*, **23**, 800–816.

European Environmental Agency (2008) *Impacts of Europe's Changing Climate: 2008 Indicator-Based Assessment.* Joint EEA-JRC-WHO report, EEA Report No. 4/2008. European Environment Agency, Copenhagen.

Fausch, K.D. & Bestgen, K.R. (1997) Ecology of fishes indigenous to the central and southwestern Great Plains. In: *Ecology and Conservation of Great Plains Vertebrates* (eds F.L. Knopf & F.B. Sampson), pp. 131–166. Springer-Verlag, New York.

Feld, C.K. (2004) Identification and measure of hydromorphological degradation in Central European lowland streams. *Hydrobiologia*, **516**, 69–90.

Ficke, A.D., Myrick, C.A. & Hansen, L.J. (2007). Potential impacts of global climate change on freshwater fisheries. *Reviews in Fish Biology and Fisheries*, **17**, 581–613.

Frissell, C.A., Liss, W.J., Warren, C.E. & Hurley, M.D. (1986) A hierarchical approach to classifying stream habitat features: Viewing streams in a watershed context. *Environmental Management*, **10**, 199–214.

Fritz, K.M. & Dodds, W.K. (2004) Resistance and resilience of macroinvertebrate assemblages to drying and flood in a tallgrass prairie stream system. *Hydrobiologia*, **527**, 99–112.

Gasith, A. & Resh, V.H. (1999) Streams in Mediterranean climate regions: Abiotic influences and biotic responses to predictable seasonal events. *Annual Review of Ecology and Systematics*, **30**, 51–81.

Gerhard, M. & Reich, M. (2000) Restoration of streams with large wood: Effects of accumulated and built-in wood on channel morphology, habitat diversity and aquatic fauna. *International Review of Hydrobiology*, 85 (1), 123–137.

Gippel, C.J. & Stewardson, M.J. (1996). Use of wetted perimeter in defining minimum environmental flows. In: *Ecohydraulics 2000, Proceedings of the 2nd International Symposium on Habitat Hydraulics* (eds M. Leclerc, H. Capra, S. Valentin, A. Boudreault & Y. Côté), pp. A571–A582. INRS-Eau, Québec.

Gordon, N.D., McMahon, T.A., Finlayson, B.L., Gippel, C.J. & Nathan, R.J. (2004) *Stream Hydrology: An Introduction for Ecologists*. John Wiley & Sons Ltd., Chichester.

Gore, J.A., Layzer, J.B. & Mead, J. (2001) Macroinvertebrate instream flow studies after 20 years: A role in stream management and restoration. *Regulated Rivers: Research & Management*, **17**, 527–542.

Hansen, H.O., Boon, P.J., Madsen, B.L. & Iversen, T.M. (1998) River restoration. The physical dimension. A series of papers presented at the International Conference River Restoration '96, organized by the European Centre for River Restoration, Silkeborg, Denmark. *Aquatic Conservation: Marine and Freshwater Ecosystems,* 8 (1), 1–264.

Harrison, A.D. (1966) Recolonisation of a Rhodesian stream after drought. *Archiv für Hydrobiologie*, **62**, 405–421.

Hart, D.D. & Finelli, C.M. (1999) Physical-biological coupling in streams the pervasive effects of flow on benthic organisms. *Annual Review of Ecology and Systematics*, **30**, 363–395.

Helliwell, R.C., Lilly, A. & Bell, J. (2007) The development, distribution and properties of soils in the Lochnagar catchment and influence on surface water chemistry. In: *Lochnagar: The Natural History of a Mountain Lake* (ed. N.L. Rose), pp. 93–120. Springer, Dordrecht.

Hillbricht-Ilkowska, A. (2002) Nutrient loading and retention in lakes of the Jorka river system (Masurian lakeland, Poland): Seasonal and long-term variation. *Polish Journal of Ecology*, 50 (4), 459–474.

Holomuzki, J.R. & Biggs, B.J.F. (2003) Sediment texture mediates high-flow effects on lotic macroinvertebrates. *Journal of the North American Benthological Society*, 22 (4), 542–553.

Hughes, J. (2007). Constraints on recovery: Using molecular methods to study connectivity of aquatic biota in rivers and streams. *Freshwater Biology*, **52**, 616–631.

Imbert, J.B., Gonzalez, J.M., Basaguren, A. & Pozo, J. (2005) Influence of inorganic substrata size, leaf litter and woody debris removal on benthic invertebrates resistance to floods in two contrasting headwater streams. *International Review of Hydrobiology*, 90 (1), 51–70.

Jenkins, K.M. & Boulton A.J. (2007) Detecting impacts and setting restoration targets in arid-zone rivers: Aquatic micro-invertebrate responses to reduced floodplain inundation. *Journal of Applied Ecology*, **44**, 823–832.

Jennings, M.J., Emmons, E.E., Hatzenbeler, G.R., Edwards, C. & Bozek, M.A. (2003) Is littoral habitat affected by residential development and land use in watersheds of Wisconsin lakes? *Lake and Reservoir Management*, **19**, 272–279.

Jowett, I.G. & Duncan, M.J. (1990) Flow variability in New Zealand rivers and its relationship to in-stream habitat and biota. *New Zealand Journal of Marine and Freshwater Research*, **24**, 305–317.

Kern, K. (ed.) (1994) *Grundlagen naturnaher Gewässergestaltung – Geomorphologische Entwicklung von Fließgewässern*. Springer, Berlin.

Knox, J.C. (1987) Historical valley floor sedimentation in the Upper Mississippi valley. *Annals of the Association of American Geographers*, **77**, 224–244.

Kristensen, P. & Hansen, H.O. (eds) (1994) *European Rivers and Lakes: Assessment of their Environmental State*. European Environment Agency Environmental Monographs 1, Copenhagen.

Ladle, M. & Bass, J.A.B. (1981) The ecology of a small chalk stream and its responses to drying during drought conditions. *Archiv für Hydrobiologie*, 90, 448–466.

Lake, P.S. (2000). Disturbance, patchiness, and diversity in streams. *Journal of the North American Benthological Society*, 19 (4), 573–592.

Layzer, J.B., Nehus, T.J., Pennington, W., Gore, J.A. & Nestler, J.M. (1989) Seasonal variation in the composition of drift below a peaking hydroelectric project. *Regulated Rivers: Research & Management*, 3, 305–317.

Littlewood, I.G. (2002) Improved unit hydrograph characterisation of daily flow regime (including low flows) for the R. Teifi, Wales: Towards better rainfall-streamflow models for regionalisation. *Hydrology and Earth System Sciences*, 6, 899–911.

Lytle, D.A. & Poff, N.L. (2004) Adaptation to natural flow regimes. *Trends in Ecology and Evolution*, 19 (2), 94–100.

Mérigoux, S. & Dolédec, S. (2004). Hydraulic requirements of stream communities: A case study on invertebrates. *Freshwater Biology*, 49, 600–613.

Meyer, J.L., Sale, M.J., Mulholland, P.J. & Poff, N.L. (1999) Impacts of climate change on aquatic ecosystem functioning and health. *Journal of the American Water Resources Association*, 35 (6), 1373–1386.

Milne, B.T. (1991) Lessons from applying fractal models to landscape patterns. In: *Quantitative Methods in Landscape Ecology: The Analysis and Interpretation of Landscape Heterogeneity* (eds M.G. Turner & R.H. Gardner), pp. 199–235. Springer-Verlag, Berlin.

Mooij, W.M.G., Hülsmann, S., De Senerpont, L.N., *et al.* (2005) The impact of climate change on lakes in the Netherlands: A review. *Journal of Aquatic Ecology*, 39, 1386–2588.

Munn, M.D. & Brusven, M.A. (1991) Benthic invertebrate communities in nonregulated and regulated waters of the Clearwater River, Idaho, USA. *Regulated Rivers: Research & Management*, 6, 1–11.

Nõges, P., Kägu, M. & Nõges, T. (2007) Role of climate and agricultural practice in determining matter discharge into large, shallow Lake Võrtsjärv, Estonia. *Hydrobiologia*, 581 (1), 125–134.

Olden, J.D. & Poff, N.L. (2003) Redundancy and the choice of hydrologic indices for characterizing streamflow regimes. *River Research and Applications*, 19, 101–121.

Olsson, T. & Söderström, O. (1978) Springtime migration and growth of *Parameletus chelifer* (Ephemeroptera) in a temporary stream in northern Sweden. *Oikos*, 31, 284–289.

Palmer, M.A. & Poff, N.L. (1997) The influence of environmental heterogeneity on patterns and processes in streams. *Journal of the North American Benthological Society*, 16, 169–173.

Palmer, M.A., Arenburger, P., Botts, P.S., Hakenkamp, C.C. & Reid, J.W. (1995) Disturbance and the community structure of stream invertebrates: Patch-specific effects and the role of refugia. *Freshwater Biology*, 34, 343–356.

Palmer, M.A., Ambrose, R.F. & Poff, L.N. (1997) Ecological theory and community restoration ecology. *Restoration Ecology*, 5, 291–300.

Pedersen, M.L. & Friberg, N. (2009). Influence of disturbance on habitats and biological communities in lowland streams. *Fundamental and Applied Limnology*, 174, 27–41.

Petts, G.E., Gurnell, A.M., Gerrard, A.J., *et al.* (2000) Longitudinal variations in exposed riverine sediments: A context for the ecology of the Fiume Tagliamento. *Aquatic Conservation: Marine and Freshwater Ecosystems*, 10, 249–266.

Poff, N.L. (2002) Ecological response to and management of increased flooding caused by climate change. *Philosophical Transactions of the Royal Society A*, 360, 1497–1510.

Poff, N.L. & Allan, J.D. (1995) Functional-organization of stream fish assemblages in relation to hydrological variability. *Ecology*, 76, 606–627.

Poff, N.L. & Ward, J.V. (1989) Implications of streamflow variability and predictability for lotic community structure: A regional analysis of streamflow patterns. *Canadian Journal of Fisheries and Aquatic Sciences*, 46, 1805–1818.

Poff, N.L., Allan, J.D., Bain, M.B., *et al.* (1997) The natural flow regime: A paradigm for river conservation and restoration. *Bioscience*, 47, 769–784.

Pusey, B.J., Arthington, A.H. & Read, M.G. (1993) Spatial and temporal variation in fish assemblage structure in the Mary River, south-east Queensland: The influence of habitat structure. *Environmental Biology of Fishes*, 37, 355–380.

Radomski, P. & Goeman, T.J. (2001) Consequences of human lakeshore development on emergent and floating-leaf vegetation abundance. *North American Journal of Fisheries Management*, 21, 46–61.

Räisänen, J., Hansson, U., Ullerstig, A., *et al.* (2003) GCM driven simulations of recent and future climate with the Rossby Centre coupled atmosphere – Baltic Sea regional climate model RCAO. *SMHI Reports Meteorology and Climatology (Norrkoping, Sweden)*, **101**, S-60176.

Reid, L.M. & Page, M.J. (2003) Magnitude and frequency of landsliding in a large New Zealand catchment. *Geomorphology*, **49** (1–2), 71.

Rørslett, B., Mjelde, M. & Johansen, S.W. (1989) Effects of hydropower development on aquatic macrophytes in Norwegian Rivers: Present state of knowledge and some case studies. *Regulated Rivers: Research & Management*, **3**, 19–28.

Rose, N.L., Morley, D., Appleby, P.G., *et al.* (2010) Sediment accumulation rates in European lakes since AD 1850: Trends, reference conditions and exceedence. *Journal of Paleolimnology*, doi:10.1007/s10933-010-9424-6.

Sand-Jensen, K. & Madsen, T.V. (1992). Patch dynamics of the stream macrophyte, *Callitriche cophocarpa*. *Freshwater Biology*, **27**, 277–282.

Sandin, L. (2009). The relationship between land-use, hydromorphology and river biota at different spatial and temporal scales: A synthesis of seven case studies. *Fundamental and Applied Limnology*, **174** (1), 1–5.

Scheuerell, M.D. & Schindler, D.E. (2004) Changes in the spatial distribution of fishes in lakes along a residential development gradient. *Ecosystems*, **7**, 98–106.

Schindler, D.W., Beaty, K.G., Fee, E.J., *et al.* (1990) Effects of climatic warming on lakes of the Central Boreal Forest. *Science*, **250**, 967–970.

Schindler, D.W., Bayley, S.E., Parker, B.R., *et al.* (1996) The effects of climatic warming on the properties of boreal lakes and streams at the experimental Lakes Area, northwestern Ontario. *Limnology and Oceanography*, **41**, 1004–1017.

Schlosser, I.J. (1995) Dispersal, boundary processes, and trophic-level interactions in streams adjacent to beaver ponds. *Ecology*, **76**, 908–925.

Schröter, D., Cramer, W., Leemans, R., *et al.* (2005) Ecosystem service supply and vulnerability to global change in Europe. *Science*, **310**, 1333–1337.

Smith, W.R.E. & Pearson, G.R. (1985) Survival of *Sclerocyphon bicolor* (Coleoptera: Psephenidae) in an intermittent stream in North Queensland (Australia). *Journal of the Australian Entomological Society*, **24** (2), 101–102.

Solomon, S., Qin, D., Manning, M., *et al.* (eds) (2007) *Climate Change 2007: The Physical Science Basis, Contribution of Working Group I to the Fourth Assessment Report of the Intergovernmental Panel on Climate Change*. Cambridge University Press, Cambridge.

Statzner, B. & Holm, T.F. (1982) Morphological adaptations of benthic invertebrates to stream flow. An old question studied by means of a new technique (Laser Doppler Anemometry). *Oecologia*, **53**, 290–292.

Strommer, J.L. & Smock, L.A. (1989) Vertical distribution and abundance of invertebrates within the sandy substrate of a low-gradient headwater stream. *Freshwater Biology*, **22**, 263–274.

Syrovátka, V., Schenková, J. & Brabec, K. (2009) The distribution of chironomid larvae and oligochaetes within a stony–bottomed river stretch: The role of substrate and hydraulic characteristics. *Fundamental and Applied Limnology*, **174** (1), 43–62.

Townsend, C.R. & Hildrew, A.G. (1994) Species traits in relation to a habitat template for river systems. *Freshwater Biology*, **31**, 265–275.

Townsend, C.R., Scarsbrook, M.R. & Dolédec, S. (1997) Quantifying disturbance in streams: Alternative measures of disturbance in relation to macroinvertebrate species traits and species richness. *Journal of the North American Benthological Society*, **16** (3), 531–544.

Townsend, C.R., Downes, B.J., Peacock, K. & Arbuckle, C. (2004) Scale and the detection of land use effects on morphology, vegetation and macroinvertebrate communities of grassland streams. *Freshwater Biology*, **49**, 448–462.

Verburg, P.H., Eickhout, B. & van Meijl, H. (2008). A multi-scale, multi-model approach for analyzing the future dynamics of European land use. *Annals of Regional Science*, **42**, 57–77.

Verdonschot, P.F.M. & Nijboer, R.C. (2002) Towards a decision support system for stream restoration in the Netherlands: An overview of restoration projects and future needs. *Hydrobiologia*, **478**, 131–148.

Walker, K.F., Boulton, A.J., Thoms, M.C. & Sheldon, F. (1994) Effects of water-level changes induced by weirs on the distribution of littoral plants along the River Murray, South Australia. *Australian Journal of Marine and Freshwater Research*, **45**, 1421–1438.

Wantzen, K.M., Rothhaupt K-O., Mörtl, M., Cantonati, M., Tóth, L.G. & Fischer, P. (2008) Ecological effects of water-level fluctuations in lakes: An urgent issue. *Hydrobiologia*, **613**, 1–4.

Ward, J.V. (1989) The four dimensional nature of lotic ecosystems. *Journal of the North American Benthological Society*, **8** (1), 2–8.

Wiberg-Larsen, P., Brodersen, K.P., Birkholm, S., Gron, P.N, & Skriver, J. (2000). Species richness and assemblage structure of Trichoptera in Danish streams. *Freshwater Biology*, **43**, 633–647.

Williams, D.D. (1987) *The Ecology of Temporary Waters*. The Blackburn Press, Caldwell, NJ.

Wissmar, R.C. & Craig, S.D. (2004) Factors affecting habitat selection by a small spawning charr population, bull trout, *Salvelinus confluentus*: Implications for recovery of an endangered species. *Fisheries Management and Ecology*, **11** (1), 23–31.

Wood, P.J. & Armitage, P.D. (1997) Biological effects of fine sediment in the lotic environment. *Environmental Management*, **21** (2), 203–217.

Wright, J.F., Clarke, R.T., Gunn, R.J.M., Winder, J.M., Kneebone, N.T. & Davy-Bowker, J. (2003) Response of the flora and macroinvertebrate fauna of a chalk stream site to changes in management. *Freshwater Biology*, **48**, 894–911.

Wright, J.F., Clarke, R.T., Gunn, R.J.M., Kneebone, N.T. & Davy-Bowker, J. (2004) Impact of major changes in flow regime on the macroinvertebrate assemblages of four chalk stream sites, 1997–2001. *River Research and Applications*, **20**, 775–794.

5

Monitoring the Responses of Freshwater Ecosystems to Climate Change

Daniel Hering, Alexandra Haidekker, Astrid Schmidt-Kloiber, Tom Barker, Laetitia Buisson, Wolfram Graf, Gäel Grenouillet, Armin Lorenz, Leonard Sandin and Sonja Stendera

Introduction

Since 1970, freshwater biodiversity has decreased more drastically than marine or terrestrial biodiversity (Loh & Wackernagel 2004). This is the result of a complex mix of stressors and impacts (Stanner & Bordeau 1995; Malmquist & Rundle 2002). The major drivers can be summarized as multiple use (such as fisheries, navigation and water abstraction), nutrient enrichment, organic and toxic pollution, acidification and habitat degradation. Climate change is adding further stresses (temperature increase, hydrological changes) and interacts in complex ways with existing ones (Travis 2003; Wrona *et al*. 2006; Durance & Ormerod 2007; Huber *et al*. 2008).

As with other stressors, climate change will result in complex cause–effect chains, the link between them provided by many interacting environmental parameters, which are directly or indirectly influenced by temperature and precipitation. The response of the biota is, therefore, less predictable than the response of chemical or hydrological variables. On the other hand, biotic parameters such as species richness, community composition or functional diversity integrate the complex effects of many stressors on freshwater ecosystems, including those directly or indirectly associated with climate change. This is the reason for using biotic communities (such as phytoplankton, invertebrates or fish) for monitoring the ecological integrity of European surface waters, as stipulated by the EU Water Framework Directive (Heiskanen *et al*. 2004). The recently established Europe-wide monitoring programmes, however, are mainly targeted at detecting the effects of those stressors that have been dominant in the past, such as eutrophication, organic pollution, acidification or hydromorphological degradation. Climate change is not specifically targeted by the Water Framework Directive, though there is likely to be greater pressure on European aquatic ecosystems in future because of it.

Climate Change Impacts on Freshwater Ecosystems. First edition. Edited by M. Kernan, R. Battarbee and B. Moss. © 2010 Blackwell Publishing Ltd.

The direct and indirect effects of climate change on the biota of lakes, rivers and wetlands will depend on ecoregion, ecosystem type and other stressors affecting the water body. Owing to the natural variability of surface waters and the effects of many other stressors, no simple dose–response relationships among climate change and biotic effects can be expected; the linkages between climate change and biodiversity patterns cannot be understood without the overall, complex picture.

The purpose of this chapter is to suggest indicators for the effects of climate change on lake, river and wetland ecosystems that reflect the direction of their pathways, relative importance, and magnitude of change. The term 'indicator' is used here in a broad sense, i.e. a simple detectable sign of a complex process that can be used as an early warning of ecosystem change. Indicators may be chemical, hydrological, morphological, biological or functional parameters, which reflect key processes influenced by climate change and are relatively simple to monitor. There is a focus on parameters that are already used for monitoring programmes under the Water Framework Directive, but other indicators (e.g. hydrological parameters) are considered as well.

The selection of indicators is based on a literature review (up to 2007), in which we first categorize and describe potential direct and indirect climate change effects on lakes, rivers and wetlands. From this description a selection of parameters judged to most clearly reflect the effects of climate change on freshwater ecosystems is made. In two case studies, the susceptibility of selected taxonomic groups to climate change effects is analysed.

Climate change impacts on the biota of lakes

Monitoring and assessing the effects of stress on lakes

Monitoring of lake ecosystems in Europe has changed significantly in recent years. While formerly the use of physicochemical and selected biological variables (such as chlorophyll) was most widespread, the EU Water Framework Directive places emphasis on biological indicators. A variety of organism groups (phytoplankton, macrophytes, benthic invertebrates and fish) have now to be monitored, supplemented by hydromorphological and physicochemical measurements.

Most biological assessment systems, developed for the purpose of the Water Framework Directive, aim to reflect the deviation of the observed assemblage from an undisturbed reference state, thus providing an integrated appraisal of a water body's ecological quality. The current assessment systems are therefore reflecting the impact of a variety of stressors. The impact of climate change has rarely been considered in assessment systems and, as noted above, is not specifically addressed by the Water Framework Directive. However, almost all indices used to monitor the ecological status of European lakes will be affected by climate change.

Climate change impacts will be among the most important stressors on freshwaters in future, and will initiate chains of processes that are complex and difficult to classify. Simple indicators are required to judge how advanced these processes already are, and which of them could be integrated into presently applied assessment systems. Given the different types of impact in cold, temperate and warm ecoregions, different sets of indicators may be required. The theoretical background for indicator selection is described below and in earlier chapters, while a potential set of indicators for lakes in cold, temperate and warm ecoregions, respectively, is given in Table 5.1. Most indicators could be easily incorporated into routine monitoring programmes, e.g. by adding indices reflecting the impact of temperature increase on phytoplankton already monitored for the purpose of the Water Framework Directive. In many cases, simple physicochemical measurements are most appropriate, as they are easy to make, are often included in routine monitoring programmes and are located at the starting points of cause–effect chains, thus giving context to subsequent changes.

Hydrology and physicochemistry

The timing of ice cover is directly dependent on winter and spring temperatures, and therefore is one of the early indicators for climate change in lakes in cold and temperate regions as are the nature and duration of summer stratification. They have been discussed in Chapters 3 and 4. Likewise, there are many consequent chemical features of these that have already been discussed and provide strong initial indicators.

Primary production

Climate-sensitive physicochemical and hydrologic conditions can be major determinants for primary production in lakes. Phytoplankton community composition may be altered by changes in winter and spring temperatures, depending on lake type and location (Anneville *et al.* 2002; Christoffersen *et al.* 2006; Elliott *et al.* 2006). The phytoplankton assemblages of shallow cold-water ecosystems seem to be especially sensitive to temperature changes (Schindler *et al.* 1990; Findlay *et al.* 2001).

A shift towards dominance of cyanophytes in warmer water, with possible implications for water quality, is widely predicted and may lead to a progressive loss of phytoplankton biodiversity (Chapter 6). Models suggest that cyanobacterial dominance will be greatest if high water temperatures are combined with high nutrient loads. At low nutrient levels, the effect of water temperature change is reduced considerably (Anneville *et al.* 2004; Elliott *et al.* 2005, 2006).

Generally, increased phytoplankton productivity and biomass are correlated with higher spring water temperatures as well as changes in hydrochemistry such as increased nutrient availability (Schindler *et al.* 1996; Straile & Geller 1998; Findlay *et al.* 2001). Earlier stratification and deeper thermoclines may have an opposite effect, however. Furthermore, improved light conditions affect phytoplankton biomass. The better light conditions during warmer winters with shorter ice cover and less snow promote phytoplankton growth in winter, even doubling chlorophyll *a* levels (Pettersson *et al.* 2003).

Table 5.1 Direct and indirect impacts of climate change on lakes. c, t, w: variable relevant in cold (c), temperature (t) or warm (w) ecoregions

Category	Response	Indicator	Justification of indicator	c	t	w
Ice cover	Higher air, and thus higher water temperature, leads to a shorter ice cover period. The relationship between air temperature and timing of lake ice break-up shows an arc cosine function. This nonlinearity results in marked differences in the response of ice break-up timing to changes in air temperature between colder and warmer regions.	Ice-cover duration, timing of ice break-up, ice thickness	Ice cover duration is simple to monitor, e.g. by remote sensing.	x	(x)	
Hydrology — Stratification	Higher temperatures result in earlier onset and prolongation of summer stratification. As a result, changing mixing processes occur and systems may change from dimictic to warm monomictic. A lack of full turnover in winter might lead to a permanent thermocline in deeper regions.	Duration of summer stratification as reflected by water temperature	Water temperature reflects the status of lake stratification.	x	x	(x)
Water level	Increased temperature and decreased precipitation in conjunction with intensive water use will decrease water volumes. This will lead to water level imbalances and, in many cases, to the complete loss of water bodies.	Lake surface	Easy to monitor by remote sensing.	(x)	(x)	x

Table 5.1 (Cont'd)

Category	Response	Indicator	Justification of indicator	c	t	w
Oxygen depletion	High temperatures will stimulate phytoplankton growth, which will lead to oxygen depletion of profundal habitats.	Oxygen concentration of the bottom water in summer	The parameter is easy to record and often incorporated into routine water chemistry monitoring.	(x)	(x)	x
Sulphate concentration	With less precipitation in El Nino years and resulting droughts, stored reduced S in anoxic zones (wetlands) is oxidized during drought, with subsequently high sulphate export rates. Elevated sulphate concentrations in lakes will be the result.	Sulphate concentration	Directly reflecting the responding parameter; often incorporated into routine water quality monitoring	x	(x)	
Physicochemistry						
DOC	Rising temperatures in combination with declining acid deposition cause increasing DOC concentrations.	DOC	Incorporated into routine water quality monitoring	x	(x)	
Acidification effects on phytoplankton	Acidification pulses occur due to drought (El Nino). Acidification pulses will cause changes in phytoplankton richness and biomass.	pH; biotic acid indices	pH is easy to record and often incorporated in water chemistry monitoring. As pH varies seasonally and daily, biotic indices are often more stable.	x		

Parameter	Description	Indicator	Notes			
Salinity	Warmer winters cause extreme rainstorms and heavy sea-salt deposition, which might affect water chemistry.	Acidifying substances	These parameters are easy to record and often incorporated into routine water chemistry monitoring.	x	x	
Total Organic Carbon (TOC) runoff patterns	Warmer winters produce higher levels of runoff TOC release with subsequently increasing TOC water concentrations.	TOC levels and/or absorbance (water colour)	Water TOC concentrations reflect changes in runoff and input of allochthonous material.	x		
Water temperature effects on phytoplankton	Increasing water temperatures lead to shifts from a dominance of diatoms and cryptophytes to cyanobacteria. This effect is especially pronounced at temperatures > 20°C, since cyanobacteria (especially large, filamentous) and green algae are favoured at higher temperatures.	Phytoplankton biomass and composition, cyanobacterial algal blooms	The shift in community composition gives information about the response of biota to changed lake characteristics as water temperatures. Phytoplankton community composition is routinely monitored for the Water Framework Directive.	x	x	(x)
Water temperature effects on macrophytes	Inter-annual variation in water temperature results in deeper macrophyte colonization, greater wet weight biomass, and an increase in whole lake biomass.	Water temperature	The parameter is easy to record and often incorporated into routine monitoring programmes.	x		

Primary production

Table 5.1 *(Cont'd)*

Category	Response	Indicator	Justification of indicator	c	t	w
Water temperature effects on zooplankton	Higher water temperature leads to shifts in zooplankton community composition. Higher, earlier population growth rates of *Daphnia* and earlier summer decline occur due to higher spring temperatures. As a result, higher *Daphnia* biomass leads to earlier phytoplankton suppression and a shift from a dominance of large-bodied to smaller species.	Zooplankton biomass and composition, size classes	The response of zooplankton (although not monitored for the Water Framework Directive) might be a good indicator for changes in food web dynamics due to temperature increase.	(x)	x	(x)
Secondary production	Water temperature effects on cold water fish	Higher water temperatures (especially in the epilimnion) lead to the progressive reduction of thermal habitats for, e.g. *Salvelinus namaycush*. As a result, cold-water species will disappear from littoral areas in spring and summer. Furthermore, higher water temperatures will reduce reproduction success of cold-water species and increase parasitic and predator pressure on the egg and young life stages.	Summer water temperature or air temperature	Water temperature is easy to measure, but even air temperature reflects warming up of mixed layer temperature.	x	x	

Category	Indicator	Description				
	Spread of alien species	Higher temperatures often favour alien fish, macrophyte or macroinvertebrate species.	Share of alien species in the community	This parameter can often be inferred from routine monitoring for the Water Framework Directive.	(x)	x	x
Food webs	Water temperature effects on food webs	Increased water temperature generates principal shifts in food webs. As cyprinid planktivorous fish species are supported, large zooplankton species are suppressed and grazing intensity is reduced.	Proportion of planktivorous and piscivorous fish species; proportion of large and small zooplankton species	Food web structure is well reflected by these two parameters. The share of large zooplankton species determines the effects on phytoplankton, the share of planktivorous species determines the effects on zooplankton.	x	x	x

Category: ecosystem component being affected by direct or indirect climate change effects. Response: describes how the variables change under the stressor considered. Indicator: a judgemental selection of the variables that most clearly reflect climate change.

With increasing spring water temperature and light availability, spring phytoplankton species may grow earlier (Chen & Folt 1996; Müller-Navarra *et al.* 1997; Elliott *et al.* 2006; Adrian *et al.* 2006). The spring peak may also be more heavily grazed (Chen & Folt 1996; Müller-Navarra *et al.* 1997; Straile 2000). In summer, nutrient limitation may occur earlier. For example, the summer cyanobacteria peak may decline earlier because of nutrient limitation from increased spring growth at higher water temperatures (Bleckner *et al.* 2002; Elliott *et al.* 2005). Larger growths can be expected in autumn and winter, benefitting from higher temperatures and delayed light-limitation before ice forms (Bleckner *et al.* 2002).

Increased temperatures may result in greater macrophyte biomass or changes in macrophyte community composition (Rooney & Kalff 2000; McKee *et al.* 2002; Feuchtmayr *et al.* 2007).

Secondary production

Trends in average temperature sometimes correlate significantly with changes in zooplankton community composition, even over comparatively short periods of 10–15 years (Burgmer *et al.* 2007). Water temperature increase may be associated with a shift of zooplankton assemblages from larger to smaller bodied forms, a shift in food availability to inedible Cyanobacteria and possibly a heightened sensitivity to algal toxins (Moore *et al.* 1996). Individual taxa have very different threshold or maximum temperatures for growth that are species specific rather than functional group specific. A shift from the dominance of large-bodied *Daphnia galeata* to smaller *D. cucullata* has been observed with higher spring temperatures (Adrian & Deneke 1996). Slow-growing summer zooplankton with more complex life cycles responded specifically to seasonal warming, depending on its timing (Adrian *et al.* 2006). Changes in the vertical temperature gradient of a lake may affect zooplankton vertical migration. In warmer months, zooplankton occurred closer to the surface (Helland *et al.* 2007). There are, however, powerful effects of predation on zooplankton communities that may mask any direct climate effects.

Community composition and species richness in lake fish communities is strongly related to air temperature. In a study in third-order catchments, the presence or absence of 33 out of 61 fish species was related to temperature, as well as to geographic factors (Minns & Moore 1995). For many species, the relationships with temperature have long been investigated: for instance, *Coregonus albula*, an autumn spawning fish species, is vulnerable to spring temperature increases because of a timing mismatch of hatching and the spring development of zooplankton, in combination with higher rates of predation by warm-water predatory fish species (Nyberg *et al.* 2001). Another threat for this species is a general reduction in occurrence of the cold, oxygen-rich hypolimnetic conditions that it requires in summer (George *et al.* 2006). Generally, higher water temperatures increase growth and production for warm water fish and inhibit growth and production for fish at or above their thermal optimum (DeStasio *et al.* 1996; Petchey *et al.* 1999; Mackenzie-Grieve & Post 2006).

Changes in precipitation and temperature may have opposing effects on fish populations. In Norway, higher winter precipitation, leading to more accumulated snow in April, can be detrimental to the recruitment of brown trout (*Salmo trutta*), whereas warmer summers increase recruitment to levels that may lead to overpopulation and to the establishment of brown trout populations at higher elevations (Borgstrom & Museth 2005). Exceeding water temperature thresholds may limit the survival of fish in lakes. More fish may die from oxygen deficiency or physiological stress in warmer water (DeStasio *et al.* 1996). On the other hand, in shallow, eutrophic lakes, winter fish-kills caused by low dissolved oxygen under ice will be reduced or eliminated (Fang & Stefan 2000). Increased hypolimnetic temperatures may lead to a loss of juvenile fish requiring cool water as a summer refuge; thus, climate change can eliminate fish populations at the margins of their range (Gunn 2002). The effect of climate change at different latitudes has been modelled based on temperature and the minimum oxygen requirements of cold-, cool- and warm-water fish (Stefan *et al.* 1996). Zoogeographical boundaries could move significantly north (Carpenter *et al.* 1992; Petchey *et al.* 1999), a problem for several Salmonidae (DeStasio *et al.* 1996; Jansen & Hesslein 2004), whereas Percidae and Cyprinidae may benefit from an increased thermal habitat in the case of moderate warming (Jansen & Hesslein 2004). Hydrologic changes, together with changes in temperature, will probably favour invasive species over rare and threatened native species (Rogers & McCarty 2000).

As lake fish communities are well correlated with temperature patterns through the variety of pathways described above, water (or air) temperature is an indicator for changes in fish communities, supported by composition measures, such as the proportion of alien species in the fish community.

Food webs

The principal alteration in lake food webs caused by global warming may be a reduction in zooplankton grazing intensity, leading to eutrophication effects (see Chapter 6 for detailed discussion). This is caused by a variety of factors: the density of planktivorous cyprinid fish species is enhanced (Jansen & Hesslein 2004) and the density of piscivorous species reduced, which may lead to a strong top-down control of large zooplankton species (Jeppesen *et al.* 2005). Warmer spring temperatures may disrupt food web linkages between phytoplankton and zooplankton because of different sensitivities to warming. The timing of thermal stratification and spring diatom growth may advance significantly with increasing spring temperatures. Thus, a long-term decline in *Daphnia* populations, frequently a keystone herbivore, may be associated with an expanding temporal mismatch with the spring diatom bloom (Winder & Schindler 2004). A timing mismatch between phytoplankton maxima and the peak abundance of *Daphnia* may lead to the absence of a clear water phase (De Senerpont Domis *et al.* 2007) that is a feature of many lakes in late spring. Even modest warming (less than 2°C) during a short but critical seasonal period may induce changes in whole lake food webs and thus alter entire ecosystems (Strecker *et al.* 2004; Hampton *et al.* 2006; Wagner & Benndorf 2007).

Climate change impacts on the biota of rivers

Monitoring and assessing the effects of stress on rivers

As with lakes, monitoring of European rivers has changed with the introduction of the Water Framework Directive. Formerly, the use of physicochemical variables and selected hydromorphological and biological indices using primarily macroinvertebrates (such as Average Score per Taxon (ASPT) or Saprobic Indices) was most widespread, while now several organism groups (phytoplankton for large rivers, benthic algae, macrophytes, benthic invertebrates and fish) are being monitored, supplemented by hydromorphological and physicochemical variables.

The biological assessment systems for the Water Framework Directive reflect the deviation of the observed assemblage from an undisturbed reference state; in the case of rivers, the deviation is mainly caused by organic pollution, hydromorphological degradation, eutrophication and acidification. While organic pollution was formerly the most widespread stressor, hydromorphological degradation is now a main concern, particularly in Central Europe. Acidification is mainly restricted to north-west Europe, some Alpine regions and a number of other upland areas, although eutrophication universally affects the lowland reaches of rivers in Europe.

Table 5.2 gives a selection of potential indicators for climate change impacts for small rivers in cold, temperate and warm ecoregions of Europe. As for lakes, many of the biotic indicators could be easily incorporated into routine monitoring programmes, e.g. by adding indices reflecting the impact of temperature change on benthic invertebrates, which are already monitored for the purposes of the Water Framework Directive. However, physicochemical variables are most appropriate as early warning indicators.

Hydrology

Changes in hydrology have been discussed in Chapters 3 and 4.

Primary production

Changed runoff and water temperature are expected to cause changes in riparian vegetation (Hauer *et al.* 1997; Primack 2000) and may enhance macrophyte and algal growth. Lower water levels and increased nutrient availability will lead to a greater proportion of terrestrial plant species in floodplains (Hudon 2004). In Fennoscandia, macrophyte species richness decreases with latitude and altitude, mainly due to decreased July temperature. Thus, macrophyte biodiversity is expected to respond strongly to climate change (Heino 2002).

Secondary production

Water temperature, flow regime, channel morphology and sedimentation, which are all subject to impacts from climate change, are decisive factors for river invertebrates and fish. Key variables are summer and maximum temperatures.

Table 5.2 Direct and indirect impacts of climate change on rivers. c, t, w: variable relevant in cold (c), temperate (t) or warm (w) ecoregions

Category	Response	Indicator	Justification of indicator	c	t	w
Hydrology						
Decrease in ice cover duration	Higher temperatures will reduce ice cover duration.	Ice cover duration	Ice cover is a key factor for the productivity of boreal aquatic ecosystems and easy to monitor.	x	(x)	
Increase in drought frequency and duration	Decreased summer precipitation and increasing air temperature in some parts of Central, Eastern and Southern Europe change the character of several small streams from permanent to temporary.	Drought periods	As gauging stations are not installed in most small headwater streams, drought periods can be easily recorded by visiting the respective streams.	x	x	x
Change of permanent to intermittent regime	Due to less precipitation and increased demand for freshwater, higher temperatures, and higher transpirations, many small rivers will become intermittent with long dry phases in summer.	Drought periods	As gauging stations are not installed in most small headwater streams, drought periods can be easily recorded by visiting the respective streams.	(x)	(x)	x
Morphology						
Increased fine sediment entry	Extreme precipitation events increase surface runoff and lead to large amounts of fine sediments entering the streams; sediments accumulate and clog the bottom interstitial.	Number and discharge or flood events in unusual seasons (recorded by gauging stations)	Extreme precipitation events will wash out fine sediments from adjacent cropland and other land-use types. They are well reflected by the discharge of a river.	(x)	x	(x)

Table 5.2 (Cont'd)

Category	Response	Indicator	Justification of indicator	c	t	w
	Increase of eutrophicating substances	N flux in the runoff and decomposition of soil organic matter increases with temperature, which increases nutrient concentrations. Eutrophication is further promoted by high water retention time through low discharge, while denitrification counteracts this effect.	Nitrate, total N, phosphate	Nutrients are routinely monitored in most European countries.	x	x	x
Physicochemistry	Reduced water quality	Increasing water temperatures enhance production and decomposition intensity, thus leading to oxygen depletion, particularly at night.	Saprobic indices	Saprobic indices reflect the organic load in streams and eventually the oxygen content. Species with a high oxygen demand (typical for low saprobic indices) will disappear while species with a low oxygen demand (typical for high saprobic indices) will benefit.	(x)	x	x

Process / effect	Description	Indicators	Explanation			
'Potamalization' – effects on nutrients	Higher water temperatures lead to a more rapid mineralization of organic matter (leaves, wood) and thus to eutrophication effects. As a result, small streams ('rhithral') will change character and resemble larger rivers ('potamal').	Water temperature (maximum monthly values)	The response of communities is mainly determined by extremes; secondary effects (e.g. oxygen depletion at night times) are most extreme in summer.	(x)	x	(x)
Acidification	Increased precipitation increases acid runoff from borealic coniferous forests leading to cascading acidification effects on aquatic biota.	pH, invertebrate-based acid-indices	pH-values decrease with increasing acid deposition. Since these events are of short duration, community based indices are often better at reflecting acidification.	x		
Primary production — Increased macrophyte/algae growth	Higher water temperatures and lower discharge enhance macrophyte and algae growth. Furthermore, higher temperatures increase mineralization processes and deliver more nutrients for macrophyte and algae growth.	Water temperature (mean monthly values), macrophyte coverage	Mean monthly temperatures indicate overall temperature increase. Macrophyte coverage is simple to record and well correlated with biomass.	x	x	x
Secondary production and food webs — Increase of respiration rate	The metabolic rates of bacteria and fungi and the metabolic rates of detritivorous species will rise with increasing temperatures. The proportion between primary production and respiration will decrease.	Percentage of collectors in the invertebrate community	Collectors gather organic material. If this food source increases, the percentage of collectors rises to about 40%.	x	x	(x)

Table 5.2 (Cont'd)

Category	Response	Indicator	Justification of indicator	c	t	w
Reduced availability of leaves	Processing rates of leaves and wood increase with temperature. Floods in winter cause more than 50% of leaf inputs to be exported, leaving little detrital material available for invertebrate consumption.	Share of the feeding type 'shredder' in the invertebrate community	Benthic invertebrates are routinely monitored in most European countries. If the availability of leaves decreases, the share of shredders will decrease, too; in small headwater streams, shredders should typically account for 30%–40% of the invertebrate community.	x	x	(x)
Replacement of cold water species (fish, macroinvertebrates)	Many fish and invertebrate species in cold regions are highly adapted to cold water temperatures (cold stenotherms) and vanish with higher temperatures.	Water temperature (maximum monthly values)	Physiologic barriers are mainly determined by extremes. For cold water species, these are too warm temperatures in crucial phases of their life cycle.	x	x	(x)
Increase or decrease of species number	Low temperatures are a migration and physiological barrier for many aquatic species. With temperature increase, several species can invade rivers in cold ecoregions. In contrast, in temperate and warm regions, increasing water temperatures lead to the extinction of cold stenothermic taxa.	Number of species (e.g. fish, selected invertebrate groups)	The increase of species numbers is best evaluated by a simple richness index, e.g. the number of species that can be easily inferred from routine monitoring results.	x	x	x

Indicator	Description	Measure	Detail			
Increase of r-strategists	Invertebrate r-strategists benefit from unpredictable flood events, e.g. in summer, which remove most invertebrates and thus favour species rapidly colonizing the competition-free space.	Number and discharge of flood events in unusual seasons (recorded by gauging stations)	Unusual hydrological events, e.g. floods in summer, cause catastrophic drifts of invertebrate species and favour r-strategists.	(x)	x	x
Changes in life strategies	If small rivers become intermittent, species with a bivoltine or semivoltine life cycle cannot survive and the community will change to univoltine species with an early emergence period.	Drought periods	As gauging stations are not installed in most small headwater streams, drought periods can be easily recorded by visiting the respective streams.	(x)	x	
'Potamalization' – effects on invertebrates	Higher water temperature leads to the disappearance of species adapted to cold water temperature and the associated high oxygen content, e.g. several stonefly (Plecoptera) species. They are replaced by species typical for warmer water previously colonizing more downstream reaches. Thus, invertebrate species typical for small streams ('rhithral') will be replaced by species from larger rivers ('potamal').	Share of invertebrate taxa preferring the metarhithral (trout zone)	Benthic invertebrates are routinely monitored in most European countries. The response of the invertebrate community to temperature increase is reflected by their longitudinal zonation preference. The share of 'metarhithral taxa' in an unimpacted small stream differs between ecoregions, but should typically be around 50%.	(x)	x	(x)

Table 5.2 (Cont'd)

Category	Response	Indicator	Justification of indicator	c	t	w
Replacement of salmonid by cyprinid fish species	Higher water temperatures will reduce reproductive success of salmonid species and increase parasitic and predator pressure on the egg and young larval stages. Warm water cyprinid species will invade in cold water regions.	Water temperature (maximum and minimum monthly values); fish species composition	The eggs of salmonid species need high oxygen concentrations, which will be reduced by higher water temperatures. Parasites and fungi benefit from high temperatures.	x	x	(x)
Standing stock of cold water fish	Brook trout populations could either benefit from increased growth rates in spring and fall or suffer from shrinking habitat and reduced growth rates in summer, depending on the magnitude of temperature change and on food availability.	Abundance and biomass of brook trout	Brook trout is a keystone species in most Northern European countries.	x	x	(x)
Spread of alien species	Higher temperatures often favour alien species that increasingly colonize small streams. These could be alien fish, macrophyte or macroinvertebrate species.	Water temperature (maximum and minimum monthly values); share of alien species in the community	The survival and reproduction of several alien species in temperate ecoregions is controlled by minimum temperatures. The second parameter can often be inferred from routine monitoring for the Water Framework Directive.	(x)	x	x

Category: ecosystem component being affected by direct or indirect climate change effects. Response: describes how the variables change under the stressor considered. Indicator: a judgemental selection of the variables that most clearly reflect climate change.

Invertebrates

In studies that compared invertebrate communities throughout Europe, community composition was shown to change with maximum stream temperature, mean July temperature, and with decreasing latitude and altitude (Lake *et al.* 2000; Heino 2002; Xenopoulos & Lodge 2006). Temperature affects invertebrate community composition by influencing species-specific developmental rates and overall assemblage phenology and by excluding taxa unable to tolerate certain temperature ranges (Hawkins *et al.* 1997; Haidekker & Hering 2008). Higher summer water temperature and low flow result in the progressive replacement of upstream, cold-water invertebrate taxa by downstream, thermophilic invertebrate taxa (Daufresne *et al.* 2004). Benthic invertebrate abundance and diversity (particularly of Ephemeroptera, Plecoptera and Trichoptera) is predicted to decrease as a result of changes in water temperature, flow regime, increased sedimentation and changes in channel morphology, and thus changes in habitat availability (Lake *et al.* 2000). Flow regime is another key factor for invertebrate assemblage structure. From a set of hydrological variables, those associated with flow had the highest correlation with macroinvertebrate community metrics for sites in England and Wales (Monk *et al.* 2006). Leaf litter quality and quantity, major food sources in small streams, respond also to climate change with coarse particulate organic matter availability, decreasing with flood frequency. Buzby & Perry (2000) showed that more than 50% of leaf inputs were exported, leaving only sparse leaf litter available to invertebrates. Many of these changes are reflected in the invertebrate community, e.g. by the composition of feeding types and life history traits (Table 5.2).

The vulnerability of freshwater organisms to the direct and indirect effects of climate change can be estimated by the ecological preferences of species. Species may be classified as follows:

- Species with limited distribution ('endemic species') are characterized by a restricted ecological niche and limited dispersal capacity, and are thus more affected by climate change than widely distributed species (Malcolm *et al.* 2006; Brown *et al.* 2007).
- Species inhabiting large rivers characterized by relatively high water temperatures are generally physiologically adaptive and may react to globally rising temperatures by colonizing upstream river reaches; species inhabiting springs cannot move further upstream and are thus more threatened (Fossa *et al.* 2004).
- Species adapted to low water temperatures ('cold-stenothermic species') are threatened by climate change more than eurythermic species (compare Schindler 2001).

Insect species potentially endangered by climate change are unevenly distributed in Europe, but there are also differences between individual insect orders. Three insect orders provide a case study of the above classification. A database on the distribution and ecological preferences of European freshwater mayflies (Ephemeroptera), stoneflies (Plecoptera) and caddisflies (Trichoptera) (www.freshwaterecology.info)

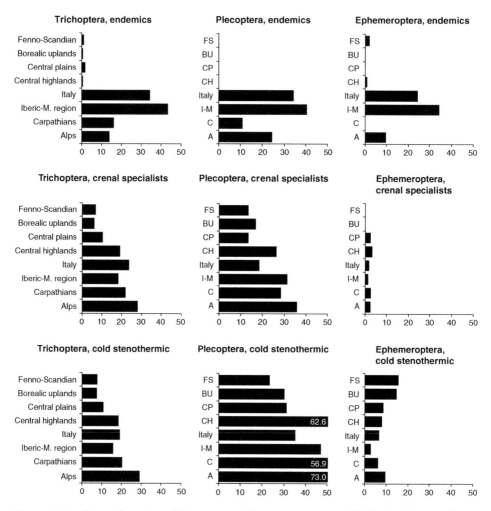

Figure 5.1   Share of species of three aquatic insect groups, caddisflies (Trichoptera), stoneflies (Plecoptera) and mayflies (Ephemeroptera), potentially endangered by climate change in selected European ecoregions according to Illies (1978) (two Northern European ecoregions, two Central European ecoregions, two Mediterranean ecoregions and two high mountain areas). Data based on 1134 Trichoptera taxa, 563 Plecoptera taxa and 344 Ephemeroptera taxa.

was used to calculate the fraction of species in each of the above groups. The data are based on an intensive literature survey and cover all European taxa of the selected organism groups (see Graf *et al.* (2008) for Trichoptera, Graf *et al.* (2009) for Plecoptera and Buffagni *et al.* (2009) for Ephemeroptera).

For all three orders, a high proportion of species in the Mediterranean ecoregions and in high mountain areas are highly vulnerable to climate change (Fig. 5.1). Most Central and Northern European species, however, are widely distributed and likely to be less affected by climate change.

Patterns of endemism are similar for the three insect orders. Up to 45% of species occurring in the Iberic-Macaronesian region are endemic and high fractions of endemic species are also found in Italy, the Balkan ecoregions and the high mountain ecoregions, such as the Alps and the Carpathians. Species restricted to springs (crenal zone) are common among the Trichoptera and Plecoptera, but rare among Ephemeroptera. Most crenal specialists are found in the Mediterranean and in the high mountain areas, while only very few of such species occur in Northern Europe. Cold-stenothermic species of Trichoptera and Plecoptera mainly occur in high mountain areas, while relatively few such species are distributed in Scandinavia. This is different from Ephemeroptera, which have generally a low number of cold stenothermic species, predominantly occurring in Northern Europe.

In general, a south-north gradient in species richness of aquatic insects can be observed. Similar patterns are found for endemic species and (to a lesser degree) for crenal specialists and cold-stenothermic species. These patterns are mainly a result of fluctuations in continental ice cover during the Pleistocene, which, in turn, caused several range extensions and regressions of species (Malicky 2000; Pauls *et al.* 2006). While glaciers covered most of Northern Europe, species retreated to Southern Europe or to ice-free parts of high mountain areas. This isolation of populations resulted in many new species and increased diversity in these areas. Most aquatic insect species occurring in Northern Europe live in Central or Southern Europe too. Mainly, generalists and species with a high dispersal capacity recolonized Northern Europe after the last ice age, while specialist species and those with limited dispersal capacities extended their range only slightly. In consequence, most of the species occurring in Northern Europe are likely to be capable of resisting climate change impacts, since they are generalists or able to rapidly colonize other areas.

Fish

Temperature and hydrological factors are major environmental determinants for fish communities, thus alterations induced by climate change are expected to modify fish assemblage structure (Poff & Allan 1995; Heino 2002). Changing river discharge causes a reduction in community richness and life cycle changes (Schindler 2001; Xenopoulos *et al.* 2005). Water temperature increase may be a threat to cold stenothermic fish species because habitats for cold stenotherms decline, isolating them in increasingly confined headwaters (Eaton & Scheller 1996; Hauer *et al.* 1997; Schindler 2001). As a result, fish assemblages are expected to shift to fewer cold-water taxa and more warm-water taxa, as well as fewer northern taxa and more southern taxa (Daufresne *et al.* 2004). As for invertebrates, fish responses to water-temperature increase are highly species-specific and depend on individual thresholds. For example, the size of Atlantic salmon (*Salmo salar*) is negatively correlated with spring air and water temperatures and with discharge and precipitation (Swansburg *et al.* 2002). Increases in winter temperature and ice break-up may affect winter survival significantly, particularly in northern populations. Because energetic deficiencies are assumed to be an important cause of winter mortality, strong negative effects on the energy budget can be expected (Finstad *et al*. 2004).

In the river Rhone (France), analysis of long-term fish data revealed that the variability of fish abundance was correlated with discharge and temperature during the reproduction period (April–June). Low flows and high temperatures coincided with high fish abundance. In line with temperature increase, southern, thermophilic fish species, e.g. chub (*Leuciscus cephalus*), and barbel (*Barbus* sp.) progressively replaced northern, cold-water fish species, e.g. dace (*Leuciscus leuciscus*) (Daufresne *et al*. 2004). With increasing water temperature the incidence of proliferative kidney disease increased in Switzerland, and populations declined by up to 66% for brown trout (Hari *et al*. 2006). High temperatures may induce an upward migration of fish at their thermal limits: e.g. brown trout (Heggenes & Dokk 2001; Hari *et al*. 2006), charr (*Salvelinus* sp.), and salmon (Carpenter *et al*. 1992) or an alteration in their migration behaviour, increasing the use of thermal refugia in cooler tributaries (chinook salmon, Cole *et al*. 1991). The timing of migration and spawning is also closely linked to thermal state (Salinger & Anderson 2006). Water temperature increase and earlier snowmelt causes an earlier return of adult salmon and alewives, and also affects the timing of migration and spawning of brown trout (Huntington *et al*. 2003). Several metrics related to the composition of fish communities and key species are suited as indicators of these changes (Table 5.2).

In a study of climate change and fish biodiversity, a list of river fish species occurring in 152 European river basins was collected from an intensive literature survey. In all, the fish database contained 306 fish species. To describe fish autecology, 21 fish species traits were considered. Species traits were diverse but most of them concerned reproduction (e.g. fecundity), feeding (e.g. trophic group) or morphology (e.g. body length). Each trait was coded in several categories ('modalities') totalling 72 trait modalities. The 306 fish species present in the database were assessed for these attributes based on the Fish-based Assessment Method for the Ecological status of European rivers (FAME) database (Schmutz *et al*. 2007) or the existing unpublished and published literature and were completed by expert judgement. The trait composition of fish assemblages was thus available for each of the 152 basins. First, 9 ecoregions (Illies 1978) describing a latitudinal gradient were selected. Only the river basins contained within a single ecoregion were retained. Fish trait composition was then compared among the different ecoregions (Fig. 5.2). A north-south gradient was observed for fish body length and traits related to reproduction, but not for traits describing habitat and feeding. In the southern ecoregions, fish assemblages are characterized by small fish, early maturation, small eggs with a short incubation period and the absence of a strategy for egg protection. Those traits are biologically concordant and counter to strong investment in reproduction, which is one of the main features of the more northern cold-water species.

Fish trait composition thus appeared to vary strongly along the latitudinal gradient. To understand more precisely this latitudinal structure, the fish trait composition was related to environmental variables that are good descriptors of the latitudinal gradient. These environmental data were extracted from 0.5°× 0.5°grid data (CIESIN 2005)

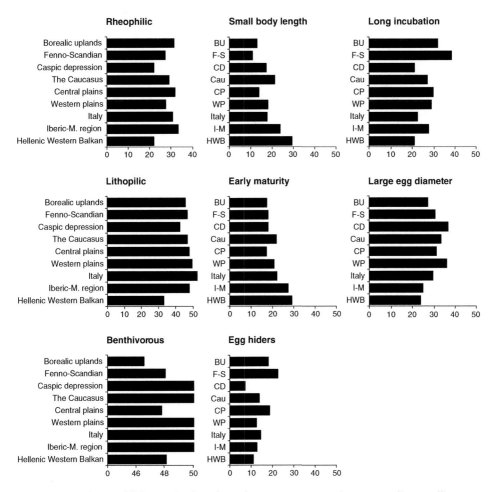

Figure 5.2   Share of fish species in selected European ecoregions according to Illies (1978). Data based on 306 fish species.

and the Atlas of Biosphere (SAGE 2002) and were thus available at the river basin scale. Climate data (temperature and precipitation) were related to fish trait modalities using Generalized Additive Models (GAMs). Among the 62 most common trait modalities, 44 and 48 responded significantly ($p < 0.05$) to mean annual temperature and precipitation, respectively (Fig. 5.3). For instance, with an increasing temperature, there is an increase in the proportion of benthivorous species but a decrease in the proportion of species laying large eggs and having long incubation periods. With an increase in precipitation, more rheophilic species are observed and these species mature early, few of them exhibiting parental care. Only 5 of the 62 modalities tested were not influenced by at least one of the two climatic variables. These relationships explain the differences in fish trait composition observed between ecoregions and suggest that Northern and Southern European fish species will not be affected equally by future climate change.

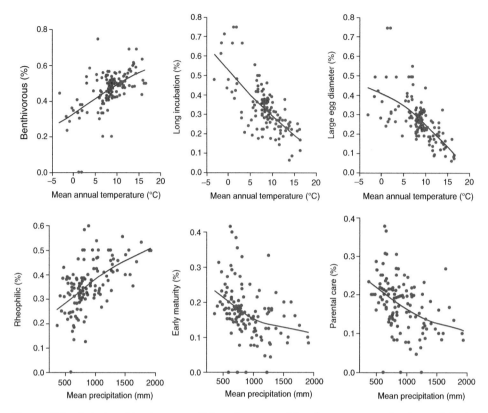

Figure 5.3 Relationships between fish trait composition (percentage of trait modality) and two climatic variables (mean annual temperature and mean precipitation) for 152 European river basins. The line was fitted using a generalized additive model (GAM).

Climate change impacts on the biota of wetlands

Monitoring and assessing the effects of stress on wetlands

Wetlands are not directly included in the Water Framework Directive, but any logical interpretation of the pristine nature of floodplain rivers must include them as key components. There is a limited number of more regional assessment systems, while Europe-wide approaches are less common. Several variables may reflect the effects of climate change on wetland ecosystems, particularly on processes and functioning. Table 5.3 lists suitable indicators for climate change impacts on wetlands.

Hydrology

Wetland hydrological regimes face impacts from both land use and climate. A rise in ambient temperature will result in increased water abstraction, and, with more

Table 5.3 Direct and indirect impacts of climate change on wetlands

Category		Response	Indicator	Justification of indicator
	Ice cover duration	Increased air temperatures lead to a later thaw.	Date of ice break-up	Indicates direct temperature effects. Influences length of season.
	Retention of flood water	Increased temperature may lead to increased rates of evaporation.	Water table height	Retention of flood water will be enhanced if the water table is lowered but reduced if the water table is higher.
Hydrology	Recharge of groundwaters	Ability to recharge aquifers is affected by desiccation.	Water table height	If the water table is high, the rate of recharge of groundwater (if any) will be increased.
	Retention of sediment	Scouring of sediments by extreme weather events.	Frequency and severity of storms	Storms and associated flash floods and spates may wash away sediments and detritus, reducing their retention.
	Acquisition of carbon I	Warm and wet conditions together increase carbon acquisition.	temperature and precipitation	Indicates gross C dynamics.
Physicochemistry	Acquisition of carbon II	Early snow melt followed by wet and warm conditions lead to high carbon acquisition through photosynthesis.	Date of snow melt, temperature and precipitation	A combination of early spring plus wet and warm weather promotes vegetation growth. Photosynthesis sequesters carbon, while rates of respiration remain comparatively stable year to year. Thus, carbon is accumulated in these conditions.

Table 5.3 (Cont'd)

Category	Response	Indicator	Justification of indicator
Export of organic carbon	Organic carbon is provided for downstream ecosystems in runoff water.	Water table height	Organic carbon can be released as dissolved organic carbon (DOC) in runoff water.
Release of CH_4	Lowered water table reduces CH_4 emission.	Water table height	Water table height and greenhouse gas emission are directly correlated.
Retention of carbon	Retention of carbonaceous material will be enhanced if warmer temperatures increase primary production while water availability is sufficient, but will be reduced if runoff events increase in frequency.	Rate of primary production	Retention of carbon in vegetation and detritus will be enhanced by increased production.
Mineralization	Lowered water table stimulates enzyme activity leading to increased mineralization.	Rates of enzyme activity	Indicates carbon loss from peat.
Export of nutrients	Increased production leads to greater provision of organic detritus, which is then available for export downstream.	Production of litter	Export of nutrients will increase if detritus production is high.
Retention of nutrients	Longer season and warmer temperatures can lead to increased primary production.	Rate of primary production	Retention of nutrients is potentially greater where production is high.

	Indicator	Description	Measure	Interpretation
Primary production	Tree survival	Increase in flooding can lead to progressive replacement of forest with bogs.	Water table height	Water table height and tree survival are directly correlated.
	Vegetation assemblages	Elevated water table leads to increase in bryophytes and reduction of shrubs in the bog, but increase in graminoids and forbs in the fen. Lowered water table leads to increased proportion of dicotyledonous plant species.	Water table height, vegetation assemblages	Water table height and vegetation assemblages are directly correlated.
	Vegetation type	Lowered water table leads to succession to forest-type vegetation from the graminoids and mosses occurring in pristine conditions.	Water table height, occurrence of higher plant species	Water table height and vegetation assemblages are directly correlated.
Secondary production	Insect species	Milder winters and hot summers are important factors in the survival of temperature-sensitive species. This will probably alter the tolerable ranges of some species, including pest species, and may lead to increased invasions into new areas by exotic species.	Taxonomic composition and abundance of insect species, especially butterflies and aquatic insects	Indicates impacts on habitat integrity.

Table 5.3 (Cont'd)

Category	Response	Indicator	Justification of indicator
Bird migration	Spring migrations start earlier with warming. This is more pronounced early in the season and with terrestrial and wetland birds than with waterfowl.	Beginning of spring migration period	
Ecosystem support	Ecosystems potentially will suffer if detritus and species are lost due to severe flooding and runoff events, and if drought levels exceed the tolerance limits of species.	Frequency and severity of storms	Reduction of biomass and species due to wash out.
Food webs	Drought and flooding both contribute to mineralization and release of nutrients from organic matter. This can increase the build up of plant-available nutrients in the sediments, which are readily washed into water courses and wetlands in runoff.	Support of food webs	Frequency and severity of storms	Increase in eutrophication may result from large runoff events.

Category: ecosystem component being affected by direct or indirect climate change effects. Response: describes how the variables change under the stressor considered. Indicator: a judgemental selection of the variables that most clearly reflect climate change. As less data than for lakes and rivers is available we do not distinguish between wetlands in different climatic regions.

frequent droughts predicted, irrigation demands will increase. Changes in land use may require the construction of dams, with barrier effects for hydrological conditions. This has implications for the size and spatial distribution of wetlands (Brinson & Malvarez 2002; Pyke 2004; Perotti *et al.* 2005). Additionally, nutrient enrichment and pollution are possible consequences of land-use changes from intensified crop growth, prolonged growing seasons and increasing urbanization and industrialization (van Breemen *et al.* 1998; Hudon 2004). Changes in precipitation, evaporation and temperature determine the groundwater level, which influences the wetland cover cycle, the transition between permanent and temporary wetlands and hydrochemical variables (e.g. Johnson *et al.* 2004; Lischeid *et al.* 2007). Water table height reflects the influences of climate change in many wetland types.

Physicochemistry

Mineralization and release of nutrients are determined by hydrology and temperature. Moisture and temperature influence microbial enzyme activity and decomposition rates. Drier, warmer conditions could stimulate nutrient mineralization and enhanced release from sediments to runoff water (Freeman *et al.* 1996; Fenner *et al.* 2005). Mineralization of C, N, and P may differ significantly among wetland types. In bog peats, nitrogen mineralization and CO_2 production may decrease with increasing ambient temperatures and lower water tables, whereas in fen peats, nitrogen mineralization may decrease and methane production may increase with higher water tables (Keller *et al.* 2004), but the generality of these findings is questionable. Stored reduced sulphur in anoxic zones of wetlands oxidizes during drought periods. Owing to the subsequent efflux into streams and lakes, sulphate concentrations and acidity can increase after droughts (Dillon *et al.* 1997; Aherne *et al.* 2004). Carbon acquisition may increase under warmer and wetter conditions, whereas in warmer and drier years, wetlands experience significant carbon losses (Carroll & Crill 1997; Griffis & Rouse 2001). Climate change is predicted to cause a doubling of net total C loss rates in wetlands (Clair *et al.* 2001). Maximum soil temperatures are correlated with maximum CH_4 emission values, whereas reduced water table levels suppress CH_4 emissions. Thus, long-term climatic changes with less precipitation and decreased water tables may reduce the incidence of CH_4 release from wetlands (Moore *et al.* 1998; Gedney & Cox 2003; Werner *et al.* 2003). Nonetheless, lowered groundwater changes due to climate change may lead to increased nitrous oxide fluxes in natural peatland soils (Regina *et al.* 1999). Under extreme drought, emissions may increase exponentially with a linear decrease in the water table (Dowrick *et al.* 1999).

Primary production

A key driver of changes in the community composition of wetland vegetation is the altered hydrological regime. Drought results in a proportional loss of native species, an influx of invasives, and a community change towards dicotyledonous species in northern wetlands (Hogenbirk & Wein 1991). In bogs, increased temperature, together with an experimentally lowered water table, caused a 50% increase in the cover of shrubs, and a 50% decrease in the cover of graminoids (Weltzin *et al.* 2003). More frequent droughts cause wetland vegetation to become both woodier and drier. Pond-meadow wetlands may acquire more species, particularly nonwetland species (Hudon 2004; Mulhouse *et al.* 2005).

Soil temperature rise causes shifts in the productivity of plant communities, e.g. an increase in shrub productivity and decreased forb (herbaceous flowering plant) productivity. A higher water table caused bryophyte productivity to increase in bog samples, while shrub productivity was lower (Weltzin *et al.* 2000). Altered riparian vegetation (herb vegetation and trees), and an altered biomass and productivity, affects detritivores and results in lower decomposition rates (Carpenter *et al.* 1992). Decomposition in river marginal wetlands is highly dependent on precipitation, whereas climate change or river flow management could disrupt floodplain nutrient dynamics, i.e. the periodic processes of organic matter retention, breakdown, mineralization and release (Andersen & Nelson 2006). Species' sensitivity to climate change is dependent on plant traits and niche properties (Thuiller *et al.* 2005). Besides water table height, which reflects many of these changes to some degree, the occurrences of certain species and vegetation assemblages may be used as indicators for climate change in plant communities (Table 5.3).

Secondary production

Lower water tables, increased temperatures and more frequent droughts lead to a loss of habitat for obligate wetland species. Invertebrates are affected directly by changing water table or temperature, but also indirectly by shifts in nutrient availability. Invasive, exotic plant species and changed nutritional quality of litter affect detritivores (Carpenter *et al.* 1992; Andersen & Nelson 2006). Other taxa affected by increased drought, weaker spring flows and reduced inundation are fish, amphibians, waterfowl and muskrat (Schindler 2001; Diamond *et al.* 2002).

Climate change advances the spring arrival of migrating birds, in both short- and long-distance migrants (Zalakevicius & Zalakeviciute 2001). It also changes the winter distribution of shorebirds (Gillings *et al.* 2006). Furthermore, global climate change alters the ranges and population state of different breeding bird species and populations. The impact of global warming on terrestrial and wetland birds is more evident than upon waterfowl (Zalakevicius & Zalakeviciute 2001). Variability of precipitation in wetlands affects population and community dynamics of wetland birds owing to egg and nestling predation, which was negatively correlated with water levels in wetlands (Fletcher & Koford 2004). Suitable indicators might include the beginning of the bird's spring migration period and metrics related to taxonomic composition of indicator taxa groups (Table. 5.3).

Conclusions

Climate change will lead to complex cause–effect chains between temperature/ precipitation and ecosystem response. As climate change will be a main stressor for aquatic ecosystems in Europe and beyond, the effects need to be specifically targeted in monitoring programmes, e.g. those implemented for the EU Water Framework Directive. Possible indicators are to some degree specific for ecoregions and freshwater ecosystem types and can be selected from a wide range of chemical, functional and biological parameters. Dose–response relationships between drivers and indicators need to be defined in future.

References

Adrian, R. & Deneke, R. (1996) Possible impact of mild winters on zooplankton succession in eutrophic lakes of the Atlantic European area. *Freshwater Biology*, **36**, 757–770.

Adrian, R., Wilhelm, S. & Gerten, D. (2006) Life-history traits of lake plankton species may govern their phenological response to climate warming. *Global Change Biology*, **12** (4), 652–661.

Aherne, J., Larssen, T., Dillon, P.J. & Cosby, B.J. (2004) Effects of climate events on environmental fluxes from forested catchments in Ontario, Canada: Modelling drought-induced redox processes. *Water, Air and Soil Pollution: Focus*, **4**, 37–48.

Andersen, D.C. & Nelson, S.M. (2006) Flood pattern and weather determine Populus leaf litter breakdown and nitrogen dynamics on a cold desert floodplain. *Journal of Arid Environments*, **64** (4), 626–650.

Anneville, O., Ginot, V., Druart, J.C. & Angeli, N. (2002) Long-term study (1974–1998) of seasonal changes in the phytoplankton in Lake Geneva: A multi-table approach. *Journal of Plankton Research*, **24** (10), 993–1007.

Anneville, O., Souissi, S., Gammeter, S. & Straile, D. (2004) Seasonal and inter-annual scales of variability in phytoplankton assemblages: Comparison of phytoplankton dynamics in three peri-alpine lakes over a period of 28 years. *Freshwater Biology*, **49** (1), 98–115.

Bleckner, T., Omstedt, A. & Rummukainen, M. (2002) A Swedish case study of contemporary and possible future consequences of climate change on lake function. *Aquatic Sciences*, **64**, 171–184.

Borgstrom, R. & Museth, J. (2005) Accumulated snow and summer temperature – Critical factors for recruitment to high mountain populations of brown trout (*Salmo trutta* L.). *Ecology of Freshwater Fish*, **14** (4), 375–384.

van Breemen, N., Jenkins, A., Wright, R.F., *et al.* (1998) Impacts of elevated carbon dioxide and temperature on a boreal forest ecosystem (CLIMEX project). *Ecosystems*, **1**, 345–351.

Brinson, M.M. & Malvarez, A.I. (2002) Temperate freshwater wetlands: Types, status, and threats. *Environmental Conservation*, **29** (2), 115–133.

Brown, L.E., Hannah, D.M. & Milner, A.M. (2007) Vulnerability of alpine stream biodiversity to shrinking glaciers and snowpacks. *Global Change Biology*, **13** (5), 958–966.

Buffagni, A., Cazzola, M., López-Rodríguez, M.J., Alba-Tercedor, J. & Armanini, D.G. (2009) Ephemeroptera. In: *Distribution and Ecological Preferences of European Freshwater Organisms*, Vol. 3 (eds A. Schmidt-Kloiber & D. Hering), 254 pp. Pensoft Publishers, Sofia, Bulgaria.

Burgmer, T., Hillebrand, H. & Pfenninger, M. (2007) Effects of climate-driven temperature changes on the diversity of freshwater macroinvertebrates. *Oecologia*, **151** (1), 93–103.

Buzby, K.M. & Perry, S.A. (2000) Modeling the potential effects of climate change on leaf pack processing in central Appalachian streams. *Canadian Journal of Fisheries and Aquatic Sciences*, **57** (9), 1773–1783.

Carpenter, S.R., Fisher, S.G., Grimm, N.B. & Kitchell, J.F. (1992) Global change and freshwater ecosystems. *Annual Review of Ecology and Systematics*, **23**, 119–139.

Carroll, P. & Crill, P.M. (1997) Carbon balance of a temperate poor fen. *Global Biogeochemical Cycles*, **11** (3), 349–356.

Center for International Earth Science Information Network (CIESIN) (2005) Gridded Gross Domestic Product (GDP). http://islscp2.sesda.com/ISLSCP2_1/html_pages/groups/soc/gdp_xdeg.html

Chen, C.Y. & Folt, C.L. (1996) Consequences of fall warming for zooplankton overwintering success. *Limnology and Oceanography*, **41**, 1077–1086.

Christoffersen, K., Andersen, N., Søndergaard, M., Liboriussen, L. & Jeppesen, E. (2006) Implications of climate-enforced temperature increases on freshwater pico- and nanoplankton populations studied in artificial ponds during 16 months. *Hydrobiologia*, **560**, 259–266.

Clair, T.A., Arp, P., Moore, T.R., Dalva, M. & Meng, F.R. (2001) Gaseous carbon dioxide and methane, as well as dissolved organic carbon losses from a small temperate wetland under a changing climate. *Environmental Pollution*, **116** (Suppl. 1), S143–S148.

Cole, J.A., Slade, S., Jones, P.D. & Gregory, J.M. (1991) Reliable yield of reservoirs and possible effects of climatic change. *Hydrological Sciences*, **36** (6), 579–598.

Daufresne, M., Roger, M.C., Capra, H. & Lamouroux, N. (2004) Long-term changes within the invertebrate and fish communities of the upper Rhône river: Effects of climatic factors. *Global Change Biology*, **10**, 124–140.

De Senerpont Domis, L.N., Mooij, W.M., Hülsmann, S., van Nes, E.H. & Scheffer, M. (2007) Can overwintering versus diapausing strategy in Daphnia determine match-mismatch events in zooplankton-algae interactions? *Oecologia*, **150** (4), 682–698.

DeStasio, B.T., Hill, D.K., Kleinhans, J.M., Nibbelink, N.P. & Magnuson, J.J. (1996) Potential effects of global climate change on small north-temperate lakes: Physics, fish and plankton. *Limnology and Oceanography*, **41** (5), 1136–1149.

Diamond, S.A., Peterson, G.S., Tietge, J.E. & Ankley, G.T. (2002) Assessment of the risk of solar ultraviolet radiation to amphibians. III. Prediction of impacts in selected northern midwestern wetlands. *Environmental Science & Technology*, **36** (13), 2866–2874.

Dillon, P.J., Molot, L.A. & Futter, M. (1997) The effect of El Niño-related drought on the recovery of acidified lakes. *Environmental Monitoring and Assessment*, **46**, 105–111.

Dowrick, D.J., Hughes, S., Freeman, C., Lock, M.A., Reynolds, B. & Hudson, J.A. (1999) Nitrous oxide emissions from a gully mire in mid-Wales, UK, under simulated summer drought. *Biogeochemistry*, **44** (2), 151–162.

Durance, I. & Ormerod, S.J. (2007) Climate change effects on upland stream macroinvertebrates over a 25 year period. *Global Change Biology*, **13**, 942–957.

Eaton, J.G. & Scheller, R.M. (1996) Effects of climate warming on fish habitat in streams of the United States. *Limnology and Oceanography*, **41** (5), 1109–1115.

Elliott, J.A., Thackeray, S.J., Huntingford, C. & Jones, R.G. (2005) Combining a regional climate model with a phytoplankton community model to predict future changes in phytoplankton in lakes. *Freshwater Biology*, **50** (8), 1404–1411.

Elliott, J.A., Jones, I.D. & Thackeray, S.J. (2006) Testing the sensitivity of phytoplankton communities to changes in water temperature and nutrient load, in a temperate lake. *Hydrobiologia*, **559**, 401–411.

Fang, X. & Stefan, H.G. (2000) Projected climate change effects on winterkill in shallow lakes in the northern United States. *Environmental Management*, **25** (3), 291–304.

Fenner, N., Freeman, C. & Reynolds, B. (2005) Observations of a seasonally shifting thermal optimum in peatland carbon-cycling processes; implications for the global carbon cycle and soil enzyme methodologies. *Soil Biology and Biochemistry*, **37** (10), 1814–1821.

Feuchtmayr, H., McKee, D., Harvey, I., Atkinson, D. & Moss, B. (2007) Response of macroinvertebrates to warming, nutrient addition and predation in large-scale mesocosm tanks. *Hydrobiologia*, **584**, 425–432.

Findlay, D.L., Kasian, S.E.M., Stainton, M.P., Beaty, K. & Lyng, M. (2001) Climatic influences on algal populations of boreal forest lakes in the Experimental Lakes Area. *Limnology and Oceanography*, **46** (7), 1784–1793.

Finstad, A.G., Forseth, T., Naesje, T.F. & Ugedal, O. (2004) The importance of ice cover for energy turnover in juvenile Atlantic salmon. *Journal of Animal Ecology*, **73** (5), 959–966.

Fletcher, R.J. & Koford, R.R. (2004) Consequences of rainfall variation for breeding wetland blackbirds. *Canadian Journal of Zoology*, **82** (8), 1316–1325.

Fossa, A.M., Sykes, M.T., Lawesson, J.E. & Gaard, M. (2004) Potential effects of climate change on plant species in the Faroe Islands. *Global Ecology and Biogeography*, **13** (5), 427–437.

Freeman, C., Liska, G., Ostle, N.J., Lock, M.A., Reynolds, B. & Hudson, J. (1996) Microbial activity and enzymic decomposition processes following peatland water table drawdown. *Plant and Soil*, **180** (1), 121–127.

Gedney, N. & Cox, P.M. (2003) The sensitivity of global climate model simulations to the representation of soil moisture heterogeneity. *Journal of Hydrometeorology*, **4** (6), 1265–1275.

George, D.G., Bell, V.A., Parker, J. & Moore, R.J. (2006) Using a 1-D mixing model to assess the potential impact of year-to-year changes in weather on the habitat of vendace (*Coregonus albula*) in Bassenthwaite Lake, Cumbria. *Freshwater Biology*, **51** (8), 1407–1416.

Gillings, S., Austin, G.E., Fuller, R. & Sutherland, W.J. (2006) Distribution shifts in wintering golden plover *Pluvialis apricaria* and lapwing *Vanellus vanellus* in Britain. *Bird Study*, **53**, 274–284.

Graf, W., Murphy, J., Dahl, J., Zamora-Munoz, C. & López-Rodríguez, M.J. (2008) Trichoptera. In: *Distribution and Ecological Preferences of European Freshwater Organisms*, Vol. 1 (eds A. Schmidt-Kloiber & D. Hering), 388 pp. Pensoft Publishers, Sofia, Bulgaria.

Graf, W., Lorenz, A.W., Tierno de Figueroa, J.M., Lücke, S., López-Rodríguez, M.J. & Davies, C. (2009) Plecoptera. In: *Distribution and Ecological Preferences of European Freshwater Organisms*, Vol. 2 (eds A. Schmidt-Kloiber & D. Hering), 262 pp. Pensoft Publishers, Sofia, Bulgaria.

Griffis, T.J. & Rouse, W.R. (2001) Modelling the interannual variability of net ecosystem CO_2 exchange at a subarctic sedge fen. *Global Change Biology*, **7** (5), 511–530.

Gunn, J.M. (2002) Impact of the 1998 El Nino event on a lake charr, *Salvelinus namaycush*, population recovering from acidification. *Environmental Biology of Fishes*, **64** (1–3), 343–351.

Haidekker, A. & Hering, D. (2008) Relationship between benthic insects (Ephemeroptera, Plecoptera, Coleoptera, Trichoptera) and temperature in small and medium-sized streams in Germany: A multivariate study. *Aquatic Ecology*, **42**, 463–481.

Hampton, S.E., Romare, P. & Seiler, D.E. (2006) Environmentally controlled Daphnia spring increase with implications for sockeye salmon fry in Lake Washington, USA. *Journal of Plankton Research*, **28** (4), 399–406.

Hari, R.E., Livingstone, D.M., Siber, R., Burkhardt-Holm, P. & Güttinger, H. (2006) Consequences of climatic change for water temperature and brown trout populations in Alpine rivers and streams. *Global Change Biology*, **12** (1), 10–26.

Hauer, F.R., Baron, J.S., Campbell, D.H., *et al.* (1997) Assessment of climate change and freshwater ecosystems of the Rocky Mountains, USA and Canada. *Hydrological Processes*, **11**, 903–924.

Hawkins, C.P., Hogue, J.N., Decker, L.M. & Feminella, J.W. (1997) Channel morphology, water temperature, and assemblage structure of stream insects. *Journal of the North American Benthological Society*, **16** (4), 728–749.

Heggenes, J. & Dokk, J.G. (2001) Contrasting temperatures, waterflows, and light: Seasonal habitat selection by young Atlantic salmon and brown trout in a Boreonemoral River, Regul. *Rivers: Research & Management*, **17**, 623–635.

Heino, J. (2002) Concordance of species richness patterns among multiple freshwater taxa: A regional perspective. *Biodiversity and Conservation*, **11** (1), 137–147.

Heiskanen, A.S., van de Bund, W., Cardoso, A.C. & Noges, P. (2004) Towards good ecological status of surface waters in Europe – Interpretation and harmonisation of the concept. *Water Science and Technology*, **49**, 169–177.

Helland, I.P., Freyhof, J., Kasprazak, P. & Mehner, T. (2007) Temperature sensitivity of vertical distributions of zooplankton and planktivorous fish in a stratified lake. *Oecologia*, **151** (2), 322–330.

Hogenbirk, J.C. & Wein, R.W. (1991) Fire and drought experiments in northern wetlands – A climate change analog. *Canadian Journal of Botany*, **69** (9), 1991–1997.

Huber, V., Adrian, R. & Gerten, D. (2008) Phytoplankton response to climate warming modified by trophic state. *Limnology and Oceanography*, **53** (1), 1–13.

Hudon, C. (2004) Shift in wetland plant composition and biomass following low-level episodes in the St. Lawrence River: Looking into the future. *Canadian Journal of Fisheries and Aquatic Sciences*, **61** (4), 603–617.

Huntington, T.G., Hodgkins, G.A. & Dudley, R. (2003) Historical trend in river ice thickness and coherence in hydroclimatological trends in Maine. *Climatic Change*, **61**, 217–236.

Illies, J. (ed.) (1978) *Limnofauna Europaea*. Gustav Fischer Verlag, Stuttgart.

Jansen, W. & Hesslein, R.H. (2004) Potential effects of climate warming on fish habitats in temperate zone lakes with special reference to Lake 239 of the experimental lakes area (ELA), north-western Ontario. *Environmental Biology of Fishes*, **70**, 1–22.

Jeppesen, E., Sondergaard, M., Jensen, J.P., *et al.* (2005) Lake responses to reduced nutrient loading – An analysis of contemporary long-term data from 35 case studies. *Freshwater Biology*, **50**, 1747–1771.

Johnson, W.C., Boettcher, S.E., Poiani, K.A. & Gunterspergen, G. (2004) Influence of weather extremes on the water levels of glaciated prairie wetlands. *Wetlands*, **24** (2), 385–398.

Keller, J.K., White, J.R., Bridgham, S.D. & Pastor, J. (2004) Climate change effects on carbon and nitrogen mineralization in peatlands through changes in soil quality. *Global Change Biology*, **10**, 1053–1064.

Lake, P.S., Palmer, M.A., Biro, P., *et al.* (2000) Global change and the biodiversity of freshwater ecosystems: Impacts on linkages between above-sediment and sediment biota. *BioScience*, **50** (12), 1099–1106.

Lischeid, G., Kolb, A., Alewell, C. & Paul, S. (2007) Impact of redox and transport processes in a riparian wetland on stream water quality in the Fichtelgebirge region, southern Germany. *Hydrological Processes*, **21** (1), 123–132.

Loh, J. & Wackernagel, M. (2004) *Living Planet Report 2004*, 40 pp. WWF International, Gland, Switzerland.

Mackenzie-Grieve, J. L. & Post, J.R. (2006) Projected impacts of climate warming on production of lake trout (*Salvelinus namaycush*) in southern Yukon lakes. *Canadian Journal of Fisheries and Aquatic Sciences*, **63** (4), 788–797.

Malcolm, J.R., Liu, C., Neilson, R.P., Hansen, L. & Hannah, L. (2006) Global warming and extinctions of endemic species from biodiversity hotspots. *Conservation Biology*, **20** (2), 538–548.

Malicky, H. (2000) Arealdynamik und Biomgrundtypen am Beispiel der Köcherfliegen (Trichoptera). *Entomologica Basiliensia*, **22**, 235–259.

Malmqvist, B. & Rundle, S. (2002) Threats to the running water ecosystems of the world. *Environmental Conservation*, **29**, 134–153.

McKee, D., Hatton, K., Eaton, J.W., *et al.* (2002) Effects of simulated climate warming on macrophytes in freshwater microcosm communities. *Aquatic Botany*, **74**, 71–83.

Minns, C.K. & Moore, J.E. (1995) Factors limiting the distributions of Ontario's freshwater fishes: The role of climate and other variables, and the potential impacts of climate change. *Canadian Journal of Fisheries and Aquatic Sciences Special Publications*, **121**, 137–160.

Monk, W.A., Wood, P.J., Hannah, D.M., Wilson, D.A., Extence, C.A. & Chadd, R.P. (2006) Flow variability and macroinvertebrate community response within riverine systems. *River Research and Applications*, **22** (5), 595–615.

Moore, M.V., Folt, C.L. & Stemberger, R.S. (1996) Consequences of elevated temperatures for zooplankton assemblages in temperate lakes. *Archiv für Hydrobiologie*, **135** (3), 289–319.

Moore, T.R., Roulet, N.T. & Waddington, J.M. (1998) Uncertainty in predicting the effect of climatic change on the carbon cycling of Canadian peatlands. *Climatic Change*, **40**, 229–245.

Müller-Navarra, D., Güss, S. & von Storch, H. (1997) Interannual variability of seasonal succession events in a temperate lake and its relation to temperature variability. *Global Change Biology*, **3**, 429–438.

Mulhouse, J.M., De Steven, D., Lide, R.F. & Sharitz, R.R. (2005) Effects of dominant species on vegetation change in Carolina bay wetlands following a multi-year drought. *Journal of the Torrey Botanical Society*, **132** (3), 411–420.

Nyberg, P., Bergstrand, E., Degerman, E. & Enderlein, O. (2001) Recruitment of pelagic fish in an unstable climate: Studies in Sweden's four largest lakes. *Ambio*, **30** (8), 559–564.

Pauls, S.U., Lumbsch, H.T. & Haase, P. (2006) Phylogeography of the montane caddisfly *Drusus discolor*: Evidence for multiple refugia and periglacial survival. *Molecular Ecology*, **15** (8), 2153–2169.

Perotti, M.G., Dieguez, M.C. & Jara, F.G. (2005) State of the knowledge of north Patagonian wetlands (Argentina): Major aspects and importance for regional biodiversity conservation. *Revista Chilena De Historia Natural*, **78** (4), 723–737.

Petchey, O.L., McPhearson, P.T., Casey, T.M. & Morin, P.J. (1999) Environmental warming alters food-web structure and ecosystem function. *Nature*, **402**, 69–72.

Pettersson, K., Grust, K., Weyhenmeyer, G. & Blenckner, T. (2003) Seasonality of chlorophyll and nutrients in Lake Erken – Effects of weather conditions. *Hydrobiologia*, **506–509**, 75–81.

Poff, N.L. & Allan, J.D. (1995) Functional-organization of stream fish assemblages in relation to hydrological variability. *Ecology*, 76 (2), 606–627.

Primack, A.G.B. (2000) Simulation of climate-change effects on riparian vegetation in the Pere Marquette River, Michigan. *Wetlands*, 20 (3), 538–547.

Pyke, C.R. (2004) Habitat loss confounds climate change impacts. *Frontiers in Ecology and the Environment*, 2 (4), 178–182.

Regina, K., Silvola, J. & Martikainen, P.J. (1999) Short-term effects of changing water table on N_2O fluxes from peat monoliths from natural and drained boreal peatlands. *Global Change Biology*, 5, 183–189.

Rogers, C.E. & McCarty, J.P. (2000) Climate change and ecosystems of the mid-Atlantic region. *Climate Research*, 14, 235–244.

Rooney, N. & J. Kalff (2000) Inter-annual variation in submerged macrophyte community biomass and distribution: The influence of temperature and lake morphometry. *Aquatic Botany*, 68, 321–335.

SAGE (2002) *Atlas of the Biosphere*. Center for Sustainability and the Global Environment, Madison. http://www.sage.wisc.edu/.

Salinger, D.H. & Anderson, J.J. (2006) Effects of water temperature and flow on adult salmon migration swim speed and delay. *Transactions of the American Fisheries Society*, 135 (1), 188–199.

Schindler, D.W. (2001) The cumulative effects of climate warming and other human stresses on Canadian freshwaters in the new millennium. *Canadian Journal of Fisheries and Aquatic Sciences*, 58 (1), 18–29.

Schindler, D.W., Beaty, K.G., Fee, E.J., *et al.* (1990) Effects of climatic warming on lakes of the Central Boreal Forest. *Science*, 250, 967–970.

Schindler, D.W., Bayley, S.E., Parker, B.R., *et al.* (1996) The effects of climatic warming on the properties of boreal lakes and streams at the Experimental Lakes Area, northwestern Ontario. *Limnology and Oceanography*, 41 (5), 1004–1017.

Schmutz, S., Cowx, I.G., Haidvogl, G. & Pont, D. (2007) Fish-based methods for assessing European running waters: A synthesis. *Fisheries Management and Ecology*, 14, 369–380.

Stanner, D. & Bordeau, P. (eds) (1995) *Europe's Environment. The Dobris Assessment*. European Environment Agency, Copenhagen.

Stefan, H.G., Hondzo, M., Fang, X., Eaton, J.G. & McCormick, J.H. (1996) Simulated long-term temperature and dissolved oxygen characteristics of lakes in the north-central United States and associated fish habitat limits. *Limnology and Oceanography*, 41 (5), 1124–1135.

Straile, D. (2000) Meteorological forcing of plankton dynamics in a large and deep continental European lake. *Oecologia*, 122, 44–50.

Straile, D. & Geller, W. (1998) The response of Daphnia to changes in trophic status and weather patterns: A case study from Lake Constance. *ICES Journal of Marine Science*, 55, 775–782.

Strecker, A.L., Cobb, T.P. & Vinebrooke, R.D. (2004) Effects of experimental greenhouse warming on phytoplankton and zooplankton communities in fishless alpine ponds. *Limnology and Oceanography*, 49 (4), 1182–1190.

Swansburg, E., Chaput, G., Moore, D., Caissie, D. & El-Jabi, N. (2002) Size variability of juvenile Atlantic salmon: Links to environmental conditions. *Journal of Fish Biology*, 61 (3), 661–683.

Thuiller, W., Lavorel, S. & Araujo, M.B. (2005) Niche properties and geographical extent as predictors of species sensitivity to climate change. *Global Ecology and Biogeography*, 14 (4), 347–357.

Travis, J.M.J. (2003) Climate change and habitat destruction: A deadly anthropogenic cocktail. *Proceedings of the Royal Society: Biological Sciences*, 270 (1514), 467–473.

Wagner, A. & Benndorf, J. (2007) Climate-driven warming during spring destabilises a *Daphnia* population: A mechanistic food web approach. *Oecologia*, 151 (2), 351–364.

Weltzin, J.F., Pastor, J., Harth, C., Bridgham, S.D., Updegraff, K. & Chapin, C.T. (2000) Response of bog and fen plant communities to warming and water-table manipulations. *Ecology*, 8 (12), 3464–3478.

Weltzin, J.F., Bridgham, S.D., Pastor, J., Chen, J. & Harth, C. (2003) Potential effects of warming and drying on peatland plant community composition. *Global Change Biology*, 9, 141–151.

Werner, C., Davis, K., Bakwin, P., Yi, C., Hurst, D. & Lock, L. (2003) Regional-scale measurements of CH_4 exchange from a tall tower over a mixed temperate/boreal lowland and wetland forest. *Global Change Biology*, 9, 1251–1261.

Winder, M. & Schindler, D.E. (2004) Climatic effects on the phenology of lake processes. *Global Change Biology*, **10** (11), 1844–1856.

Wrona, F.J., Prowse, T.D. & Reist, J.D. (2006) Climate change impacts on Arctic freshwater ecosystems and fisheries. *Ambio*, **35**, 325–325.

Xenopoulos, M.A. & Lodge, D.M. (2006) Going with the flow: Using species-discharge relationships to forecast losses in fish biodiversity. *Ecology*, **87** (8), 1907–1914.

Xenopoulos, M.A., Lodge, D.M., Alcamo, J., Märker, M., Schulze, K. & Van Vuuren, D.P. (2005) Scenarios of freshwater fish extinctions from climate change and water withdrawal. *Global Change Biology*, **11** (10), 1557–1564.

Zalakevicius, M. & Zalakeviciute, R. (2001) Global climate change impact on birds: A review of research in Lithuania. *Folia Zoologica*, **50** (1), 1–17.

6

Interaction of Climate Change and Eutrophication

Erik Jeppesen, Brian Moss, Helen Bennion, Laurence Carvalho, Luc DeMeester, Heidrun Feuchtmayr, Nikolai Friberg, Mark O. Gessner, Mariet Hefting, Torben L. Lauridsen, Lone Liboriussen, Hilmar J. Malmquist, Linda May, Mariana Meerhoff, Jon S. Olafsson, Merel B. Soons and Jos T.A. Verhoeven

Introduction

The conversion of much of the earth's land surface to agriculture or urban land has had major effects on nutrient flows from the local to the global scales. Natural ecosystems generally conserve nutrients and store organic matter effectively. In all agricultural systems, such conservation mechanisms are weakened, and management increases nutrient inputs over natural levels. These altered features of ecosystems together result in massive leakage of nutrients into waterways. Humans also concentrate nutrients from food and release excreta that generate immense nutrient flows in areas of high population density, even where modern sewage treatment technology is applied. The consequent eutrophication of many of the world's waters has become a widespread and expensive problem for water supply, human health and amenity, as well as for nature conservation. Eutrophication profoundly affects aquatic ecosystems by altering biodiversity, trophic structure and biogeochemical cycling.

Biological processes are temperature sensitive. Increasing temperature, expected under future climate change, might (i) change growth and respiration of organisms, potentially leading to lower net primary production, (ii) enhance oxygen consumption (and, as warm water holds less oxygen, the risk of oxygen depletion increases with problems for sensitive species) and increase nutrient release from sediments; (iii) affect life history in the direction of shorter lifespans and earlier reproduction; and (iv) change phenology and trophic dynamics, which potentially may result in a mismatch between consumer demand and their prey availability. Such temperature-induced changes are expected to interact strongly with existing increased nutrient flows and create new problems for

Climate Change Impacts on Freshwater Ecosystems. First edition. Edited by M. Kernan, R. Battarbee and B. Moss. © 2010 Blackwell Publishing Ltd.

freshwater biota or exacerbate existing ones (Mooij *et al.* 2005; Blenckner *et al.* 2006; Jeppesen *et al.* 2007a, 2009).

Attempts to adapt to or mitigate the effects of climate change as well as to cope with a rapidly increasing world population needing to be fed are likely to lead to intensified crop production in the temperate zone and, for reasons of water scarcity, less intensive production in the Mediterranean (Olesen & Bindi 2002; Alcamo *et al.* 2007). Intensification of agriculture in the temperate zone will probably cause increased nutrient export from land to water as a result of increased run-off (Olesen & Bindi 2002), now that the P and N cycles have been massively accelerated by human activities (Galloway *et al.* 2008; Smil 2000), unless additional nutrient retention measures are applied. However, the control of diffuse losses of nutrients from land has proven very difficult, not least because nitrate is extremely mobile and cultivation inevitably leads to erosion and downslope transport of soil particles containing phosphorus. Changes in temperature and rainfall will also lead to alterations in agricultural practices, including changes in soil cultivation and in the rates and timing of fertilization. These changes will, in turn, have cascading effects on nutrient flows (IPCC 2007). Extreme rainfall events and periodic droughts (through aeration, mineralization and subsequent washout of mineralized nutrients) will also enhance the risk of nutrient losses from catchments to freshwaters.

At the higher temperatures of Southern Europe, the predicted decrease in precipitation and higher evaporation will result in less run-off and, as a result, possibly lower nutrient loading to freshwaters (Jeppesen *et al.* 2009; Özen *et al.* 2010). However, this reduction is not expected to compensate for the negative consequences of water deficits that will lead to a concentration of nutrients from point sources and a reinforcement of eutrophication symptoms in aquatic ecosystems (Beklioglu *et al.* 2007; Beklioglu & Tan 2008; Özen *et al.* 2010). Moreover, in Southern Europe, drought and reduced discharge into inland waters will result in greater salinization exacerbated by increased evaporation and greater use of water for irrigation (Zalidis *et al.* 2002). For example, a decrease in hydraulic loading has been found to lead to a threefold increase in salinity in two shallow Mediterranean lakes (Beklioglu & Tan 2008; Beklioglu & Özen 2008). Salinities of a few parts per thousand, which are not uncommon in such inland basins, lead to reduced species richness (Williams 2001; Barker *et al.* 2008; Brucet *et al.* 2009), community shifts and a higher risk of clear water switching to a phytoplankton-dominated turbid state (Jeppesen *et al.* 2007b).

There is a substantial body of literature on inland water eutrophication, which has been summarized in several reviews (e.g. Carpenter *et al.* 1998; Smith 2003; Schindler 2006; Dodds 2007). It is perhaps the most widespread problem in freshwater ecosystems now that aerial deposition of nitrogen fertilizes ecosystems at the global scale (e.g. Galloway *et al.* 2008). Similarly, the anticipated effects of global warming on freshwater ecosystems have been repeatedly reviewed (e.g. Firth & Fisher 1992; Mooij 2005; Jeppesen *et al.* 2009). In this chapter, we will describe new results on the interaction between eutrophication and climate change based on studies carried out as part of the Euro-limpacs project during the period 2004–8.

Studies of climate–nutrient interactions in lakes, streams and wetlands in the Euro-limpacs project and elsewhere have shown that warming is likely to exacerbate some symptoms of eutrophication in freshwaters, but not others. For wetlands and streams, this conclusion is mainly based on experiments in paired sites in Iceland, supplemented with space-for-time cross-comparisons of systems from different climate regions. For lakes, we have used several approaches, including space-for-time comparisons, experiments, palaeolimnology and modelling. The key questions asked were

- Do nutrients structure ecosystems in different ways under current and anticipated future climatic conditions?
- Will changing climate interact with increased nutrient supply to alter ecosystem processes?
- Will changing climate aggravate eutrophication symptoms?
- Can effects of climate change be distinguished from those of eutrophication?
- Are there lessons to be learnt from the past to understand future problems better?
- Can we mitigate negative effects of climate change on ecosystems in terms of enhanced eutrophication?

These questions are addressed in the following four sections on the basis of the different approaches used, and the final section attempts to provide some general conclusions.

Changes in trophic structure

Space-for-time analyses of large data sets from lakes

Eutrophication was formerly seen largely in terms of nutrient flows entering rivers and lakes from the catchment, and this is still fundamentally the basis of assessing eutrophication effects. However, once nutrients have entered a basin, the way that they are apportioned may have different consequences. For example, the absolute and relative amounts of vascular plants and algae produced are dependent on the structure of the food web (also called trophic structure). Internal processes, such as the recycling of nutrients from sediments, may be altered, sometimes exacerbating the effects of increased external loads (Jeppesen *et al.* 2009).

Based on comparisons across latitudinal gradients, and presuming free migration of species northwards, major changes in trophic structure are to be expected in a future warmer European climate (Jeppesen *et al.* 2007; 2010). The fish community will change, with higher dominance of zooplanktivorous and omnivorous fish, including the thermo-tolerant carp (*Cyprinus carpio*) (Lehtonen 1996), implying increased predation on zooplankton and, consequently, less grazing on phytoplankton (less top-down control), and a higher algal biomass (and chlorophyll *a*) per unit of phosphorus. A study of 84 shallow European lakes from northern Sweden to Spain showed that the ratio of fish biomass (expressed as catch per net-night in multi-mesh sized gillnets) to zooplankton biomass increased southwards whilst the zooplankton:phytoplankton biomass ratio decreased in the same direction, both substantially (Gyllström *et al.* 2005).

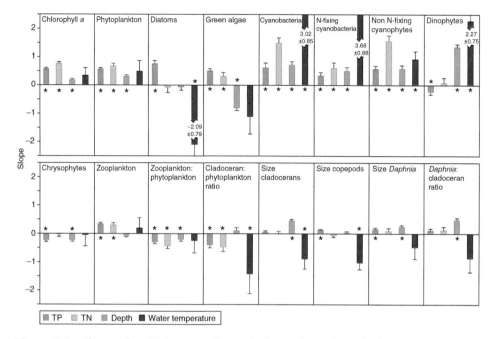

Figure 6.1 Slopes of multiple regressions relating various phytoplankton and zooplankton variables (log transformed) to concentrations of total phosphorus (TP) and total nitrogen (TN) in surface water, mean lake depth and surface water temperature (all log transformed) measured in August in 250 lakes and over 800 lake-years in Denmark. Significant ($p < 0.05$) slopes are indicated by asterisks. (From Jeppesen *et al.* 2009.)

Size structure of fish populations also changes latitudinally, with a higher proportion of small fish in warm lakes. Higher-latitude fish species are often not only larger but grow more slowly, mature later, have longer lifespans and allocate more energy to reproduction than populations at lower latitudes (Blanck & Lamouroux 2007). Even within species, such changes can be seen along the latitudinal gradient (Blanck & Lamouroux 2007).

Fish abundance, and thus predation on zooplankton and benthos, may also be influenced by changes in the duration of ice cover (see also Chapter 3). Comparative studies of Danish coastal lakes and continental Canadian lakes with similar summer temperatures, but major differences during winter, have shown much lower chlorophyll:TP ratios and higher zooplankton:phytoplankton ratios in the winter-cold Canadian lakes, perhaps due to lower winter survival of zooplanktivorous fish under ice (Jackson *et al.* 2007). Thus, reduced ice cover in winter should enhance fish survival, which might cascade down the food web and also reinforce symptoms of eutrophication. The effects of warming on fish survival may be contradictory, however, as discussed below.

Enhanced predation on zooplankton at higher temperatures is also evident from analyses of data from Danish lakes, showing a decrease in the average size of cladocerans and copepods with increasing temperature (Fig. 6.1, Jeppesen *et al.* 2009). Tendencies for decreasing zooplankton:phytoplankton biomass ratios and lower proportions of *Daphnia* among the cladocerans point in the same

(a)   (b)

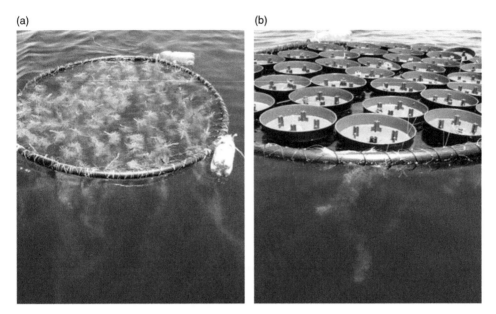

Figure 6.2   Experimental set-up showing (a) submerged and (b) free-floating artificial plant beds. During summer 2005, four replicate plant beds of each type were introduced to 10 lakes in Uruguay and 10 similar lakes in Denmark.

direction. With fewer large-sized *Daphnia* and a lower average size of zooplankton, grazing on phytoplankton is likely to decline.

Further evidence comes from subtropical lakes. Here, the typical cladoceran community comprises small-bodied genera (Meerhoff *et al.* 2007a; Iglesias *et al.* 2007), and experimental studies have shown that high fish predation is the key factor for the observed patterns in the structure of the littoral zooplankton communities (Iglesias *et al.* 2008). Comparative experimental studies with artificial plants in lakes in Uruguay (warm temperate to subtropical) and Denmark (temperate), described below, provide further evidence of lower grazer control of algae in warm lakes compared with otherwise similar temperate lakes (Meerhoff *et al.* 2007a, b).

Field experiments in Uruguay and Denmark

To test the hypotheses that in a warmer climate (i) fish impact the littoral trophic structure more strongly; (ii) trophic cascades are more truncated; and (iii) the key role of submerged plants in maintaining clear water may be negatively affected, we carried out experiments in subtropical Uruguay (30–35 °S) and temperate Denmark (55–57 °N). We introduced artificial plastic plant beds mimicking submerged and free-floating macrophytes (Fig. 6.2) to 10 shallow lakes, paired in terms of size, nutrient concentrations and water turbidity (Secchi depth), and compared the structure of the communities associated with the plant beds.

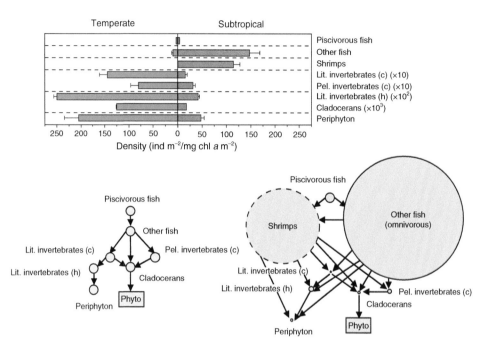

Figure 6.3   Structure of the littoral communities in artificial plant beds in temperate (Denmark, left) and subtropical (Uruguay, right) lakes. Above: Density of potentially piscivorous fish (numbers of individuals m^{-2}, all other fish, shrimps, littoral carnivorous macro-invertebrates ($\times 10$), pelagic carnivorous invertebrates ($\times 10$), littoral herbivorous macro-invertebrates ($\times 10^2$), pelagic herbivorous cladocerans ($\times 10^3$) and biomass of periphyton (as chlorophyll *a*)). Below: Simplified schematic depiction of trophic interactions among the same trophic groups as above, showing densities in the subtropics relative to those in temperate lakes. Shrimps were absent in the temperate lakes, and except for fish, the same taxa received the same trophic classification in both climate zones. Abbreviations: (c), carnivorous; (h), herbivorous; Pel., pelagic; Lit., littoral, phyto, phytoplankton. Data are sample means ($\pm$ 1 SE) of five lakes paired between climate regions in terms of phytoplankton biomass and physico-chemical and morphometric characteristics (adapted from Meerhoff *et al.* 2007b).

Notable differences were found between the two climate zones, regardless of other variables, supporting our first two hypotheses (Meerhoff *et al.* 2007a, b). The littoral food web was more complex and less hierarchically structured in the subtropical lakes (Fig. 6.3). Their fish communities had higher diversity, greater density (on average 11-fold) and biomass, less piscivory and widespread omnivory, smaller individual size (90% of fish smaller than 3.0 cm standard nose to tail-fork length) and stronger association with the submerged plants than in the temperate lakes. Fish communities in the warm lakes were dominated by omnivorous cyprinodontids, whereas somewhat larger cyprinids and percids dominated in the temperate lakes (Teixeira de Mello *et al.* 2009). More fish species co-occurred with fewer cladoceran and littoral invertebrates genera in the subtropical lakes, with the opposite in the temperate lakes.

Higher taxon richness and significantly greater densities of littoral invertebrates occurred in the temperate lakes with an about eightfold higher densities of predators, tenfold of grazers and twofold of collectors. Cladoceran species richness, density and body size were also significantly higher in the temperate lakes. However, periphyton biomass in the subtropical lakes was much lower than expected from the lower density of grazing invertebrates and the more positive environmental conditions for periphyton growth (more light and higher temperature). The observed fourfold reduction in periphyton biomass in the subtropical lakes seems likely to have been the result of periphyton feeding by fish and shrimps.

These results help to explain why the buffers that the plants provide in the temperate zone against increasing external nutrient load apparently are less strong in warmer lakes (Romo *et al.* 2005; Blenckner *et al.* 2007; Jeppesen *et al.* 2009). Danish lakes (as in most Northern European lakes) show a clear positive effect of submerged macrophytes on water clarity, but no clear difference with plant cover was found in the chlorophyll:TP or Secchi depth:TP relationships in shallow lakes in Florida (USA), irrespective of plant cover (Bachmann *et al.* 2002; Jeppesen *et al.* 2007a, E. Jeppesen *et al.* unpublished data). It has been argued that macrophyte growth will be stimulated by climate warming (Scheffer *et al.* 2001) due to higher temperature and, in the Mediterranean region, due also to reduced water depth and therefore greater light penetration to the bottom (Coops *et al.* 2003; Beklioglu *et al.* 2006). However, an analysis of data from lakes from the temperate zone to the tropics suggests a lower probability of macrophyte dominance in warm lakes (Kosten *et al.* 2009).

Moreover, even when submerged macrophytes are abundant, the positive effect of the plants on water clarity is much less pronounced at higher temperatures (Jeppesen *et al.* 2007a), probably reflecting much lower zooplankton grazing on phytoplankton due to a higher fish predation on zooplankton (Jeppesen *et al.* 2007b) by small omnivorous fish. In addition, higher dominance of benthivorous fish foraging in the sediment may enhance internal nutrient loading when lake temperature increases. This may accelerate eutrophication and lead to higher and prolonged dominance of nuisance algae, especially cyanobacteria. Cyanobacteria, most notably the N-fixing forms, are highly sensitive to increases in temperature (Fig. 6.1). Likewise, dinophytes, which may also cause problems for human water supplies, may become more important with higher temperatures, whereas diatom proportion, in particular, diminishes (Jeppesen *et al.* 2009; Fig. 6.1).

Mesocosm experiments

The question of whether warming intensifies symptoms of eutrophication was also addressed using controlled mesocosm experiments. These included studies in tanks in the United Kingdom, a flow-through pond system in Denmark and *in situ* enclosure experiments in a littoral reed stand of a lake in Switzerland (Fig. 6.4). In such experiments, temperature was controlled at desired values above ambient, and additional treatments included differential nutrient loading or a change in the fish community or both.

Denmark   The United Kingdom   Switzerland

Figure 6.4 Experimental mesocosm facilities set-up in Denmark, the United Kingdom and Switzerland to study the combined effects of warming and nutrient enrichment on freshwater ecosystems.

Experimental tanks in the United Kingdom

Two experiments have been carried out in a system of forty-eight 3-m^3 tanks near Liverpool in north-west England. Mesocosms mimicking ponds and shallow lakes were warmed by a pumped hot-water system (McKee *et al.* 2000, Fig. 6.2). The first, in 1998–2000, used planted macrophyte communities, including *Lagarosiphon major, Elodea nuttallii* and *Potamogeton natans*, a 3 °C rise in temperature above ambient, rather infertile inorganic sediments and a fertilization regime in which concentrations given were 0.5 mg N l^{-1} and 0.05 mg P l^{-1}, compared with controls in which no nutrients were applied. Sticklebacks (*Gasterosteus aculeatus*) were present in half of the mesocosms, which were set up in a randomized block design with quadruple replication.

The second experiment, in 2005–7, involved a temperature increase of 4 °C above ambient, presence and absence of sticklebacks and three nutrient regimes in mesocosms that had sediments higher in organic matter and nutrients (loss on ignition 7.5%, TN 0.83 mg g^{-1} and TP 0.2 mg g^{-1}) than in the previous experiment. Controls with no additions of nutrients constituted the first regime, the second used a lower (0.25 mg N l^{-1}) level and the third a higher (2.5 mg N l^{-1}) level than in the first experiment, in line with the loadings on waters in regions with intensive agriculture. Only a small phosphorus dose was given, because the tanks were phosphorus rich as a result of release from sediment. The plant community was self-determined from seed and fragments and dominated by the submerged *E. nuttallii* and *Ceratophyllum demersum* and the free-floating *Lemna trisulca*, together with other floating lemnids (*Lemna minor, Spirodela polyrhiza*). Absolute maximum ambient temperatures in the summers of 1999 and 2000 were 23.7 °C and 24.8 °C, respectively. Ambient summer means were 16.4 °C and 15.5 °C (McKee *et al.* 2003). In the second experiment, in 2007, the summer ambient maximum was 24.9 °C and the summer mean, 15.5 °C.

In the first experiment (McKee *et al.* 2002, 2003), warming did not change phytoplankton biomass, even with increased nutrients and fish presence. Macrophyte communities remained dominant. Total plant abundance remained relatively high and was unaffected by warming, but the proportion of an exotic species, *L. major*, introduced from South Africa, increased. Warming had no influence on *E. nuttallii*, a result observed again in the later experiment.

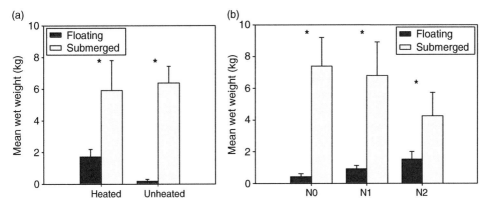

Figure 6.5 Effects of (a) warming by 4 °C above ambient and (b) increased nutrient loading (N0, N1, N2) on submerged and free-floating plants in a mesocosm experiment in the United Kingdom. Values are in kg fresh biomass per mesocosm. Asterisks indicate significant differences at $p<0.05$, which occurred among treatment levels as well as between macrophyte growth forms.

Macrophytes still dominated in the second experiment (Feuchtmayr *et al.* 2009), but warming led to lower phytoplankton biomass, almost certainly because substantial floating lemnid communities developed, especially at the higher nutrient levels (Fig. 6.5). Warming increased phosphorus concentrations and conductivity, decreased pH and oxygen saturation and increased the frequency of severe deoxygenation in the first experiment. In the second experiment, raised temperature had similar effects, but these were more extreme, except for pH, which did not change.

Effects of nutrient addition and the presence of fish were independent of warming in the first experiment and tended to increase and decrease macrophyte abundance, respectively. This was not the case in the second experiment where warming and nutrients together increased the growth of floating lemnids, which had been negligible in the previous experiment. There were also much more severe effects on fish in the second experiment (Moran *et al.* 2009). Heating reduced the stickleback population by 60%, and nutrient addition reduced it by approximately 80%. A combination of nutrients and heating resulted in a complete loss of the stickleback populations. This was attributed to the increased likelihood of hypoxia in mesocosms subject to heating and nutrient addition, rather than direct thermal stress, though aquarium experiments showed reduced breeding male activity at temperatures above 17 °C in well-oxygenated conditions (K. Hopkins, B. Moss & A.B. Gill unpublished data).

Results of the two experiments thus suggest that warming exacerbates the symptoms of eutrophication, to a small extent with a temperature increase of 3 °C and low nutrient loading but with high severity with 4 °C and nutrient loadings comparable with those in many European lowlands at present. These include increased phosphorus loading owing to release from the sediments, deoxygenation, reduced fish biomass, reduced plant diversity and species richness, an increase in warm-water exotics such as *L. major* and predominance of free-floating plants such as lemnids. One common symptom of eutrophication, increased phytoplankton biomass, was not observed, probably because of shading by lemnids. Disappearance of submerged plants, seen by many *per se* as a symptom of advanced eutrophication, also did not occur. *C. demersum* increased with warming, but *L. trisulca* decreased, resulting in only a small net reduction of macrophyte biomass. It may be that *C. demersum* was more tolerant of the increased lemnid shading.

The implications of warming/deoxygenation for fish communities are severe, however, for sticklebacks are resilient fish, more tolerant of reduced oxygen concentrations than most other European fish species. In the United Kingdom, only tench (*Tinca tinca*), crucian carp (*Carassius carassius*) and the common carp (*Cyprinus carpio*) are more tolerant. Perhaps even more significantly, an analysis of the oxygen balance of the tanks in summer 2007 suggested that respiration rates increased markedly through warming compared with photosynthetic rates, largely through increased metabolism of heterotrophs. The effect also occurred in response to nutrient loading and could imply a positive feedback of warming on carbon dioxide release (Moss 2010).

Experimental ponds in Denmark

A longer-term experiment was carried out in Denmark with the aim of contrasting the effects of warming on the two alternative states in which shallow lakes typically occur: clear water, macrophyte dominated, with small zooplanktivorous fish populations and turbid, phytoplankton dominated, with relatively large zooplanktivorous fish populations. Twenty-four cylindrical outdoor mesocosms, each 2.8 m³ in volume, were used (Fig. 6.4). Groundwater was pumped in above the sediment and drained through an outlet at the water surface. The theoretical water retention time was 2.5 months, and the water was heated by electrical elements and continuously mixed by paddles. The system was run at low and high nutrient concentrations, the latter obtained by weekly dosing of N and P, and at three temperatures, ambient and elevated according to the IPPC climate scenarios A2 and A2 + 50%, downscaled to local 25 × 25 km grid cells. There were four replicates of each treatment. The modelled temperature difference for the A2 scenario is generally higher in August to January (max. 4.4 °C in September) than during the rest of the year (min. 2.5 °C in June).

There was high inter-annual variability within and among mesocosms and among replicates during the season. Cumulative data for the 4-year period, however, showed a clear pattern. As expected, chlorophyll *a* was higher overall in the high-nutrient mesocosms (Fig. 6.6). However, the effects of warming differed. At low nutrient loading, warming had a substantial effect on chlorophyll *a*, which was approximately 2.5 times higher in the heated mesocosms than in the controls. Chlorophyll *a* was also 50% higher in the A2 scenario than in the control mesocosms at high nutrient loading, while it was much lower than in the control mesocosms at A2 + 50%, although still double that in the low-nutrient mesocosms run at the same temperature scenario (Fig. 6.6).

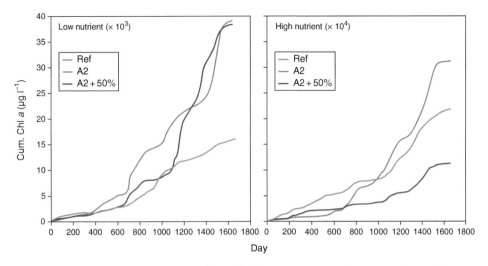

Figure 6.6 Cumulative changes in chlorophyll *a* concentration (average of 4 replicate mesocosms) during the first 4 years (number of days) of the Danish mesocosm experiment involving two levels of nutrient loading (left, low; right, high). Green line gives data for ambient temperature, blue line for the A2 scenario of IPCC (2007) and red line for the A2+50% scenario.

Changes in seasonality were also evident. In the low-nutrient mesocosms, two plant species, *Potamogeton crispus* and *Elodea canadensis*, became dominant at the raised temperature. Growth of *P. crispus* started earlier in the heated mesocosms than in the controls (Fig. 6.7) in all four growth seasons studied and reached a peak earlier in the heated mesocosms (T.L. Lauridsen *et al.* unpublished data). Controlled growth experiments in pots confirmed this pattern. However, there was a complicated pattern of effects. For *E. canadensis*, growth in pots was independent of temperature in both spring and summer. Growth potential in autumn, in contrast, was higher under the A2+ scenario at both low and high nutrient concentrations compared with the control mesocosms. In winter, growth was almost zero, except in the A2+ mesocosms where a slight growth was recorded. Higher temperatures also increased the likelihood of filamentous algae becoming dominant in the experimental ponds. Finally, earlier emergence of flying insects and earlier reproduction of snails were recorded (T.B. Kristensen unpublished data).

Oxygen in the mesocosms was continuously measured, which allowed the calculation of gross and net primary production (GPP and NPP, respectively) and ecosystem respiration (ER) from changes in concentrations over 24-hour cycles. Net ecosystem production (NEP) was calculated as the difference between GPP and ER. Seasonal variation in GPP and in daytime ER followed the variations in light and temperature, being about 10 times higher during summer than in winter. At ambient temperatures, the addition of extra nutrients had only a minor effect on GPP and ER, and the overall seasonal dynamics remained similar at low and high nutrient loading. Only during summer did nutrient addition lead to higher rates. That GPP and ER were not greatly affected by nutrient loading

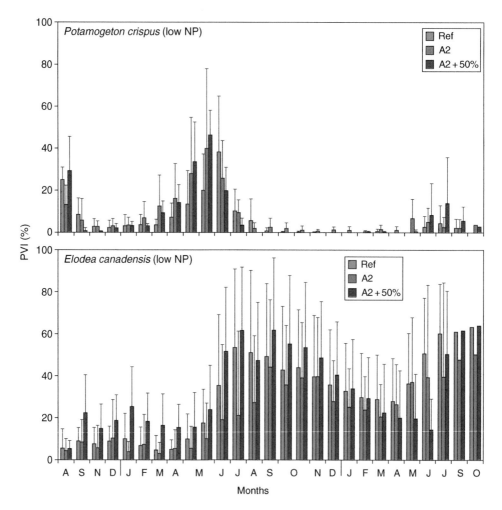

Figure 6.7   Seasonal variation in per cent plant volume inhabited (PVI) (mean ± SE) of the two dominant macrophytes in mesocosms receiving low nutrient loading at three different temperature regimes in the Danish mesocosm experiment.

confirms earlier studies (Vadeboncoeur *et al.* 2003; Liboriussen & Jeppesen 2003) of shallow lakes, where a shift was observed from benthic to pelagic production with increasing loading but only minor changes in total primary productivity. Warming affected GPP and daytime ER differently at the two nutrient levels. At low levels, warming had no effect in summer but stimulated both processes during winter. The situation is different when changes over 24 hours are examined. Net ecosystem production on an annual basis was always negative (probably due to respiration of organic matter in the sediments) and became more negative with warming at low nutrient loading, while in the warmed tanks with high loading, NEP first increased from the control to the A2 regime and then decreased again in the A2+50 regime back to the control level, partly following the pattern shown by chlorophyll *a*.

As for the UK experiments, the Danish mesocosm results suggest that warming exacerbates symptoms of eutrophication, in this case with substantially higher phytoplankton biomass occurring both at low nutrient level and in the A2 scenario with high nutrient levels. A shift to stronger dominance of filamentous algae or macrophytes was observed in the A2 scenario with high nutrient levels. The season with macrophyte growth was, however, prolonged at higher temperature, and the biomass of submerged macrophytes did not decline at elevated temperature in these mesocosms. In fact, higher temperature gave *P. crispus*, commonly a dominant species in spring, a window of opportunity that led to a higher average biomass, whereas no clear effect was seen for *E. canadensis* (T.L. Lauridsen *et al.* unpublished data). The preliminary results on the oxygen balance indicate a daily GPP and ER that were higher overall with warming but hint at lower NEP when 24-hour periods are considered. The long-term responses of these variables are yet to be evaluated.

Of great interest is the fact that although warming noticeably increased many symptoms of eutrophication in both the Danish and the UK experiments, it sometimes had no effects and occasionally reduced them. There was great variety in the details of the responses, even under the well-controlled conditions in the experimental ponds compared with natural ponds and lakes. Starting conditions, intrinsic chaotic dynamics and interactions of nutrient loading and temperature increase combine to make precise predictions of the future a challenging task.

Mesocosms in a littoral reed stand in Switzerland

Experiments with mesocosms in Switzerland (Fig. 6.4) addressed impacts of warming and eutrophication on heterotrophic processes of the carbon cycle in freshwater marshes, an important wetland type with large potential for feedback effects on climate (e.g. Brix *et al.* 2001). In contrast to the experiments conducted in Denmark and the United Kingdom, the mesocosms were not isolated tanks but were placed within the littoral zone of a lake and included unfenced control plots (Flury *et al.* 2010). Emphasis of the Swiss experiments was on plant litter decomposition, although responses of whole-system metabolism and several other variables were also assessed (Flury 2008; Hammrich 2008). The mesocosms were arranged in a randomized block design in a monospecific stand of common reed, *Phragmites australis*. Half of them were continuously heated to a target temperature of 4°C above ambient (typically >2°C achieved) and/or periodically fertilized with $Ca(NO_3)_2$ to increase dissolved nitrogen concentration fivefold (in practice threefold to ninefold) above the ambient concentration (annual load of approximately $10\,g\,N\,m^{-2}$).

Experiments conducted in these mesocosms produced notable and sometimes unexpected results. Raised temperature stimulated microbial respiration associated with decomposing leaves in the short term, as anticipated, but had no effect in the long run, that is, respiration rates on litter from heated and control mesocosms did not differ when measured at a standard temperature in the laboratory (Fig. 6.8). Nitrogen enrichment, in contrast, tended to curtail respiration both in the long and in the short term, and leaf litter mass loss was also slowed in mesocosms receiving extra nitrogen. This counterintuitive inhibition of decomposition by increased NO_3^- supply has previously been observed in forests where it was

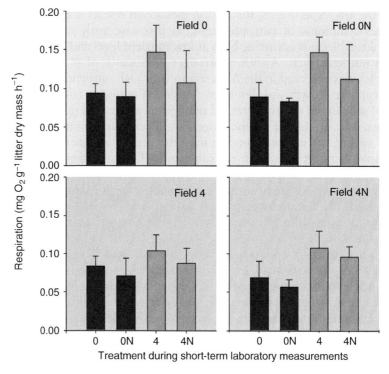

Figure 6.8   Microbial respiration rates associated with decomposing leaf litter in response to experimental warming and nitrogen enrichment. Standard litter samples were collected from control (0), experimentally heated (4), nitrogen-enriched (0N), and heated and enriched (4N) enclosures, and respiration of all these samples was measured in the laboratory at ambient lake temperature (0), at elevated temperature (4) with nitrogen added, and at elevated temperature with nitrogen added (4N). Data shown in each of the four panels refer to conditions in the field, while treatments indicated on the x-axis refer to conditions during measurements in the laboratory.

attributed to the inhibition of lignin-degrading enzymes. An alternative hypothesis is that increased N supply stimulates processes other than litter decomposition (e.g. algal growth), thereby exacerbating deprivation of litter decomposers of potentially limiting nutrients such as phosphorus. Irrespective of the underlying mechanism, the observed inhibition of heterotrophic processes by NO_3^- enrichment suggests that increased nitrogen availability, resulting from continuing high rates of atmospheric N deposition, might buffer against the acceleration of litter decomposition caused by warming in the future.

   Understanding changes in the regulation of decomposition rates under warmer and nutrient-richer conditions may be more intricate than suggested by our results on leaf decomposition. Neither warming nor nitrate enrichment affected the decomposition of stem litter of *P. australis*. This may be due to the intrinsic recalcitrance of stem litter and/or because environmental conditions other than temperature and nitrate availability were overwhelming in determining stem decomposition rate. One of these factors could be oxygen, since litter placed in unfenced control plots in the reed stand, which freely exchanged water with the open lake, clearly decomposed faster than litter in any of the mesocosms. Nonetheless, it is remarkable that the pattern of stem litter decomposition was distinct from that of leaf litter decomposing in the same place. Thus, it appears that effects of warming and nitrogen enrichment vary with litter quality, making forecasts of global change effects on litter decomposition challenging. Consequently, predictions based only on temperature-response relationships such as Q_{10} models may be of limited value.

An assessment of whole-mesocosm metabolism by monitoring oxygen concentrations over two 3-day periods on four occasions revealed that GPP, ER and NEP were characterized by invariably low activity in summer, autumn and winter. As a result, no clear differences were apparent among treatments during these periods. In spring, NEP was also unaffected by warming. However, although not statistically significant, a tendency emerged towards higher GPP and ER at this time. Metabolism in N-enriched mesocosms also tended to be lower than in unfertilized mesocosms, as had been seen for leaf decomposition and leaf-associated respiration rates. Provided this pattern turns out to be robust when substantiated by more data, it would imply that the stimulating effects of predicted global warming on GPP and ER could be offset, at least in part, by sustained nitrogen deposition. This contrasts with a finding in the UK experiments that respiration by heterotrophs was stimulated by both warming and nitrogen addition, which suggests the opposite. Distinct differences between systems should be borne in mind, however. The UK system was very rich in nutrients, dominated by submerged and free-floating plants; the Swiss system was in an emergent plant stand in a much less eutrophic lake. Extension of such results on a global basis will thus need many more experiments. However, it is clear that important implications for carbon sequestration and the balance between aerobic and anaerobic decomposition processes in freshwater systems can be expected.

Microevolution experiments in the UK and Danish mesocosms

Overall results of the three mesocosm experiments in the United Kingdom, Denmark and Switzerland point to a general exacerbation of eutrophication symptoms with warming. However, some of the observed responses were counterintuitive. Another process that needs to be considered is evolutionary adaptation to warming. We used two of the mesocosm systems of the Euro-limpacs project to investigate the possibility that such adaptations can occur on short timescales.

When local communities inhabiting a lake or pond experience a temperature increase, strong selection is imposed, to which the community may respond by changes in species composition. For example, in the UK mesocosm experiments, ostracods significantly increased with warming. In addition to species sorting, however, each individual population may respond to selection pressures by changing its trait values to become better adapted to the new conditions. These microevolutionary changes may potentially allow local populations to cope with climate change and may impact community composition by reducing the number of species replacements (see Urban *et al.* 2008).

We studied evolutionary adaptation to temperature of various cladocerans in the mesocosm experiments run in the United Kingdom and Denmark. Life-history experiments on clones isolated from the cladoceran *Simocephalus* populations inhabiting mesocosms in Denmark provided solid proof of microevolutionary responses to temperature increase in this species, in terms of both survival and key life-history traits such as age at reproduction or number of offspring. Changes at the within-species level are then buffering against changes at the among-species level (Van Doorslaer *et al.* 2007).

All mesocosms in the UK second experiment were inoculated in March 2006 with the same, genetically well-characterized *Daphnia magna* population, consisting of 150 clones hatched from the dormant egg bank of a local population. The fate of these populations was monitored until the experiment ended in September 2007. We studied dynamics of the populations establishing in the different treatments, characterized their genetic structure using neutral DNA markers and carried out life-table experiments in the laboratory using clones isolated from mesocosms exposed to different temperatures. In addition, we carried out enclosure and competition experiments in the tanks to quantify the degree to which populations in differently heated mesocosms were locally adapted to their new environment (Van Doorslaer *et al.* 2009).

Key observations were that the populations exposed to a higher temperature indeed differed genetically from control populations for life-history traits and that these differences translated into differences in competitive strength at higher temperatures. The most striking result was that adaptation to heated mesocosms made the UK genotypes better than the control genotypes at competing with clones isolated from Southern France, mimicking immigration by southern genotypes. Although the French clones were still competitively stronger at the higher temperature than even the warm-adapted UK clones, the difference was substantially smaller than when French clones competed with the initial UK populations. These results suggest that evolutionary responses may rapidly reduce the vulnerability for invasion by southern genotypes. Yet, 1 year of evolutionary change proved insufficient to increase the fitness of UK clones to a level exceeding the fitness of southern immigrant clones (Van Doorslaer *et al.* 2009).

Overall, our results indicate potential for rapid evolutionary change. This suggests that there is potential for interaction between evolutionary dynamics and ecological processes, including population dynamics, community assembly and ecosystem functioning. These interactions have hardly been explored (Urban *et al.* 2008), even though they may change our perspective on ecological responses to anthropogenic environmental changes such as climate warming. They might, for example, explain the negligible apparent effects of temperature increase on the phytoplankton communities during the first UK mesocosm experiment, where a mere two of over ninety species showed declines and a further two increased their abundance in response to warming (Moss *et al.* 2003). It must be borne in mind, however, that algae and zooplankton undergo many generations in a year. Evolutionary response may be much slower for organisms that reproduce only annually or less frequently, such as fish.

Stream and wetland experiments at paired sites

Studies in controlled mesocosms have the advantage that many factors such as temperature and nutrient availability can be experimentally controlled. The disadvantage is that the systems are relatively small compared with natural ecosystems and may reflect conditions in ponds rather than lakes (though ponds and other small water bodies dominate the world's area of freshwaters and have been unjustifiably derogated). The same arguments apply to short self-contained experimental streams, though warming experiments in such systems have yet to be attempted. Given the open nature of streams with their unidirectional flow and often extensive associated wetlands, experiments involving artificial heating are very difficult and expensive to perform (Hogg & Williams 1996), with substantial constraints on both experimental scale and replication. A compromise is to use natural settings that provide the contrasted conditions required for testing a given hypothesis. For example, to conduct experiments on temperature effects, areas of geothermal activity provide excellent opportunities for adopting this paired-site approach. Thus, we took advantage in Euro-limpacs of adjacent streams and wetlands of different temperature regimes created in the actively volcanic landscapes of Iceland.

Stream studies

There is a wealth of information on stream macro-invertebrates in relation to temperature (e.g. Ward & Stanford 1982; Ward 1992), whether effects of temperature are direct or mediated by changed hydrological regimes of streams (see Chapter 4). Recent analyses of long-term data sets have also suggested major effects of temperature on the species distributions of invertebrates and on local community structure (Daufresne *et al.* 2003; Mouton & Daufresne 2006; Burgmer *et al.* 2007). Furthermore, predictions of community change in response to rising temperatures have been supported by recent invasions of terrestrial species from lower latitudes and altitudes coinciding with global warming (Parmesan *et al.* 1999; Hickling *et al.* 2006).

   The streams at Hengill in Iceland are almost ideal model systems to investigate the foundations of such field observations, as they provide a relevant temperature gradient that is largely unconfounded by other environmental variables (Friberg *et al.* 2009).

   The Hengill geothermal area (64°03'N, 21°18'W) of south-western Iceland (Fig. 6.9) is one of the most extensive geothermal fields on the island, with a total coverage of 110 km² (Saemundsson 1967; Gunnlaugsson & Gíslason 2005). Precipitation, which seeps through the porous volcanic bedrock, is among the highest in Iceland and feeds the numerous streams emerging from the mountain ridges at an elevation of 300–500 m a.s.l. The study site where the paired-stream experiments were carried out is located in the valleys of Miðdalur and Innstidalur at 350–420 m a.s.l. These valleys have lush grassland, wetland patches and streams with a range in temperature from 6 to 42 °C. The streams are primarily groundwater fed with little or no influence from volcanic gases. As a result, pH is comparable among streams differing in water temperature, as are other chemical variables which could have a negative effect on the biota. We investigated 10 streams which drained individual sub-catchments and experienced a summer temperature range of 7–23 °C (Fig. 6.10).

   In the streams investigated, structural and functional attributes of macro-invertebrates responded to temperature changes in similar ways as observed elsewhere along nutrient enrichment gradients (e.g. Pascoal *et al.* 2003;

Figure 6.9   Two of the streams sampled at Hengill, Iceland. The stream on the right is cold (range 2–8 °C), whereas the stream to the left is geothermally heated (range 15–23 °C). Both streams are spring fed and less than 50 m long before their confluence with the main river.

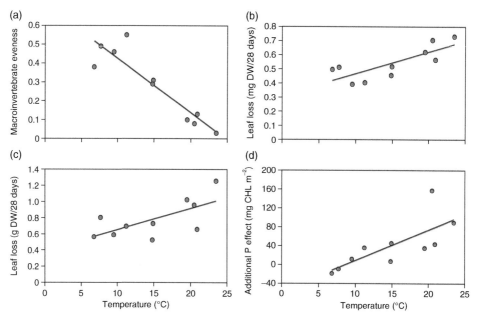

Figure 6.10   Variation in structural and functional attributes of contrasting stream ecosystems at Hengill, Iceland, in relation to differences in stream temperature.
(a) macro-invertebrate evenness; (b) loss in leaf mass from fine-mesh litter bags; (c) loss in leaf mass from coarse-mesh litter bags; and (d) additional effect on leaf mass loss of P in the N + P treatment (Friberg *et al.* 2009).

Hering *et al.* 2006). Warm streams under similar nutrient loading showed changes similar to those found in moderately eutrophic streams: dominance of a few macro-invertebrate taxa and increased productivity. Experimental nutrient additions suggested, furthermore, that the response to higher nutrient loading could increase with temperature, thereby making freshwater ecosystems more susceptible to eutrophication, similar to the responses in lakes described above.

Macro-invertebrate densities ranged from 3,000 to 16,000 individuals m^{-2} and were positively correlated with temperature. Communities had between 12 and 22 macro-invertebrate taxa and were dominated by Chironomidae (16 species out of 35 taxa in total). Chironomidae were especially abundant in the colder streams; in some, *Eukiefferiella minor* constituted up to 50% of the individuals recorded. The density of the only blackfly species found, *Simulium vittatum*, was highest in the warmer streams with only low densities recorded in colder streams. Even more pronounced was the distribution of the snail *Lymnaea peregra*. It was only found in streams with a temperature above 14 °C. There, however, it dominated the macro-invertebrate community in five of the six streams sampled, constituting up to 63% of the total number of individuals. Taxon richness tended to be unimodal with the highest number of taxa occurring at mid-range temperatures, whereas evenness was strongly negatively related to temperature (Fig. 6.10). The invertebrate species pool is limited in Iceland. Therefore, warming in the more species-rich streams of mainland Europe may show stronger effects, which may further exacerbate eutrophication symptoms.

Litter decomposition rates of Arctic downy birch (*Betula pubescens*) were used as an indicator of ecosystem functioning. Decomposition in both fine (mesh size: 200 µm) and coarse (mesh size: 1 cm) litter bags placed for 28 days in the streams was significantly correlated with temperature (Fig. 6.10). The leaf mass lost (initial weight 2.00 g DW) also differed significantly between the two types of bags, and a significant temperature effect was observed. Loss in coarse bags ranged between 0.53 (26.5%) and 1.26 g (64.5%) DW (28 days)$^{-1}$, being higher than the loss in fine mesh bags (0.39 (19.5%) to 0.71 g (35.5%) DW (28 days)$^{-1}$).

Nutrient-diffusing substrata consisting of plastic pots (surface area 20 cm^2, covered with 200-µm nylon mesh) containing either 2% agar (controls) or agar with N, P or both added were used in the same 10 streams to investigate the combined effects of temperature and nutrient additions. Algal biomass that grew on the pots was significantly higher in the N + P treatment (mean: 93 mg chl m^{-2}) compared with N alone (mean: 53 mg chl m^{-2}), and the N treatment had significantly higher algal biomass than both the P treatment and the controls, which did not differ. There were no significant relationships between temperature and algal biomass in any of the nutrient treatments or controls. However, compared with the other cold streams, the coldest stream investigated showed a deviant pattern, with high algal biomass on all substrates, including the controls, and a particularly strong response to both N and N + P additions. If this stream was excluded from the analysis, there was a clear trend in the N + P treatment towards higher algal biomass at raised temperature. Subtracting algal biomasses in the N treatments from those in the N + P treatments also revealed a clear temperature effect (Fig. 6.10), indicating that the additional effect of adding P in the N + P treatment was temperature sensitive.

Wetlands

Wetlands provide many ecosystem services that play important roles at landscape and global scales (Bobbink et al. 2006). Among other functions, wetlands play crucial roles in regional and global nutrient fluxes and are globally significant as locations of nitrogen removal by denitrification (Zedler & Kercher 2005; Verhoeven et al. 2006). This aspect of wetland function was investigated by using paired wet grassland and sedge communities near the Hengill streams presented above. Local geothermal activity heats up not only the stream water but also the soil, thereby creating naturally heated spots that closely resemble unheated ('ambient') spots nearby except for their temperature.

Denitrification removes nitrogen (as NO_3^- or NO_2^-) from ecosystems and releases it into the air as nitrous oxide (N_2O) or dinitrogen (N_2). This mechanism is of increasing importance in a world of high fertilizer use in human-dominated landscapes, because denitrification is a key mechanism through which excess nitrogen in landscapes is removed (e.g. Hefting et al. 2005). However, if denitrification does not proceed fully towards the formation of N_2, a stable greenhouse gas, N_2O, is emitted with a relative contribution to the greenhouse effect calculated to be 300 times greater than the contribution of CO_2 (Rodhe 1990).

We selected six naturally heated plots (average soil temperature at 10 cm depth was 28.7°C in August) and six very similar ambient plots (16.6°C) in the Hengill area of south-western Iceland (Fig. 6.11). At the start of the growing season, a 25×25 cm subplot in each plot was fertilized with 10 g N m^{-2} (100 kg ha^{-1}, which is equivalent to that of moderately intensive agriculture) in the form of *Agroblen* slow-release fertilizer. Just before taking gas samples from the fertilized and control subplots of each plot, we applied an additional fertilization of 12.5 g N m^{-2} as KNO_3 to the fertilized subplots. We sampled accumulated gases in each subplot using closed chambers (diameter 15.2 cm) put in place over the soil for 16 days at the end of the growing season in August. Gas samples were analysed for N_2O by gas chromatography. A great advantage of our approach was that we could combine N-enrichment and temperature differences in a two-factorial design without introducing artefacts that are normally associated with experimental heating of soils.

Nitrogen enrichment significantly increased N_2O emissions, but differences between plots of different soil temperatures were not significant (Fig. 6.12). There might have been a slight interaction effect between N-enrichment and temperature, as mean N_2O emissions from the warm plots were higher than from the cold plots, but this effect was not statistically significant. These field results were confirmed by N_2O emissions from Icelandic soil samples measured under controlled conditions in the laboratory (Fig. 6.12). We measured N_2O emissions from slurries with and without acetylene addition (acetylene inhibits enzymes involved in denitrification) and calculated N_2O/N_2 ratios following Castaldi and Smith (1998) for denitrification enzyme activity (DEA) measurements. We amended fresh soil from Icelandic soil cores with demineralized water or a solution of 10 mg NO_3^- N g^{-1} soil fresh weight and incubated them at 15°C or 30°C under anaerobic conditions. Gas samples were taken six hours after incubation and analysed in the same way as the gas samples from the field. Again, N_2O

Figure 6.11   Example of an experimental plot in a *Carex*-dominated fen at the geothermally active research site near the Hengill area in Iceland.

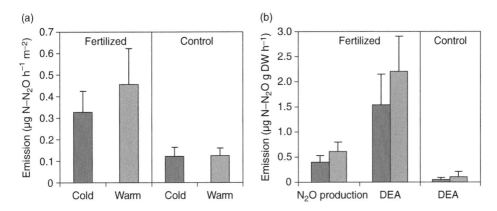

Figure 6.12 N_2O emissions measured (a) *in situ* from wetland soils of N-fertilized and control plots at both ambient and geothermally elevated temperatures (mean ± SE, $N = 6$) and (b) in wetland soil slurries under controlled conditions in the laboratory. Green bars represent incubation of soils from warm locations (30°C), blue bars from cold locations (15°C) (mean ± SE, $N = 8$). DEA, denitrification enzyme activity.

emission and DEA increased with N-enrichment but not significantly with temperature, although emissions were highest under both N-enrichment and elevated temperature (Fig. 6.12). The N_2O/N_2 ratios did not differ significantly between N-amended soils sampled from ambient and warm field sites.

   In conclusion, N-enrichment thus clearly stimulated N_2O production under both field and laboratory conditions, whereas a rise in temperature alone had only a small effect. While the effect of N-enrichment confirms results of previous studies (Velthof *et al.* 1996), the lack of a temperature effect was unexpected and might be explained by antagonistic effects of warming on the ratio of denitrification end products. The temperature difference in the study sites was around 12 °C and thus twice the estimated maximum temperature rise anticipated by the end of the 21st century (max. 6.4 °C; IPCC 2007). This result implies that a positive feedback loop between global warming and enhanced N_2O emissions from wetlands is by no means a certainty.

Palaeolimnology and modelling

To improve our understanding of how climate change might affect water quality in the future, we need a better knowledge of how it has modified lake ecosystems in the past. This information is contained within historical and palaeolimnological records, but it is difficult to extract because many of the lakes that have been affected by climate change have experienced changes in nutrient input over a similar period. In the Euro-limpacs project, several approaches have been applied to selected lake sites in an attempt to disentangle the effects of nutrient enrichment and climate change and assess interactions between the two. Statistical modelling was used on long-term monitoring data to identify trends and relationships among variables recorded seasonally (e.g. Ferguson *et al.* 2008), process-based modelling was applied to sites where historical catchment and lake data are available to provide information on interactions among variables (e.g. Elliott *et al.* 2005; Elliott & May 2008), and palaeolimnology was used to reconstruct environmental change from the remains of biota preserved in lake sediments and, thereby, to provide data over a longer period (e.g. Marchetto *et al.* 2004; Battarbee *et al.* 2005; Manca *et al.* 2007).

Insights from the application of these methods are illustrated here for Loch Leven, the largest shallow lake in Great Britain, which has been monitored fortnightly since 1968 and where significant changes in both the climate and the nutrient availability have been recorded (Carvalho & Kirika 2003). In response to severe eutrophication problems, phosphorus input from external sources, mostly a woollen mill, was reduced during the 1990s. The total phosphorus (TP) loading, which had steadily risen to 20 t yr^{-1} by 1985, was reduced to 8 t TP yr^{-1} by 1995 (Bailey-Watts & Kirika 1999). The period of eutrophication and recovery also spans a period when there has been a measurable impact of climate change on the lake. Winter ice cover has become less frequent and less extensive, spring air temperatures have increased markedly and winter rainfall has significantly increased (Ferguson *et al.* 2008).

In response to the reductions in point-source nutrient loading, there have been significant decreasing trends of soluble reactive phosphorus (SRP) concentrations in summer, autumn and winter, mainly because of decreases in the last decade (1995–2005). There has, however, been an increasing trend in winter nitrate concentrations, probably due partly to the significantly increased winter rainfall. Chlorophyll *a* concentration has decreased in spring as years have become warmer, but increased during the rest of the year (Ferguson *et al.* 2009).

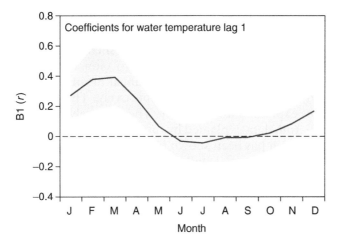

Figure 6.13 Calculated correlation coefficients for the relationship between water temperature and *Daphnia* population density at Loch Leven (the United Kingdom) over the year. Shaded region identifies ± 2 standard errors, indicating a significant positive relationship during winter and spring months, but no significant effect from May to November.

Daphnia numbers have also increased in winter and spring (Ferguson *et al.* 2007, 2009) (Fig. 6.13). Higher temperatures have direct physiological effects (e.g. increased growth rates) on plankton. However, the negative relationship at Loch Leven between chlorophyll *a* and water temperature in spring suggests an indirect response to higher spring temperatures, probably related to greater zooplankton production and grazing.

The phytoplankton community model PROTECH (Phytoplankton Responses To Environmental Change; Reynolds *et al.* 2001) was used to investigate the impacts of changing water temperature and nutrient loading upon the phytoplankton in Loch Leven (Elliott & May 2008). A range of water temperature and nutrient loading scenarios was run to examine the effect of changes in these variables on chlorophyll *a* concentration, phytoplankton species diversity and cyanobacterial abundance. Change in water temperature had relatively little effect on phytoplankton biomass and species diversity in comparison with changes in nutrient loading. However, phytoplankton varied according to the way in which nutrient loading changed. For example, increasing P load alone caused a large increase in total chlorophyll *a* concentration and in *Anabaena* (cyanobacteria) abundance. In contrast, simultaneously increasing the loads of phosphate and nitrate resulted in higher *Anabaena* densities at lower nutrient concentrations. A likely explanation for this observation is that *Anabaena*, which is a nitrogen fixer, is able to exploit the available P better than other phytoplankton species when nitrogen levels are low.

Sediment cores provide an alternative way of examining relationships over long periods. A sediment core from Loch Leven, representing the period 1940–2005, showed that changes in diatom composition and abundance were relatively subtle, yet highly variable (H. Bennion, unpublished data). Diatoms from the core were compared with physico-chemical conditions monitored in the lake over the same period, 1969–2005. Only small amounts of variation in the diatom assemblages could be attributed to climate and nutrients, with 2.2%, 3.7% and 5.7% being explained by annual mean TP, annual maximum temperature and annual total rainfall, respectively. The sum of interactions between variables explained less than 2%. The relatively higher proportion of variance in the diatom data explained by rainfall is possibly linked to its impact on flushing rate, which has been shown to influence the plankton populations of the loch (Bailey-Watts *et al.* 1990).

A comparison of the planktonic diatom composition in the palaeo-record with that in the phytoplankton records for 1996–2005 revealed a good match for the dominant taxa (J. Wischnewski & H. Bennion unpublished data) (Fig. 6.14). Whilst there were discrepancies between the two data sets for some taxa (e.g. *Aulacoseira ambigua*), all of the key species recorded in the plankton were also observed in the fossil assemblages, indicating that the sediment core almost faithfully records changes observed in the water column. Both records show a decline in *Cyclotella radiosa* and *Asterionella formosa*, and more recently in *Aulacoseira subarctica*, *A. granulata* and *Diatoma elongatum*, and a relative increase in *Fragilaria crotonensis*, *Stephanodiscus hantzschii* and *S. parvus*, taxa associated with nutrient-rich conditions. Indeed there is some evidence that water quality may have declined recently with an increase in measured annual mean TP concentration from $\sim 50\,\mu g\,l^{-1}$ in 2000 to $> 70\,\mu g\,l^{-1}$ in 2005. The degree to which this increase in nutrient concentrations can be attributed to changes in climate remains unclear. Increased water temperature, however, is associated with greater internal P loading at the site (Spears *et al.* 2008).

The research on Loch Leven to date suggests that each of the approaches above provides different, yet complementary, perspectives about lake responses to environmental change. In Loch Leven, the data sets show that the lake has been slow to recover from eutrophication and that temperature has variable effects, both positive and negative, on its plankton community. Responses, such as changes in chlorophyll *a* concentrations and *Daphnia* densities, are often interrelated, and the relationships between variables may vary over the year. Overall, the study highlights the complexity of lake responses to changes in nutrient regimes and climate and, despite advances in analytical methods, the difficulty of unravelling their effects.

Synthesis

This chapter started with a series of questions about the combined effects of warming and eutrophication. We set out to answer them by using a range of approaches and techniques, and although there will never be a complete answer to any of the questions, the results obtained in Euro-limpacs and presented above give at least partial ones.

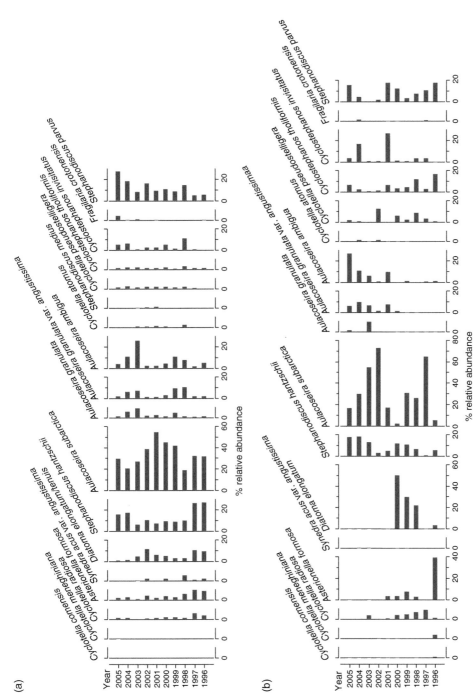

Figure 6.14 Planktonic diatom composition (expressed as annual means) for the period 1996–2005 in (a) fossil diatom assemblages of a sediment core and (b) plankton records from the water column.

Do nutrients structure ecosystems in different ways under current and anticipated future climatic conditions?

There is a clear change in trophic structure in lakes along the climate gradient from simple, often elongated, food webs in cold climates to truncated webs in warm ones with a higher degree of omnivory. Preliminary results indicate that this conclusion, well supported by data for lakes, also holds for streams.

The effect of increasing nutrient supply also differs among climate regions. In temperate lakes, a shift often occurs from high proportions of potential piscivorous fish, few and large plankti-benthivorous fish, high abundance of zooplankton and clear water often with macrophytes to a turbid state with dominance of small plankti-benthivorous fish and phytoplankton. This shift is associated with increased nutrient supply, although additional factors may be needed to trigger the switch and to maintain one state or the other. In the subtropics, lakes subject to both low and high nutrient loading are typically dominated by numerous small omnivorous fish that exert a high predation pressure on zooplankton. They leave little scope for daytime hiding of zooplankton in vegetation because the fish aggregate in the vegetation. Such systems can have clear water when nutrient loading is low. However, they are very vulnerable to increases in nutrient levels because the top-down effect of zooplankton is weaker than in temperate lakes due to the higher predation on zooplankton by fish.

Although less well studied than in lakes, the same probably holds true for streams and rivers, with a higher likelihood that benthic filamentous algae will grow abundantly on hard substrata at the bottom of small streams and dense phytoplankton communities develop in large rivers. There remains a big gap in knowledge about how wetlands will change in response to climate change and eutrophication. As many have a strong terrestrial character, analyses have focused on vegetation rather than whole food webs, and although animal communities, particularly birds, have also been extensively described in wetlands, overall trophic structures and their potential changes are not well understood.

Will changing climate interact with increased nutrient supply to alter ecosystem processes?

Whether changing climate interacts with increased nutrient supply to alter ecosystem processes is more uncertain at the moment than change of trophic structure. There is evidence that processes like deoxygenation, decomposition and denitrification are influenced by both nutrients and warming, but the interaction among factors appears to be complex and variable, and there are discrepancies about how system components and whole systems are affected. In lakes, we can expect (i) higher internal loading of phosphorus in response to higher temperatures, more prolonged stratification in deep lakes and higher sedimentation rates as phytoplankton becomes more abundant and grazing by zooplankton is reduced; (ii) possibly a higher likelihood of losing submerged macrophytes and thereby shifting shallow lakes from benthic- to pelagic-dominated systems and reducing biodiversity (although different pictures emerge from space-for-time studies and mesocosm experiments); and (iii) higher nutrient and carbon turnover and higher productivity, though perhaps unchanged or reduced NEP.

There is evidence that nitrogen, now increasingly seen as just as important as phosphorus in freshwater systems (Elser *et al.* 2007), may have counterintuitive effects, for example, by inhibiting decomposition, but sometimes also interacting with temperature to change the structure of aquatic plant communities in ways that will give more problems, such as increases in nuisance free-floating plants. There is also evidence that increased temperature will be associated with rapid expansion in exotic species of plants, already causing management problems, and fish such as carp, as more thermo-sensitive fish fail to survive. Where they have been introduced, carp are regarded as a problem to all but a section of the angling community that avidly seeks to catch large specimens. Important ecosystem processes in wetlands, such as denitrification, show complex responses to interactions between nutrient enrichment and warming. While increased loading of nitrogen increases denitrification in wetlands, antagonistic effects of warming on the ratio of denitrification end products may prevent emissions of N_2O, a potent greenhouse gas, from increasing at the same rate. This implies that warming under nutrient-enriched conditions need not necessarily lead to positive feedbacks on greenhouse gas production and emissions, though more experiments are needed to get a more clear picture.

Will changing climate aggravate eutrophication symptoms?

There is clear evidence from the space-for-time comparison of data, controlled experiments in ponds and lakes and the paired-site studies in Iceland that climate warming will exacerbate the symptoms of eutrophication. However, the effects will be complex and will vary depending on initial conditions and location. Shifts in trophic structure will lead to higher risk of eutrophication symptoms such as algal blooms and perhaps higher risk of cyanobacterial dominance in larger lakes and of filamentous algae in shallow lakes and streams. Some symptoms may be unaffected. For example, no experimental evidence to date suggests that macrophyte-dominated clear-water lakes will shift to a turbid state. Indeed, results of the mesocosm experiments provide no support for the presumption that increases in nutrient supply alone drive the displacement of aquatic plants. Other factors such as increased salinity, raised water levels, changes in fish communities, mechanical damage or damage by vertebrate grazers or toxins to invertebrate grazers or the plants themselves will generally be involved. In Icelandic wetlands, symptoms such as increased N_2O emission may even be abated following global warming, but further studies are needed to verify this finding for a wider range of wetland types.

Can effects of climate change be distinguished from those of eutrophication?

It is patently difficult to distinguish climate change effects from eutrophication effects in long-term records and palaeolimnological studies, since the symptoms accompanying both changes are rather similar and both have occurred simultaneously, at least in the last 150 years. However, the shift in trophic structure, phenology and life histories of organisms (e.g. age and size of first reproduction, longevity, seasonal dynamics) can be used as indicators of warming effects as the changes typically will be larger than expected from those of eutrophication alone. The results of controlled experiments can separate distinct effects in relatively simple though not necessarily unrealistic systems, but in more complex real ecosystems, many pressures not only act simultaneously but also interact, so that demonstration of the individual effects of possibly reduced nutrient loading and anthropogenic CO_2 emissions in the future will continue to be challenging and only give very general answers, even when advanced statistical approaches are used in the analysis of trends.

Are there lessons to be learnt from the past that can help to understand future problems better?

Our answer to the question whether useful lessons for the future can be learnt from the past is yes, particularly if several approaches are employed in concert to analyse past situations and make predictions. If we extend our records sufficiently far back in time and consider change over long timescales, we reach a situation where climate change was an important driver of ecosystem change while anthropogenic effects such as eutrophication were less pronounced or even negligible (cf. Chapter 2). Information from the past may therefore provide a clearer picture of ecosystem change driven largely by climate and can help us to extract a climate signal from contemporary data sets. Palaeolimnology holds great potential to achieve this goal.

In several lakes, there is evidence of ecological response to changes in both nutrient levels and climate, although in most cases, the eutrophication signal tends to eclipse the climate signal. Nonetheless, in Lago Maggiore of Northern Italy, for example, there is evidence of increased temporal variation in zooplankton communities following exceptional meteorological events and changes in fish predation, which might be partly attributable to increased temperatures (Manca *et al.* 2007). Furthermore, several data sets indicate that climate change is likely to have a confounding effect on recovery from eutrophication (Jeppesen *et al.* 2009), suggesting that lake responses to nutrient reduction may be somewhat slower under warmer and wetter conditions than originally envisaged based on current climate.

Can we mitigate negative effects of climate change on ecosystems in terms of enhanced eutrophication?

There is much scope for combating eutrophication symptoms that will be aggravated by global warming by taking measures to reduce external nutrient loading to freshwaters beyond those already implemented or planned. These include (i) less intensive land use in catchments with sensitive freshwaters to reduce diffuse nutrient inputs; (ii) re-establishment of riparian vegetation to buffer nutrient transfers to streams and rivers and improve in-channel structures to increase the retention of organic matter and nutrients; (iii) improved land management to reduce sediment and nutrient export from catchments; (iv) improved design of sewage works to cope with the consequences of flood events and low flows in receiving waters; and (v) more effective reduction of nutrient loading from point sources and, for N, from the atmosphere.

Conclusion

It seems clear in general terms that warming will exacerbate many, though not necessarily all, symptoms of eutrophication, but there remain three key problems in answering the questions posed here. The first is that responses are complex and varied depending on the specific context. As a result, deduction, from general principles, of specific measures to be taken in particular situations will be associated with great uncertainty in view of the limited scope of research that is likely to be achievable over the period in which climate change is occurring. Taking a precautionary approach thus remains eminently wise. The second concern is that several lines of evidence hint at biological feedback mechanisms that may result in increased respiratory production of carbon dioxide, if not of nitrous oxide (N_2O), and methane (CH_4). This might mean that the purely physical models that are the basis for climate change predictions made by the IPCC (2007) are severe underestimates. The third point is that as world population grows, pressure to grow more food, whilst simultaneously producing crops for biofuel, will probably lead to further increased nutrient inputs and an intensification of eutrophication problems in receiving fresh waters. Coupled with the inherent resistance of human society to acknowledge that the root of the climate problem is economic growth regardless of environmental consequences, and that serious alternative economic models to continued growth are lacking, the future of fresh waters has become highly uncertain.

References

Alcamo, J., Florke, M. & Marker, M. (2007) Future long-term changes in global water resources driven by socio-economic and climatic changes. *Hydrological Sciences Journal*, 52, 247–275.

Bachmann, R.W., Horsburgh, C.A., Hoyer, M.V., Mataraza, L.K. & Canfield, D.E. (2002) Relations between trophic state indicators and plant biomass in Florida lakes. *Hydrobiologia*, 470, 219–234.

Bailey-Watts, A.E. & Kirika, A. (1999) Poor water quality in Loch Leven (Scotland) in 1995, in spite of reduced phosphorus loadings since 1985: The influences of catchment management and inter-annual weather variation. *Hydrobiologia*, 403, 135–151.

Bailey-Watts, A.E., Kirika, A., May, L. & Jones, D.H. (1990) Changes in phytoplankton over various time scales in a shallow, eutrophic: The Loch Leven experience with special reference to the influence of flushing rate. *Freshwater Biology*, 23, 85–111.

Barker, T., Hatton, K., O'Connor, M., Connor, L., Bagnell, L. & Moss, B. (2008) Control of ecosystem state in a shallow, brackish lake: Implications for the conservation of stonewort communities. *Aquatic Conservation-Marine and Freshwater Ecosystems*, 18, 221–240.

Battarbee, R.W., Anderson, N.J., Jeppesen, E. & Leavitt, P.R. (2005) Combining palaeolimnological and limnological approaches in assessing lake ecosystem response to nutrient reduction. *Freshwater Biology*, 50, 1772–1780.

Beklioglu, M. & Özen, A. (2008) Ülkemiz göllerinde kuraklık etkisi ve ekolojik tepkiler. *Uluslararası Küresel İklim Değişimi ve Çevresel Etkileri Konferansı*, 1, s.299–s.306.

Beklioglu, M. & Tan, C.O. (2008) Drought complicated restoration of a Mediterranean shallow lake by biomanipulation. *Archive für Hydrobiologie/Fundamentals of Applied Limnology*, 171, 105–118.

Beklioglu, M., Altınayar, G. & Tan, C.T. (2006) Water level control over submerged macrophyte development in five Mediterranean Turkey. *Archive für Hydrobiologie*, 166, 535–556.

Beklioglu, M., Romo, S., Kagalou, I., Quintana, X. & Bécares, E. (2007) State of the art in the functioning of shallow Mediterranean lakes: Workshop conclusions. *Hydrobiologia*, 584, 317–326.

Blanck, A. & Lamouroux, N. (2007) Large-scale intraspecific variation in life-history traits of European freshwater fish. *Journal of Biogeography*, **34**, 862–875.

Blenckner, T., Malmaeus, J.M. & Pettersson, K. (2006) Climatic change and the risk of lake eutrophication. *International Association of Theoretical and Applied Limnology*, **29**, 1837–1840.

Blenckner, T., Adrian, R., Livingstone, D.M., *et al.* (2007) Large-scale climatic signatures in lakes across Europe: A meta-analysis. *Global Change Biology*, **13**, 1314–1326.

Bobbink, R., Beltman, B., Verhoeven, J.T.A. & Whigham, D.F. (eds) (2006) *Wetlands: Functioning, Biodiversity, Conservation and Restoration*. Springer-Verlag, Heidelberg.

Brix, H., Sorrell, B.K. & Lorenzen, B. (2001) Are Phragmites-dominated wetlands a net source or net sink of greenhouse gases? *Aquatic Botany*, **69**, 313–324.

Brucet, S., Boix, D., Gascan, S., *et al.* (2009) Species richness of crustacean zooplankton and trophic structure of brackish lagoons in contrasting climate zones: North temperate Denmark and Mediterranean Catalonia (Spain). *Ecography*, **32**, 692–702.

Burgmer, T., Hillebrand, H. & Pfenninger M. (2007) Effects of climate-driven temperature changes on the diversity of freshwater macroinvertebrates. *Oecologia*, **151**, 93–103.

Carpenter, S.R., Caraco, N.F., Correll, D.L., Howarth, R.W., Sharpley, A.N. & Smith, V.H. (1998) Nonpoint pollution of surface waters with phosphorus and nitrogen. *Ecological Applications*, **8**, 559–568.

Carvalho, L. & Kirika, A. (2003) Changes in shallow lake functioning: Response to climate change and nutrient reduction. *Hydrobiology*, **506/509**, 789–796.

Castaldi, S. & Smith, K.A. (1998) The effect of different N substrates on biological N_2O production from forest and agricultural light textured soils. *Plant and Soil*, **2**, 229–238.

Coops, H., Beklioglu, M. & Crisman, T.L. (2003) The role of water-level fluctuations in shallow lake ecosystems – workshop conclusions. *Hydrobiologia*, **506/509**, 23–27.

Daufresne, M., Roger, M.C., Capra, H. & Lamoroux, N. (2003) Long-term changes within the invertebrate and fish communities of the Upper Rhone River: Effects of climatic factors. *Global Change Biology*, **10**, 124–140.

Dodds, W.K. (2007) Trophic state, eutrophication, and nutrient criteria in streams. *Trends in Ecology and Evolution*, **22**, 669–676.

Elliott, J.A. & May, L. (2008) The sensitivity of phytoplankton in Loch Leven (U.K.) to changes in nutrient load and water temperature. *Freshwater Biology*, **53**, 32–41.

Elliott, J.A., Thackeray, S.J., Huntingford, C. & Jones, R.G. (2005) Combining a regional climate model with a phytoplankton community model to predict future changes in phytoplankton in lakes. *Freshwater Biology*, **50**, 1404–1411.

Elser, J.J., Bracken, M.E.S., Cleland, E.E., *et al.* (2007) Global analysis of nitrogen and phosphorus limitation of primary producers in freshwater, marine and terrestrial ecosystems. *Ecology Letters*, **10**, 1135–1142.

Ferguson, C.A., Scott, E.M., Bowman, A.W. & Carvalho, L. (2007) Model comparison for a complex ecological system. *Journal of the Royal Statistical Society Series A*, **170**, 691–711.

Ferguson, C.A., Carvalho, L., Scott, E.M., Bowman, A.W. & Kirika, A. (2008) Assessing ecological responses to environmental change using statistical models. *Journal of Applied Ecology*, **45**, 193–203.

Ferguson, C.A., Bowman, A.W., Scott, E.M. & Carvalho, L. (2009) Multivariate varying-coefficient models for ecological systems. *Environmetrics*, **20**, 460–476.

Feuchtmayr, H., Moran, R., Hatton, K., *et al.* (2009) Global warming and eutrophication: Effects on water chemistry and autotrophic communities in experimental hypertrophic shallow lake mesocosms. *Journal of Applied Ecology*, **46**, 713–723.

Firth, P. & Fisher, S.G. (1992) *Global Climate Change and Freshwater Ecosystems*. Springer-Verlag, New York.

Flury, S. (2008) *Carbon fluxes in a freshwater wetland under simulated global change: Litter decomposition, microbes and methane emission*. PhD dissertation, ETH Zurich, Zurich.

Flury, S., McGinnis, D.F. & Gessner, M.O. (2010) Methane emissions from a freshwater marsh in response to experimentally simulated global warming and nitrogen enrichment. *Journal of Geophysical Research-Biogeosciences*, **115**, G01007, doi: 10.1029/2009JG001079.

Friberg, N., Dybkjær, J.B., Olagsson, J.S., Gislason, G.M., Larsen, S.E. & Lauridsen, T.L. (2009) Relationships between structure and function in streams contrasting in temperature. *Freshwater Biology*, **54**, 2051.

Galloway, J.N., Townsend, A.R., Erisman, J.W., *et al.* (2008) Transformation of the nitrogen cycle: Recent trends, questions, and potential solutions. *Science*, 320, 889–892.

Gunnlaugsson, E. & Gíslason, G. (2005) Preparation for a new power plant in the Hengill geothermal area, Iceland. *Proceedings of the World Geothermal Congress 2005*, Antalya, Turkey, 24–29 April 2005.

Gyllström, M., Hansson, L.-A., Jeppesen, E., *et al.* (2005) The role of climate in shaping zooplankton communities of shallow lakes. *Limnology and Oceanography*, 50, 2008–2021.

Hammrich, A. (2008) *Effects of warming and nitrogen enrichment on carbon turnover in a littoral wetland*. PhD dissertation, ETH Zurich, Zurich.

Hefting, M.M., Clement, J.-C., Bienkowski, P., *et al.* (2005) The role of vegetation and litter in the nitrogen dynamics of riparian buffer zones in Europe. *Ecological Engineering*, 24, 465–482.

Hering, D., Johnson, R.K., Kramm, S., Schmutz, S., Szoszkiewicz, K. & Verdonschot, P.F.M. (2006) Assessment of European streams with diatoms, macrophytes, macroinvertebrates and fish: A comparative metric-based analysis of organism response to stress. *Freshwater Biology*, 51, 1757–1785.

Hickling, R., Roy, R.B., Hill, J.K., Fox, R. & Thomas, C.D. (2006) The distributions of a wide range of taxonomic groups are expanding polewards. *Global Change Biology*, 12, 450–455.

Hogg, I.D. & Williams, D.D. (1996) Response of stream invertebrates to a global-warming thermal regime: An ecosystem-level manipulation. *Ecology*, 77, 395–407.

Iglesias, C., Goyenola, G., Mazzeo, N., Meerhoff, M., Rodó, E. & Jeppesen, E. (2007) Horizontal dynamics of zooplankton in subtropical Lake Blanca (Uruguay) hosting multiple zooplankton predators and aquatic plant refuges. *Hydrobiologia*, 584, 179–189.

Iglesias, C., Mazzeo, N., Goyenola, G., *et al.* (2008) Field and experimental evidence of the effect of *Jenynsia multidentata*, a small omnivorous–planktivorous fish, on the size distribution of zooplankton in subtropical lakes. *Freshwater Biology*, 53, 1797–1807.

IPCC (Intergovernmental Panel on Climate Change) (2007) Summary for policymakers l. In: *Climate Change 2007: The Physical Science Basis. Contribution of Working Group I to the Fourth Assessment Report of the Intergovernmental Panel on Climate Change* (eds S. Solomon, M. Manning, Z. Chen, *et al.*). Cambridge University Press, Cambridge and New York.

Jackson, L.J., Søndergaard, M., Lauridsen, T.L. & Jeppesen, E. (2007) Patterns, processes, and contrast of macrophyte-dominated and turbid Danish and Canadian shallow lakes, and implications of climate change. *Freshwater Biology*, 52, 1782–1792.

Jeppesen, E., Søndergaard, M., Meerhoff, M., Lauridsen, T.L. & Jensen, J.P. (2007a) Shallow lake restoration by nutrient loading reduction – Some recent findings and challenges ahead. *Hydrobiologia*, 584, 239–252.

Jeppesen, E., Søndergaard, M., Pedersen, A.R., *et al.* (2007b) Salinity induced regime shift in shallow brackish lagoons. *Ecosystems*, 10, 47–57.

Jeppesen, E., Kronvang, B., Meerhoff, M., *et al.* (2009) Climate change effects on runoff, catchment phosphorus loading and lake ecological state, and potential adaptations. *Journal of Environmental Quality*, 48, 1930–1941.

Jeppesen, E., Søndergaard, M., Holmgren, K., *et al.* (2010) Impacts of climate warming on lake fish community structure and potential effects on ecosystem function. *Hydrobiologia*, 646, 73–90.

Kosten, S., Kamarainen, A., Jeppesen, E., *et al.* (2009) Likelihood of abundant submerged vegetation growth in shallow lakes differs across climate zones. *Global Change Biology*, 15, 2503–2517.

Lehtonen, H. (1996) Potential effects of global warming on northern European freshwater fish and fisheries. *Fisheries Management Ecology*, 3, 59–71.

Liboriussen, L. & Jeppesen, E. (2003) Temporal dynamics in epipelic, pelagic and epiphytic algal production in a clear and a turbid shallow lake. *Freshwater Biology*, 48, 418–431.

Manca, M., Torretta, B., Comoli, P., Amsinck, S.L. & Jeppesen, E. (2007) Major changes in the trophic dynamics in large, deep subalpine Lake Maggiore from 1943 to 2002: A high resolution comparative palaeo-neolimnological study. *Freshwater Biology*, 52, 2256–2269.

Marchetto, A., Lami, A., Musazzi, S., Massaferro, J., Langone, L. & Guilizzoni, P. (2004) Lake Maggiore (N. Italy) trophic history: Fossil diatom, plant pigments, and chironomids, and comparison with long-term limnological data. *Quaternary International*, 13, 97–110.

McKee, D., Atkinson, D., Collings, S.E., *et al.* (2000) Heated aquatic microcosms for climate change experiments. *Freshwater Forum*, 14, 51–58.

McKee, D., Hatton, K., Eaton, J.W., *et al.* (2002) Effects of simulated climate warming on macrophytes in freshwater microcosm communities. *Aquatic Botany*, 74, 71–83.

McKee, D., Atkinson, D., Collings, S.E., *et al.* (2003) Response of freshwater microcosm communities to nutrients, fish and elevated temperature during winter and summer. *Limnology and Oceanography*, **48**, 707–722.

Meerhoff, M., Iglesias, C., Teixeira de Mello, F., *et al.* (2007a) Effects of contrasting climates and habitat complexity on community structure and predator avoidance behaviour of zooplankton in the shallow lake littoral. *Freshwater Biology*, **52**, 1009–1021.

Meerhoff, M., Clemente, J.M., de Mello, F.T., Iglesias, C., Pedersen, A.R. & Jeppesen, E. (2007b) Can warm climate-related structure of littoral predator assemblies weaken the clear water state in shallow lakes? *Global Change Biology*, **13**, 1888–1897.

Mooij, W.M., Hulsmann, S., De Senerpont Domis, L.N., *et al.* (2005) The impact of climate change on lakes in the Netherlands: A review. *Aquatic Ecology*, **39**, 381–400.

Moran, R., Harvey, I., Moss, B., *et al.* (2009) Influence of simulated climate change and eutrophication on three-spined stickleback populations: A large scale mesocosm experiment. *Freshwater Biology*, **55**, 315.

Moss, B. (2010) Climate change, nutrient pollution and the bargain of Dr Faustus. *Freshwater Biology*, **55** (1), 175–187.

Moss, B., McKee, D., Atkinson, D., *et al.* (2003) How important is climate? Effects of warming, nutrient addition and fish on phytoplankton in shallow lake microcosms. *Journal of Applied Ecology*, **40**, 782–792.

Mouton, J. & Daufresne, M. (2006) Effects of the 2003 heatwave and climatic warming on mollusc communities of the Saone: A large lowland river and its two main tributaries (France). *Global Change Biology*, **12**, 441–449.

Olesen, J.E. & Bindi, M. (2002) Consequences of climate change for European agricultural productivity, land use and policy. *European Journal of Agronomy*, **16**, 239–262.

Özen, A., Karapınar, B., Kucuk, I., Jeppesen, E. & Beklioglu, M. (2010) Drought-induced changes in nutrient concentrations and retention in two shallow Mediterranean lakes subjected to different degrees of management. *Hydrobiologia*, **646**, 61–72.

Parmesan, C., Ryrholm, N., Steanescu, C., *et al.* (1999) Poleward shifts in geographical ranges of butterfly species associated with regional warming. *Nature*, **399**, 579–583.

Pascoal, C., Pinho, M., Cassio, F. & Gomes, P. (2003) Assessing structural and functional ecosystem condition using leaf breakdown: Studies on a polluted river. *Freshwater Biology*, **48**, 2033–2044.

Reynolds, C.S., Irish, A.E. & Elliott, J.A. (2001) The ecological basis for simulating phytoplankton responses to environmental change (PROTECH). *Ecological Modelling*, **140**, 271–291.

Rodhe, H. (1990) A comparison of the greenhouse contribution of various gases to the greenhouse effect. *Science*, **248**, 1217–1219.

Romo, S., Villena, M.-J., Sahuquillo, M., *et al.* (2005) Response of a shallow Mediterranean lake to nutrient diversion: Does it follow similar patterns as in northern shallow lakes? *Freshwater Biology*, **50**, 1706–1717.

Saemundsson, K. (1967) Vulkanismus und tectonic des Hegnillgebietes in Sudwest-Island. *Acta Naturalia Islandica*, **II** (7) 1–105.

Scheffer, M., Straile, D., van Nes, E.H. & Hosper, H. (2001) Climatic warming causes regime shifts in lake food webs. *Limnology and Oceanography*, **46**, 1780–1783.

Schindler, D.W. (2006) Recent advances in the understanding and management of eutrophication. *Limnology and Oceanography*, **51**, 356–363.

Smil, V. (2000) Phosphorus in the environment: Natural flows and human interferences. *Annual Review of Energy Environment*, **25**, 53–88.

Smith V.H. (2003) Eutrophication of freshwater and marine ecosystems: A global problem. *Environmental Science and Pollution Research*, **10**, 126–139.

Spears, B.M., Carvalho, L., Perkins, R. & Paterson, D.M. (2008) Effects of light on sediment nutrient flux and water column nutrient stoichiometry in a shallow lake. *Water Research*, **42**, 977–986.

Teixeira de Mello, F., Meerhoff, M., Pekcan-Hekim, Z. & Jeppesen, E. (2009) Substantial differences in littoral fish community structure and dynamics in subtropical and temperate shallow lakes. *Freshwater Biology*, **54**, 1202–1215.

Urban, M.C., Leibold, M.A., Amarasekare, P., *et al.* (2008) The evolutionary ecology of metacommunities. *Trends in Ecology and Evolution*, **23**, 311–317.

Vadeboncoeur, Y., Jeppesen, E., Vander Zanden, M.J., Schierup, H.-H., Christoffersen, K. & Lodge, D. (2003) From Greenland to green lakes: Cultural eutrophication and the loss of benthic pathways. *Limnology and Oceanography*, **48**, 1408–1418.

Van Doorslaer, W., Stoks, R., Jeppesen, E. & De Meester, L. (2007) Adaptive responses to simulated global warming in *Simocephalus vetulus*: A mesocosm study. *Global Change Biology*, 13, 878–886.

Van Doorslaer, W., Vanoverbeke, J., Duvivier, C., *et al.* (2009) Local adaptation to higher temperatures reduces immigration success of genotypes from a warmer region in the water flea *Daphnia*. *Global Change Biology*, 15(12), 3046–3055.

Velthof, G.L., Oenema, O., Postma, R. & Van Beusichem, M.L. (1996) Effects of type and amount of applied nitrogen fertilizer on nitrous oxide fluxes from intensively managed grassland. *Nutrient Cycling in Agroecosystems*, 46, 257–267.

Verhoeven, J.T.A., Arheimer, B., Chenqing, Y. & Hefting, M.M. (2006) Regional and global concerns over wetlands and water quality. *Trends in Ecology and Evolution*, 21, 96–103.

Ward, J.V. (1992) *Aquatic Insect Ecology. 1. Biology and Habitat*. John Wiley & Sons, New York.

Ward, J.V. & Stanford, J.A. (1982) Thermal responses in the evolutionary ecology of aquatic insects. *Annual Review of Entomology*, 27, 97–117.

Williams, W.D. (2001) Anthropogenic salinisation of inland waters. *Hydrobiologia*, 466, 329–337.

Zalidis, G., Stamatiadis, S., Takavakoglou, V., Eskridge, K. & Misopolinos, N. (2002) Impacts of agricultural practices on soil and water quality in the Mediterranean region and proposed assessment methodology. *Agricultural Ecosystems Environment*, 88, 137–146.

Zedler, J.B. & Kercher, S. (2005) Wetland resources: Status, trends, ecosystem services, and restorability. *Annual Review of Environment and Resources*, 30, 39–74.

7

Interaction of Climate Change and Acid Deposition

Richard F. Wright, Julian Aherne, Kevin Bishop, Peter J. Dillon, Martin Erlandsson, Chris D. Evans, Martin Forsius, David W. Hardekopf, Rachel C. Helliwell, Jakub Hruška, Mike Hutchins, Øyvind Kaste, Jiří Kopáček, Pavel Krám, Hjalmar Laudon, Filip Moldan, Michela Rogora, Anne Merete S. Sjøeng and Heleen A. de Wit

Introduction

Both climate change and acidification ultimately involve chemical changes in the atmosphere and, not surprisingly, the two are strongly linked. Particular focus here is on how climate change might delay recovery of acidified, and hence damaged, aquatic ecosystems. Data come from acid-impacted areas of Europe and eastern North America and involve large-scale experiments with altered local climate, conducted in small catchments and lakes. Long-term datasets (30+ years) have also been analysed to link variations in acid deposition and climate on water chemistry and biology. Together, the experimental and empirical data have been used to develop, modify and calibrate statistical and process-orientated models. These models have then provided tools to predict ecosystem changes, given the future scenarios of acid deposition and climate.

Chronic emissions of sulphur (S) and nitrogen (N) compounds to the atmosphere, long-range transport and the resultant deposition of S and N pollutants have caused acidification of freshwaters over large regions of Europe and eastern North America (Overrein *et al.* 1980; Schindler 1988; Clair *et al.* 1995; Driscoll *et al.* 2001; Jeffries *et al.* 2003). In Europe, S emissions were greatest in the late 1970s/early 1980s and have since declined strongly, while N emissions have had a much more modest decline (Fig. 7.1).

Ecological damage included loss of salmon and trout populations from thousands of water bodies and changes in the species composition of invertebrate, aquatic macrophyte and algal communities. In Scandinavia, for example, a survey

Climate Change Impacts on Freshwater Ecosystems. First edition. Edited by M. Kernan, R. Battarbee and B. Moss. © 2010 Blackwell Publishing Ltd.

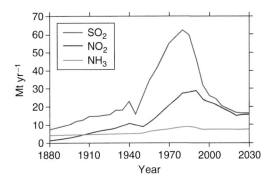

Figure 7.1   The rise and fall of emissions of sulphur and nitrogen in Europe over the period 1880–2030 as estimated by Schöpp *et al.* (2003). Units are Mt yr^{-1} of SO$_2$ (red line), NO$_2$ (blue line) and NH$_3$ (green line), respectively. Estimates for the future (2000–30) assume full implementation of current legislation (CLe scenario).

conducted in the 1990s indicated that about 10% of all fish stocks in western Scandinavia's 126,000 lakes had been affected by acidification (Tammi *et al.* 2003). The situation in Norway has been quite dramatic, with lakes in 30% of the country suffering damage to fish populations (Hesthagen *et al.* 1999). The story has been similar in other parts of Europe and in eastern North America, with extensive acidification reported from south-eastern Canada and upland areas of eastern United States. Here also, S emissions peaked in the 1980s. Inorganic aluminium and hydrogen ions are the primary agents of toxicity. The strong acid anions, sulphate, nitrate and chloride, bring these acid cations into soil solution and to runoff. The links between emissions of air pollutants, deposition of S and N, acidification of surface waters and damage to aquatic organisms thus lie in following the fates of S, N and, to a lesser extent, chloride (Cl).

Two factors are needed for surface water acidification: the water must be acid-sensitive and the area must receive sufficient amounts of acid deposition (Wright & Henriksen 1978). Acid sensitive lakes and streams are found throughout the world in catchments with weathering-resistant bedrock such as granite and quartzite and young, often poorly-developed podsolic and organic-rich soils (Skjelkvåle & Wright 1990). In these waters, the dominant inorganic anion is usually the weak-acid anion bicarbonate (HCO$_3$), whose source is respiration by plant roots and dissolution in soil water. HCO$_3$ is generally accompanied by the base cations, calcium (Ca) and magnesium (Mg), and the concentrations of all three depend on the ease by which the soil minerals can be broken down by weathering. Acid-sensitive waters have low concentrations of these ions.

The second factor is the amount of acid deposition. The most sensitive waters are affected when the rain is more acidic than about pH 4.7 and pollutant sulphate (SO$_4$) concentrations exceed about 20 µeq l^{-1}. In acidified waters, the strong-acid anion SO$_4$ is usually the dominant ion and replaces HCO$_3$. Acidified, SO$_4$-rich waters have pH often below 5 and increased concentrations of inorganic aluminium species (Al^{n+}). The acid and inorganic Al are toxic to fish and other aquatic organisms.

By the 1980s, acid rain was clearly recognized as an international problem that required an international solution. In 1979, negotiations to reduce the emissions of air pollutants began under the auspices of the United Nations Economic Commission for Europe (UNECE) with the establishment of the Convention on Long-Range Transboundary Air Pollution (LRTAP) (http://www.unece.org/env/lrtap) (Bull et al. 2001). Work under the Convention had produced a series of protocols in which countries agreed to reduce emissions of sulphur and nitrogen compounds to the atmosphere. The latest protocol, signed in 1999 at Gothenburg, Sweden, called for about 80% reduction in S and 50% reduction in N emissions in Europe by the year 2010 relative to levels in 1980. Similar agreements were reached in North America and have resulted in large decreases in S emissions in both eastern Canada and eastern United States.

The result has been dramatic reductions in S deposition in both Europe (Fig. 7.1) and North America. As these protocols began to be implemented, emissions of acidifying gases in Europe declined from their peak levels in the late 1970s/early 1980s, and by the year 2000, sulphur deposition had decreased by >50% and nitrogen deposition by about 20% (UNECE 1999). Further decreases are predicted (Fig. 7.1) for the next 20 years as the Gothenburg protocol and other, national, legislations are implemented (called the current legislation scenario – CLe).

And it has begun to work. In the 1990s, surface waters in Europe showed the first signs of recovery in response to lower levels of acid deposition; SO_4 concentrations decreased, pH and acid neutralizing capacity (ANC) increased and concentrations of Al^{n+} decreased (Stoddard et al. 1999; Evans et al. 2001). The waters are becoming less toxic for fish and other aquatic organisms (Monteith et al. 2005), but there is still a long way to go before recovery is complete.

In a modelling study of surface waters in 12 acid-sensitive areas of Europe, Wright et al. (2005) showed that while many waters have shown chemical recovery since the peak of acid deposition in the 1980s, even with full implementation of the Gothenburg protocol, many others will continue to be acidified for decades to come (Fig. 7.2).

Climate change can affect the chemical and biological recovery of freshwaters from acidification. Long-term, seasonal and episodic changes in climate all potentially affect a variety of processes in catchments and surface water bodies. For example, warming can be expected to increase rates of mineralization of soil organic matter, which, in turn, might release nutrients such as nitrogen in the catchment in runoff. This was the result of the CLIMEX (Climate Change Experiment) experiment in Norway, in which, nitrogen flux in runoff increased following a whole-catchment warming (Wright 1998). The most biologically damaging effects of acidification often occur during short acid episodes in lakes and streams. These episodes typically coincide with climatic extremes, such as droughts, storm events, snowmelt or periods of winter freezing. Acid pulses following droughts have been documented from Ontario, Canada (Dillon et al. 1997), and acid pulses following storms with high winds and inputs of sea salts have been reported from Southern Norway (Hindar et al. 1994). The prognosis for future climate in Europe is generally warmer, wetter in the north but drier in the south, and stormier in all areas with more frequent extremes. Climate change may thus offset or even reverse some of the ongoing recovery expected as S and N emissions continue to decline.

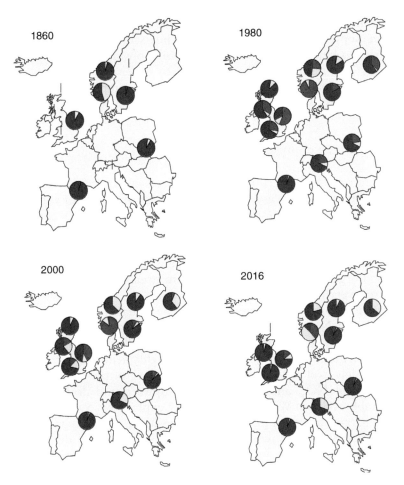

Figure 7.2 ANC concentrations (μeq l^{-1}) in surface waters for 12 regions in Europe as simulated by the acidification model MAGIC (Model of Acidification of Groundwater in Catchments) (SMART (Simulation Model for Acidification's Regional Trends) in Finland), given full implementation of the Gothenburg protocol and other legislation but not considering the confounding influence of climate change. The three ANC classes relate to probability of viable populations of brown trout and other key indicator organisms. Red: ANC < 0 barren of fish; Yellow: ANC 0–20 sparse population; Blue: ANC > 20 good population. Four key years are shown: 1860 pre-acidification (no simulations for Finland, as the SMART model was initiated in year 1960); 1980 maximum acidification; 2000 present-day; 2016 complete implementation of emission reduction protocols. (From Wright *et al.* 2005.)

Three research approaches have been used to study the effects of acid deposition and climate change on surface waters. The first is the analysis of data collected regularly at one site over time or at many sites simultaneously to record spatial variations. The second is experimental, either in the laboratory, in mesocosms or in large-scale whole ecosystem experiments in the field. And the third is by modelling, either statistical or process-orientated models that can be used to evaluate the effects of various future scenarios for acid deposition and climate. These three approaches are interrelated: trends seen in analyses of empirical data give rise to hypotheses of cause–effect relationships that can be tested by experiments. The results can then be used to develop and calibrate models. Modelling can reveal shortcomings and gaps in the empirical data, which can then form the basis for new monitoring or measurements.

Effects of climate change on nitrate leaching

The role of nitrate (NO_3) in climate change effects on recovery of surface waters from acidification is of special interest. While S concentrations in deposition have decreased by 60%–80% since the peak years in the 1980s, N concentrations continue to be high. As many terrestrial ecosystems in acid-sensitive regions are growth-limited by nitrogen, typically most of the N deposited is retained, largely in soil organic matter. If climate change entails mobilization and release of N stored in soil organic matter, the NO_3 levels in streams and lakes could increase and delay recovery or even cause re-acidification.

Climate effects on NO_3 concentrations can be revealed by analysing long data records of climate, N deposition and streamwater NO_3 concentrations. de Wit *et al.* (2008) analysed 20-year records of NO_3 in four small streams in Norway, three of which (Birkenes, Langtjern, Storgama) are highly acidified with pH 4.5–5.5. Empirical models explained between 45% and 61% of the variation in weekly concentrations of NO_3, and described both upward and downward seasonal trends tolerably well (de Wit *et al.* 2008). Key explanatory variables were snow depth, discharge, temperature and N deposition. All catchments showed reductions over time in snow depth and increases in winter discharge due to warmer winters in the years since about 1990. In two inland catchments, located in moderate N deposition areas, these climatic changes appeared to drive distinct decreases in winter and spring concentrations and fluxes of NO_3.

At one of the sites, Storgama, a series of experiments was conducted in 2003–7 to test the role of snowpack in regulating NO_3 concentrations and fluxes in runoff (Kaste *et al.* 2008). Here, whole-catchment manipulations in mini-catchments included extra insulation of soils in two catchments (by means of rock wool mats) to prevent sub-zero temperatures during winter and removal of snow in two other catchments to promote soil frost (Fig. 7.3). The main results from this study show that increased soil temperatures during winter (due to heavy snow pack and/or extra insulation) increased the springtime concentrations and fluxes of NH_4 and NO_3 in runoff (Kaste *et al.* 2008). The experiments thus support the statistical analysis of the long-term record from Storgama which indicate that less snow gives colder soils and lower flux of N from the soil.

de Wit and Wright (2008) then used the statistical analyses of de Wit *et al.* (2008) (as supported by the experiments of Kaste *et al.* (2008)) to project future NO_3 concentrations and fluxes at Storgama, given several future scenarios of N deposition and climate. Two N deposition scenarios were used, CLe and maximum feasible reduction (MFR), together with four climate scenarios, two greenhouse gas emission scenarios A2 and B2 (IPCC 2007) run

Figure 7.3 View from the experimental catchments: Site 2 with insulation mats (photograph by Jarle Håvardstun) and site 5 during snow removal (photograph by Live S. Vestgarden). (From Kaste *et al.* 2008.)

with two global climate models, HadAM3 of the Hadley Centre and ECHAM4/ OPYC3 of the Max Planck Institute. The climate scenarios come from regional downscaling provided by the PRUDENCE project (http://prudence.dmi.dk) (see Chapter 3). All scenarios suggested reduced NO_3 in runoff in the future. Warmer winters mean less snow, colder soils in winter, less mobilization of soil

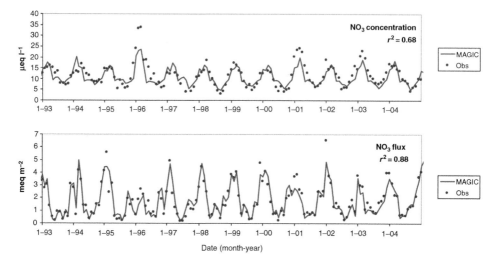

Figure 7.4   Simulated (solid line) versus observed NO$_3$ concentrations (upper panel) and fluxes (bottom panel) for the best set-up during calibration of MAGIC for Øygard. Correlation coefficients (r^2) between observed and modelled values are shown. (From Sjøeng et al. 2009a.)

N and thus lower flux of NO$_3$ in winter and spring runoff. The results must be treated with caution, however, because the projections are based on extrapolations of regressions outside the range of observations because future mean annual temperature will be much higher than the range of the year-to-year variations in the 20-year record, 1986–2005. Also, the long-term fate of stored N in the soil is unknown.

Process-orientated models offer another tool by which the effect of climate change can be assessed. Sjøeng *et al.* (2009a, b) used the MAGIC model (Model for Acidification of Groundwater in Catchments; Cosby *et al.* 1985a, b, 2001) to project future NO$_3$ concentrations in streamwater at Øygard, a small catchment in south-western Norway. MAGIC was first calibrated for monthly time-steps to the 12-year data record and then driven by four climate scenarios from PRUDENCE (A2, B2 with the Hadley and MPI models) under two different 'storylines' of assumptions involving future rates of plant processes. The calibration of the model best fits the monthly observations when the variations in precipitation and snowpack accumulation were taken into account (Fig. 7.4).

The calibrated model projected increases in future NO$_3$ concentrations under the four climate scenarios, but the magnitude of the increase was dependent on the assumptions made for future plant response to climate change (Fig. 7.5). Under storyline 1, plant growth is assumed not to change, whereas under storyline 2, the plant growth is assumed to increase (due to increased temperature) and result in more C and N in litter input to the soil. Whereas under storyline 1 the C pool of the soil decreases over time as increased soil temperature results in increased decomposition of soil organic matter, under storyline 2 the increased litter input of C holds the C pool of the soil nearly unchanged (Sjøeng *et al.* 2009b).

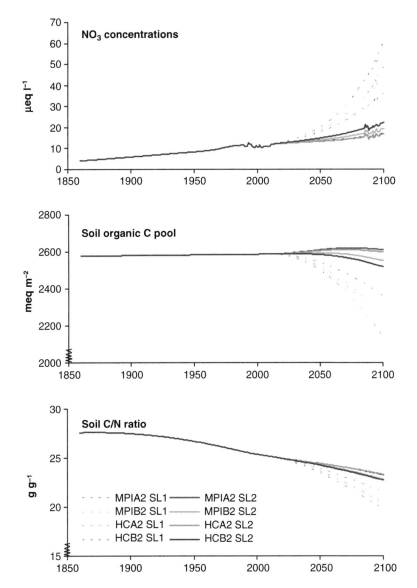

Figure 7.5 Projected NO_3 concentrations in streamwater at Øygard, Norway, under four climate scenarios and two storylines (SL1 and SL2) of plant processes. The MAGIC model was used. Also shown are soil C pool and C/N ratio of the soil organic matter. (From Sjøeng *et al.* 2009b.)

Northern Italy is a hot spot with respect to increasing concentrations of NO_3 in streams and rivers. The trends indicate increasing degree of N saturation in catchments of small and medium-sized rivers south of the Alps, but climate variations also appear to play a significant role (Rogora 2007; Rogora & Mosello 2007). The increase in NO_3 concentrations closely follows the pattern

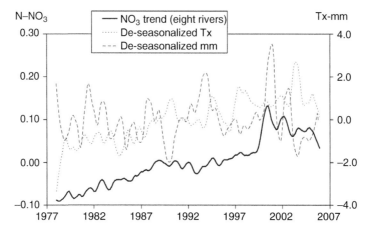

Figure 7.6 Main common trend of NO$_3$ concentration in eight rivers (unbroken line), and series of maximum temperature (Tx; dotted line) and precipitation amount (mm; dashed line). The data were first de-seasonalized and normalized (mean zero; standard deviation 1). Data are unitless owing to normalization. (From Rogora 2007.)

of maximum annual temperature, possibly reflecting higher rates of N mineralization and thus the release of N to runoff (Rogora 2007) (Fig. 7.6). The highest peaks of NO$_3$ concentration were usually detected after prolonged dry and warm periods (Rogora & Mosello 2007).

   One of the major uncertainties in all future projections of water chemistry under changed climate is the long-term fate of the N from atmospheric deposition that is retained each year in terrestrial ecosystems receiving chronic high levels of N deposition. Neither the processes of N fixation nor those of denitrification are sufficiently large in most catchment soils to account for the retention of N. The soils are slowly accumulating N. These ecosystems cannot indefinitely retain 80% or more of the incoming N without eventually releasing a larger fraction of the incoming N and even some of the stored N, most probably to runoff in the form of NO$_3$. The term 'nitrogen saturation' has been used to describe this situation (Aber *et al.* 1989). Yet, the long-term records from many sites in Europe and North America (Wright *et al.* 2001; Kaste *et al.* 2007) do not yet show major increases in NO$_3$ concentrations.

   On the contrary, some of the long-term records show declines in concentrations of NO$_3$ in runoff. At the undisturbed W6 forested catchment at Hubbard Brook, New Hampshire, the United States, for example, NO$_3$ concentrations in streamwater reached peaks in the 1970s, and much of the subsequent decline is ascribed to climatic factors (Aber *et al.* 2002). The long-term record from Lange Bramke, Germany, shows a similar decline since peak years in the 1980s; there is no explanation for this decline. In heavily acidified and damaged mountain forests in the Czech Republic, NO$_3$ concentrations in streamwater have also declined, but here the cause is ascribed to improved growth of the forests, which had previously been damaged by high levels of air pollution (Majer *et al.* 2005; Oulehle *et al.* 2008).

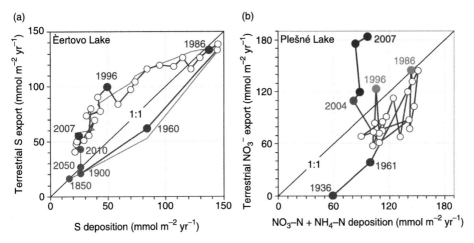

Figure 7.7 Input and output fluxes of S and inorganic N in catchments of the Bohemian Forest lakes (Czech Republic). (a) 'S leaching from' versus 'S deposition to' the catchment of Čertovo Lake. The blue points are based on measured data (Kopáček *et al.* 2006a), red points and lines are simulated by MAGIC (Majer *et al.* 2003). (b) 'NO$_3^-$ leaching from' versus 'NH$_4$–N + NO$_3$–N deposition to' the catchment of Plešné Lake. The blue points are based on measured data. Other colours represent extremely high N-exports (exceeding N deposition) due to: severe winters (green, 1986 and 1996); extremely dry and hot summer 2003 (red, 2004); and forest dieback caused by bark beetle infestation (brown, 2005–7). (Data from Kopáček *et al.* 2006b.)

The recovery of previously damaged forests, including many in central Europe with a legacy of heavy acid deposition and N saturation, can be expected to cascade down to the streams draining them and to lakes supplied by the streams. In the Bohemian Forest, Czech Republic, there has been high NO$_3$ leaching, exceeding atmospheric input of inorganic N (Fig. 7.7) to lakes, Čertovo and Plešné (Majer *et al.* 2003; Kopáček *et al.* 2006a). At Čertovo Lake, S input–output budgets show significant hysteresis compared with the 1: 1 line of export to import, with the current leaching substantially higher than that at the beginning of the 20th century at the same level of S deposition. There are apparently no anomalies related to climatic conditions. Nitrate at Plešné Lake, on the other hand, shows major changes in response to climatic variations. NO$_3$ concentrations were negligible in the 1930s and increased around the 1950s, when N deposition exceeded 70 mmol m^{-2} yr^{-1} (10 kg N ha^{-1} yr^{-1}). NO$_3$ leaching declined after the decline in N deposition, but the catchment remains N-saturated and the current NO$_3$ leaching is higher than that in the 1960s at similar N deposition rate. Extremely high N exports, exceeding N deposition, occurred in 1986 and 1996 after severe winters, in 2004 after extremely dry and hot summer 2003, and in 2005–7 after forest dieback due to bark beetle infestation in the catchment (Kopáček *et al.* 2006b).

The increased terrestrial NO$_3$ export following climatic anomalies (hot and dry summer 2003) and forest dieback (2005–8) was associated with high leaching of ionic Al and loss of base cations from the soil (Fig. 7.8). Nitrate has become

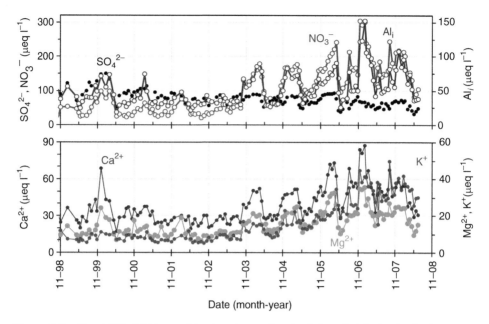

Figure 7.8 The 1998–2008 trend in chemistry of major inlet of Plešné Lake (Bohemian Forest). Elevated terrestrial export of NO_3^- results from hot and dry summer 2003 and forest dieback (2005–8) (J. Kopáček unpublished data).

the dominant strong acid anion, and its leaching governs Al toxicity of soil solutions and surface waters, depletes soil base saturation and delays soil and water recovery from acidification. The important question is to what extent the elevated Al concentration in soil solution is toxic to fine roots and mycorrhiza. A reduction of mycorrhizal fungi would leave more NH_4^+ for nitrification and result in raised NO_3 (and consequently Al) leaching, thus causing a positive feedback in terrestrial NO_3 export.

In areas receiving increasing inputs of atmospheric N, the long-term fate of N retained in catchments remains a major source of uncertainty in projecting acidification and recovery in freshwater ecosystems. Typically, catchments retain 70%–100% of the incoming N (Dise & Wright 1995; Wright *et al.* 2001). Long-term (30+ years) monitoring records do not point to dramatic changes in this % retention over time (Wright *et al.* 2001). Climate change affects the retention and loss of N in catchments. Yet, clearly, N cannot go on accumulating forever in these ecosystems. The question is, thus, when and under what circumstances will the N be leached from the terrestrial ecosystems to give acidification problems in freshwater ecosystems, and will climate change ameliorate or exacerbate the problem?

Acid episodes

Superimposed upon chronic acidification of surface waters are acid episodes that cause sudden temporary decreases of pH and ANC. Acid episodes are frequently triggered by drought, floods, rapid snowmelt and inputs of wind-borne sea salts.

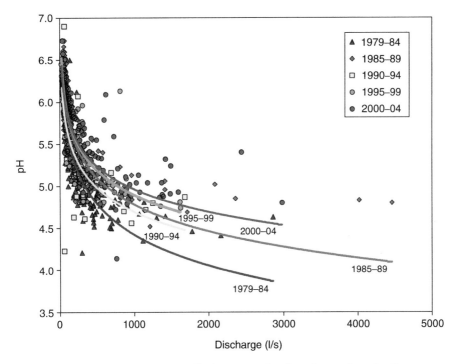

Figure 7.9   Stream pH versus mean discharge on the day of sampling, for five approximately equal time intervals. Lines show logarithmic least squares regression fit for pH versus discharge for each time interval. (From Evans *et al.* 2008.)

Acid episodes can cause acute toxicity to aquatic organisms, such as the fish kill in 1993 in Norway following a sea-salt episode (Hindar *et al.* 1994). Scenarios of future climate change entail increased frequency and severity of such climatic extremes. The implication is that these may delay biological recovery in response to decreased S and N deposition.

Evans *et al.* (2008) developed a procedure by which each acid episode could be assigned to one of four main drivers: hydrology (i.e. direct influence of precipitation), summer drought, snowmelt, and sea-salt deposition. The procedure entails quantifying the relative magnitudes of changes in base cations and the three major strong acid anions, SO_4, NO_3 and Cl, in explaining the observed decrease in ANC. At Afon Gwy, a small moorland stream in mid-Wales, Evans *et al.* (2008) found that the severity of episodes had decreased during the past 20 years, due to the decreased deposition of S and N. For example, high flow episodes in recent years have much higher pH levels than comparable flows in the early 1980s (Fig. 7.9).

A similar decrease in the severity of acid episodes has occurred at the nearby, and partially afforested, Afon Hafren catchment (M. Hutchins unpublished data). Here, the weekly data were split into four time-periods, stratified and summarized according to discharge conditions, and then the major driver was identified by means of the method of Evans *et al.* (2008). There was a strong relationship

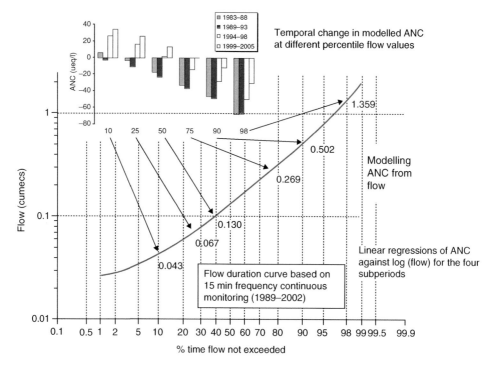

Figure 7.10 Afon Hafren, Wales. Relationships between ANC and discharge (log flow) for four different periods. As the streamwater has recovered (increasing ANC over time), the ANC at peak discharges has also increased. (From M. Hutchins, unpublished data.)

between ANC and discharge, and this relationship has changed over time due to chemical recovery (Fig. 7.10). Since the mid-1980s, there has been a relative shift from SO_4-dominated events to those dominated by sea salts. Principal component analysis of the major ionic species in Hafren streamwater suggests that about 70% of the discharge-related chemical variation is explained by sea salts. Episodic deposition of sea salts causes acid pulses due to cation exchange in the soil of Na with H^+ and Al^{n+}; the resulting streamwater has a molar Na: Cl ratio appreciably lower than that in standard sea water.

Wright (2008) used the same procedure to examine episodes in the Birkenes catchment, southernmost Norway. Here, the 30-year record of weekly (or more frequent) samples of streamwater shows gradual improvement in water quality from mean annual ANC less than $-70\,\mu eq\ l^{-1}$ in the mid-1980s to greater than $-30\,\mu eq\ l^{-1}$ in 2000–4 due to decline in S and N deposition. The severity of acid episodes has also declined. Whereas, prior to 1993, ANC during episodes decreased to less than $-100\,\mu eq\ l^{-1}$, since 2000 the ANC has not dropped to lower than $-75\,\mu eq\ l^{-1}$ (Fig. 7.11). At Birkenes, about a third of the acid episodes were driven by sea-salt inputs. As the frequency and severity of storms are projected to increase with future climate change, the expectation is that sea salt–driven episodes will become more frequent. But since the general level of acidification at Birkenes is improving, the severity of the acid pulses may decline in the future.

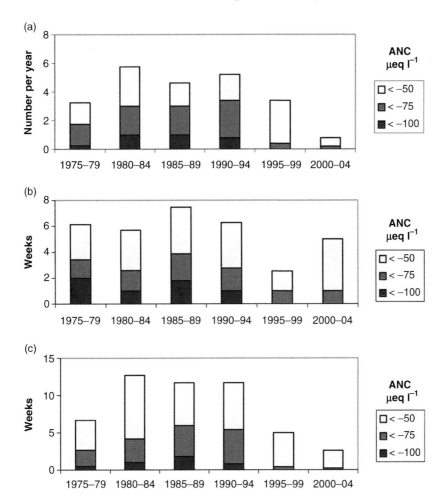

Figure 7.11 Summary of episodes with threshold ANC $< -50\,\mu$eq l^{-1} at Birkenes 1975–2004 by 5-year period. (a) Mean frequency of episodes; (b) mean duration of episode; (c) mean number of weeks per year below threshold. (From Wright 2008.)

Snowmelt is typically the period of year with lowest pH and highest toxicity of streamwater to fish and other aquatic organisms. In streams in northern Sweden, for example, the pH can be well above 5 most of the year, but below 5 and toxic during a few weeks each spring (Laudon & Bishop 1999). The severity of these snowmelt-induced acid episodes has decreased in recent years owing to reduced S and N deposition (Laudon & Bishop 2002b). Also, combined MAGIC and pBDM ('one point' Boreal Dilution Model) models for the Lysina catchment, Czech Republic, suggested that the recent recovery of pH, ANC and inorganic monomeric Al during spring snowmelt floods has been faster than the recovery of annual average chemistry (Laudon *et al.* 2005).

Acid pulses can come with the first runoff following droughts. The accompanying anion is usually SO$_4$. The mechanism is oxidation of reduced sulphur compounds

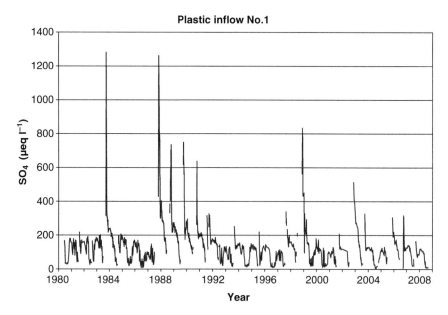

Figure 7.12 Sulphate concentrations in streamwater at Plastic Lake, Ontario, showing the pulses following drought. The breaks in the curve are due to droughts; at these times there was no flow in the stream, thus no water samples were collected. (Updated from Dillon *et al.* 1997.)

in soils and flushing of the sulphate with the first runoff. The drought causes lowering of the water table and exposes anoxic soils and peats to oxygen. The role of drought-induced acid episodes is particularly well-illustrated in the long-term record from lakes and streams in the Muskoka-Haliburton area of southern Ontario (Dillon *et al.* 1997; Eimers *et al.* 2008) (Fig. 7.12). Similar sulphate-driven acid episodes have been reported at many other sites, including Birkenes, Norway (Wright 2008), Sweden (Laudon & Bishop 2002a; Laudon 2008) and the United Kingdom (Hughes *et al.* 1997; Adamson *et al.* 2001).

Dissolved organic carbon (DOC)

Dissolved organic carbon (DOC) in surface waters gives rise to weak organic acids, complexes many toxic components and thus affects the chronic and acute toxicity of acidic water to aquatic organisms, in particular, fish. In boreal lakes and streams, DOC levels typically show a clear seasonal pattern, with low levels in winter and spring and higher levels in late summer and autumn. This pattern is climate-driven and related to biological activity in catchment vegetation and soils, as well as in-lake processes (mainly driven by UVA radiation) that degrade DOC in the water (Gennings *et al.* 2001). It can thus be expected that future climate change will affect the annual mean levels and seasonal patterns of DOC concentrations in surface waters (Laudon & Buffam 2008). This issue has already been raised in Chapter 3, but is amplified by studies from the viewpoint of acidification processes.

The decline in sulphate concentrations due to declining S deposition is believed to have resulted in a general increase in DOC concentrations in acid-sensitive lakes and streams in Europe and eastern North America since the mid-1980s, possibly due to increased solubility of some components of the organic matter under less acid conditions (Monteith *et al.* 2007). Monteith *et al.* (2007) analysed time series from 522 remote lakes and streams in Europe and North America and showed that increasing concentrations of DOC between 1990 and 2004 could be explained by atmospheric deposition chemistry and catchment acidity (Monteith *et al.* 2007). The relative change in DOC per year was strongly inversely related to the change in sulphate and chloride concentrations in atmospheric deposition. Thus, acid deposition to these ecosystems seems to have been partially buffered by changes in organic acidity, and recovery from acidification has implied an increased mobility of DOC and organic acids.

Vuorenmaa *et al.* (2006) studied trends in total organic carbon (TOC) concentrations over the period 1987–2003 in 13 small forest lakes in Finland. Recovery from acidification (reduced S deposition) and long-term changes in runoff as potential drivers for the trends were examined. Results showed that TOC concentrations have increased throughout Finland. Ten of the 13 lakes showed a significant increasing TOC trend ($p < 0.05$) and included both clear water and humic lakes. The largest annual increase in TOC occurred in lakes with the largest average concentrations. The magnitudes of the TOC trends were not significantly related to the proportion of peat soils in the catchment but catchment size was an important predictor. Decreasing S deposition and improved acid–base status in soil due to the recovery from acidification implied an increased mobilization of organic acids and TOC. There was little evidence that the long-term increasing trend in TOC concentrations was related to long-term changes in runoff.

Analysis of the long-term increase of DOC in two geochemically contrasting forested catchments in the Czech Republic did not show a clear climatic effect on DOC increases (Hruška *et al.* 2009). Between 1993 and 2007, DOC in stream water increased significantly in both catchments: the mean annual increase was 0.42 mg l^{-1} yr^{-1} ($p < 0.001$) in acidic Lysina and 0.43 mg l^{-1} yr^{-1} ($p < 0.001$) in alkaline Pluhuv Bor, resulting in cumulative increases of 64% and 65%, respectively. The long-term increase in DOC was correlated with a decrease in ionic strength (IS) at both catchments ($p < 0.001$), which resulted from declining atmospheric deposition. Only granitic Lysina was significantly acidified. Well-buffered Pluhuv Bor, with serpentinite bedrock, showed very minor changes in stream pH, as incoming SO_4^{2-} from the atmosphere was buffered by exchangeable cations in the magnesium-rich soils. Thus, the observed DOC increase at Pluhuv Bor was not due to changes in acidity. Climate change cannot be driving the increases in DOC, because none of temperature, annual precipitation or discharge (annual or weekly) showed statistically significant trends during the study period.

Results of the MAGIC simulation show that the observed DOC increases will significantly influence recovery of the Lysina catchment from acidification

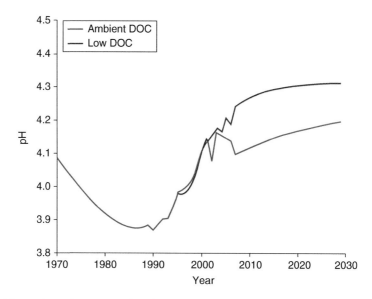

Figure 7.13 Projected recovery of pH in streamwater at Lysina catchment, Czech Republic, as simulated by the MAGIC model using two scenarios for DOC. (i) DOC was assumed constant at 1993–4 levels ('Low DOC' scenario) and (ii) DOC observed between 1993 and 2007 was used for the modelling of the period 1994–2007 and an average of period 2006–7 for the period 2008–30 ('Ambient DOC' scenario). (From Hruška *et al.* 2009.)

(Fig. 7.13). With the level of DOC measured in 1993–4, predicted pH rose close to 4.30 from the lowest value measured (3.88) and predicted for late 1980s. When the measured increase of DOC (Fig. 7.13) was incorporated into the MAGIC model, pH was predicted to rise to only 4.20 (Hruška *et al.* 2009).

Concentrations of DOC in streams typically increase during episodes of high runoff. What is less well established, however, is the role of ionic strength of the runoff on DOC. Theoretically, the higher ionic strength of the soil solution should result in lower DOC concentrations owing to the lower solubility of organic matter (Evans *et al.* 1988; Tipping & Hurley 1988). Under natural conditions, it is difficult to discern the effects of hydrology or chemistry alone since the changes in discharge and water chemistry typically occur at the same time. To investigate the potential effect of high sea-salt episodes on DOC runoff concentrations, a field-scale watering experiment was conducted in 2004 at the experimental catchment G1 at Gårdsjön, Sweden (F. Moldan unpublished data). The catchment was brought to field capacity; i.e. addition of water from the sprinkling system was approximately equal to discharge rate. After reaching field capacity, the catchment was watered for 4 days with distilled water, after which sea salt was added to the sprinkling solution.

Sea salt was added first for 2 days at a level typical for ambient throughfall and then at a high level observed during storms and periods of high winds. Experimental addition of sea salt resulted in an immediate decrease of DOC in the stream from

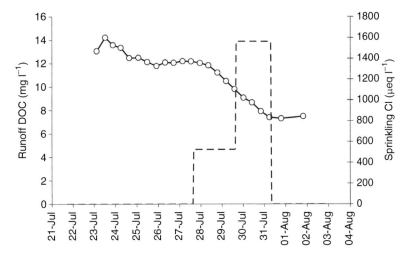

Figure 7.14 Decline of DOC (circles, left axis) at the Gårdsjön experiment during 10 days of watering and sea-salt addition. Cl⁻ concentration (dashed line, right axis) in added water (F. Moldan unpublished data.)

12 mg l⁻¹ to 8 mg l⁻¹ (Fig. 7.14). pH decreased slightly from 4.4 to 4.2 and inorganic Al increased from 3 μeq l⁻¹ to 200 μeq l⁻¹. The ionic strength of the stream water almost doubled. These results suggest that the DOC changes were driven by ionic strength or acidity increase. Climate change through enhanced sea salt input can, in coastal areas, suppress DOC runoff concentrations during hydrological episodes.

Saharan dust

Inputs of dust from deserts of Northern Africa are important sources of alkaline substances that act to neutralize acid deposition. Dust inputs are great enough to play a major role in the acid–base balance in acid-sensitive catchments in mountainous regions of Southern Europe, such as Northern Italian Alps (Rogora *et al.* 2004) and the Spanish Pyrenees (Rodà *et al.* 1993) and even as far north as the central Alps of Austria (Psenner 1999). Future climate change may mean an increase in the frequency and severity of these episodes of Saharan dust inputs; for acidified freshwaters, this means an improvement in the acidification status (Rogora *et al.* 2004).

Effects of climate warming on soil chemical processes

An increase in soil temperature should, in theory, increase the rate of chemical weathering of soil minerals. This mechanism has been postulated by Sommaruga-Wögrath *et al.* (1997) to explain recent trends in acidity of remote high elevation lakes in the Austrian Alps. This is further supported by palaeolimnological data from the last 200 years, which show a strong positive correlation between pH and mean air temperature.

Furthermore, a reduction in the extent and duration of snow cover, owing to less precipitation and higher temperatures, means a greater exposure of rocks and soils in the catchments, which enhances weathering processes. An analysis of long-term chemical data for high altitude lakes in the central Italian Alps showed an increase of solute contents in the last few years, mainly evident in lakes lying in catchments with highly soluble rocks (Rogora *et al.* 2003). At these sites, the overall effect of warming on weathering processes will be to increase ANC and accelerate recovery of lake water from acidification.

Studies of acid–base chemistry in soils and lakes in the Bohemian Forest, Czech Republic, indicate that warming results in decreased aluminium concentrations in surface waters. Vesely *et al.* (2003) hypothesize that this may be because of the temperature dependence of the equilibrium between inorganic Al and secondary minerals in the soil; higher temperature pushes the equilibrium towards lower Al concentrations. The net effect of both these processes is to increase the ANC and decrease the toxicity of runoff from soils under increased temperature.

Modelling the combined effects of climate change and acid deposition

Climate change and acid deposition have many connections, from the economic activities that give rise to emissions of gases and particles to the atmosphere to long-range transport and deposition of S and N, to their combined effects on terrestrial ecosystems and finally to their effects on aquatic ecosystems. For all these links, models have been used to evaluate future scenarios. A switch from use of fossil fuels to renewable energy sources will entail both a reduction in emissions of greenhouse gases and also a reduction in emissions of S and N compounds. A study by Mayerhofer *et al.* (2001) looked at the linkages between climate change and regional air pollution in Europe using the Integrated Model to Assess the Global Environment (IMAGE; Netherlands Environmental Assessment Agency (PBL)) for climate and the Regional Air Pollution Information and Simulation model (RAINS); International Institute for Applied Systems Analysis (IIASA) for air pollution. One of the obvious trade offs is that decreased emissions of S will result in fewer particles in the atmosphere, which, in turn, will increase solar radiation inputs through reduced cloud formation.

Climate change will also affect the transport and deposition of S and N on a regional scale. Modelling studies in Europe using regional climate models and chemical transport models indicate that the increased precipitation projected for North-West Europe will entail commensurate increased N deposition (Hole *et al.* 2008; Hole & Enghardt 2008).

There are many processes in terrestrial ecosystems that will affect the amount and chemical composition of runoff. The combined effects of these processes can be quantified by the use of process-orientated biogeochemical models. MAGIC (Cosby *et al.* 1985a, b, 2001) is such a model. It has been used to evaluate the

Table 7.1 Summary of the relative importance of factors drivers by climate change that affect soil and water acidification. (Modified from Wright *et al.* 2006.)

Factor	Potential effect on ANC	Importance	Relationship to climate
Sea salts	Decrease	Coastal sites only	Empirical links to NAO
Dust	Increase	Southern sites only	Empirical links to NAO
Runoff	Increase	Medium	Precipitation, temperature from GCMs
Weathering	Increase	Low	Temperature, from GCMs
Organic acids	No change	High	Poorly understood
Soil pCO$_2$	No change	Low	From GCMs
Forest growth	Decrease	Low	Temperature, moisture, pCO$_2$, complicated
Organic matter decomposition	Decrease	High	Temperature, moisture, complicated

potential effect of climate change on soil and water chemistry. Wright *et al.* (2006) used MAGIC to examine the relative sensitivity of eight major climate-sensitive processes on the recovery of soil and water from acidification (Table 7.1). They found that several of the factors are of minor importance, several are important only in specific areas and several are of widespread importance. The effect of climate change on production and loss of organic acids is one of the important processes, and mineralization and loss of N from soil is another.

This exercise simply tested the relative importance of various processes and did not involve the use of actual climate scenarios in projecting future response in soil and water chemistry. Hardekopf *et al.* (2008) took the next step and used the best available information to set the rates and parameter values for several of the key climate-driven processes in MAGIC to project future acidification and recovery at Litavka, a small catchment in the Czech Republic. Their simulation included the temperature dependence of weathering rate, DOC release from soil, net soil N mineralization and forest growth. They used several future climate scenarios derived from GCMs and downscaled to the Litavka site. They concluded that future climate change will have little effect on chemical recovery at this site.

Evans (2005) took another approach and used empirical statistical relationships between climate variables and observed streamwater chemistry at Afon Gwy, a small stream in Wales, the United Kingdom, to set parameter values and rates in MAGIC for evaluation of future climate scenarios. Evans used three empirical relationships: (i) discharge as a function of rainfall and temperature; (ii) sea-salt deposition as a function of the North Atlantic Oscillation Index (NAOI); and (iii) stream concentrations of DOC as a function of summer temperature and S deposition to project future soil and water acidification. He predicted that future climate change would mean increased DOC and also increased sea-salt deposition, but that these would have counteracting effects on acidification of soil and water. Increase in soil solution DOC will result in transport of base cations from the soil, thus leading to greater soil acidification but less water acidification.

Several other modelling studies have focussed on the role of DOC. Aherne *et al.* (2008a) used MAGIC coupled with INCA-C (Integrated Catchments model for Carbon), a model that simulates DOC concentrations in runoff from catchments (Futter *et al.* 2007) to project future recovery from acidification of a small tributary stream to Plastic Lake, Ontario. Downscaled climate scenarios derived from GCMs were used. At this site drought-induced pulses of sulphate play the dominant role (Fig. 7.12). As INCA-C does not provide information on soil solution DOC, the MAGIC simulations were run without any future change in soil DOC. The INCA-C model predicted increases in stream water DOC as a result of warming but does not thus fully account for the full effect of increased temperature on DOC and surface water acidity. As Futter *et al.* (2008) point out, a full understanding of surface water DOC dynamics requires incorporation of DOC processes into catchment-scale process-orientated models, such as MAGIC.

Aherne *et al.* (2008b) and Posch *et al.* (2008) expanded this type of scenario study to cover an entire lake population in Finland. They used the MAGIC model framework and extensive soil, surface water and deposition datasets for 163 Finnish forested catchments to evaluate the water chemistry response to several scenarios for acid deposition, climate change and forest harvesting. Simulations suggested that only the maximum (technically) feasible levels of reduction in emissions would result in significant recovery of soils and surface waters and would return water quality close to pre-acidification values in the studied catchments. The direct influence of climate change (temperature and runoff) had very little impact on model simulations for the sites, based on current process descriptions. However, two exploratory simple empirical DOC models indicated that changes in S deposition or temperature could have a confounding influence on the recovery of surface waters and that the corresponding increases in DOC concentrations may offset the recovery in pH due to reductions in S and N deposition.

The use of forest biomass for energy production has become an important mitigation strategy to reduce greenhouse gas emissions. To meet this new demand, future harvesting is expected to shift from stem-only to whole-tree harvesting. This increased use of forest harvest residues for biofuel production (whole tree harvest (WTH) scenario) was predicted to have a significant negative influence on the base cation budgets causing re-acidification of the study catchments (Fig. 7.15) (Aherne *et al.* 2008b). Sustainable forestry management policies must consider the combined impact of air pollution and harvesting practices. Clearly, there is a need for further emission reductions to mitigate the negative impacts of WTH, if such a policy is implemented. Additionally, increased fertilizer use, such as wood ash applications, may also be required to maintain soil nutrient status and lake water quality in these forested ecosystems.

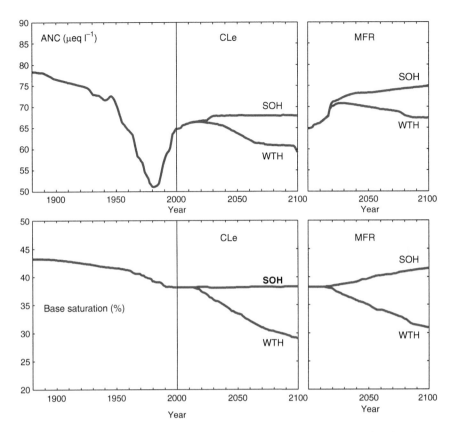

Figure 7.15 Temporal development of the median of lake water ANC (top) and the catchment soil base saturation (bottom) for the 163 Finnish study catchments under the two emission scenarios (current legislation CLe: left panel and maximum feasible reduction MFR: right panel) and two forest harvest (biomass energy) scenarios (stem-only harvest SOH and whole-tree harvest WTH). (From Aherne *et al.* 2008b.)

Effects on aquatic biota

Acid deposition affects aquatic biota because the water becomes toxic with too high concentrations of inorganic Al and H^+. Climate change interacts with acid deposition by affecting the concentrations of these toxic chemical components. But in addition, climate change can have direct effects on aquatic biota. A good example is the dependence of salmon migration on water temperature and discharge.

Hardekopf *et al.* (2008) have made one of the few studies of the interactions of acid deposition and climate change on aquatic biota. Theirs was a modelling study of a small catchment and stream in the Czech Republic. They found that under scenarios of future warmer climate, the recovery of benthic invertebrates in the stream will still be hampered by continued release of sulphate (and acid) from the soil.

The effects of drought on benthic invertebrates were investigated in a number of acid-sensitive streams in Ontario, Canada (Gilbert *et al.* 2008). Relative abundance of Ephemeroptera, Plecoptera and Trichoptera, collectively indicators of healthy benthic communities, increased the first year following drought but then decreased the second year indicating high resistance but poor resilience. Chironomids, on the other hand, showed the opposite effects. Drought acts as a disturbance mechanism that simplifies benthic community assemblages and thus reduces biodiversity.

The studies of acid episodes under future scenarios of climate change and acid deposition point to increased frequency and severity. Kroglund *et al.* (2008) point out that this implies poorer conditions for populations of salmon. A similar conclusion was reached by Kowalik *et al.* (2007) with respect to invertebrates in British streams.

There are clear thresholds at which ecosystem structure and function show dramatic changes. An obvious such threshold in response to increasing temperature is the point at which a lake is no longer ice-covered in the winter and thus subjected to circulation rather than thermal stratification. In acidified lakes, this might mean that the acid snowmelt in the spring may be mixed with the entire lake volume rather than be restricted to a 1-m thin layer beneath the ice. Such acidic layers have been shown to be the limiting factor with respect to recruitment of lake-spawning fish species, such as lake trout in Canada (Gunn & Keller 1984) and brown trout in Norway (Barlaup *et al.* 1998). Other such thresholds may be related to biogeochemical processes in the catchment, for example, a severe drought might kill the vegetation, or milder climate might promote expansion of insects and diseases with similar effect, and to species-specific events, for example, in a high elevation lake, where warmer water may result in crossing of a developmental threshold such that invertebrates can complete an entire life cycle (Borgstrøm 2001) within the year.

Conclusions

Projections of the synergistic effects of acid deposition and climate change on freshwater ecosystems are inherently fraught with the uncertainty that such projections are for climatic conditions not currently experienced. For many of the climate scenarios, the projected mean temperature in the future will be well above that observed even in extreme years during the period of observation (maximum 30 years for most ecosystems). The ecosystem responses are probably not linear; thus, extrapolation from observations, even those spanning several decades, entails going outside the range of observations.

It is acid deposition that is responsible for the widespread acidification of surface waters in sensitive areas of Europe, eastern North America and elsewhere in the world. This means that measures to reduce acidification problems can continue to be focussed on reducing emissions of S and N compounds to the atmosphere. Although reductions in emissions of S and N compounds have led to dramatic improvements and recovery in water quality in acidified freshwater ecosystems, biological recovery has lagged and the problem will remain in many areas for decades to come. Further reductions are required if the goal is to permit recovery of all impacted ecosystems.

Climate change is a confounding factor in that it can exacerbate or ameliorate the rate and degree of acidification and recovery, both with respect to chemical as well as biological effects. The absence of recovery following reduction in acid deposition, therefore, may simply be the result of the confounding influence of climatic variations. The time-scales of recovery from acid deposition are in many respects similar to those of chronic changes in climate, in part because both drivers act by affecting large pools of S, N, C and base cations in catchment soil. But extreme climatic events, such as droughts, cause extreme responses that set back the biological recovery process and slow down progress towards a stable ecosystem. The interactions are complicated and manifold, and thus the outcomes on ecosystems are difficult to predict and generalize.

Both acid deposition and climate change are caused by emissions of gases to the atmosphere and are largely due to the same types of human activities – burning of fossil fuels and other industrial processes. Clearly, there are substantial 'co-benefits' to be gained: for example, reductions in emissions of CO_2 by a switch to renewable energy sources will also bring about a reduction in S and N emissions. At the policy level, much might be gained by coordinating future emission controls, now dealt with separately under the United Nations Economic Council for Europe (UNECE) Convention on Long-Range Transboundary Air Pollution (LRTAP) and the UN Framework Convention on Climate Change.

Society will take measures to ameliorate or mitigate the effects of climate change. Some of these measures may indirectly affect the acidification of sensitive freshwaters. For example, as illustrated by the modelling example from Finland (Fig. 7.15), more intensive use of forests for biofuel may entail release of N now stored in the soil to surface waters in the form of NO_3 accompanied by acidic cations. More research is needed on the effects of adaptation and mitigation.

The mechanisms of the interactions between climate effects and acidification effects are still, however, poorly understood. Experiments, continued monitoring and analysis of long-term data series and modelling are complementary approaches that lead to new insights and knowledge on possible interactions. Research is particularly challenging in this field because the goal is to make projections for the future under climatic conditions that for many ecosystems have never been experienced previously. It is certain that climate change will have an increasing impact on freshwaters in the foreseeable future and there will certainly be effects not yet identified. Current assessments of total impact on freshwaters are probably underestimated.

References

Aber, J.D., Nadelhoffer, K.J., Steudler, P. & Melillo, J. (1989) Nitrogen saturation in northern forest ecosystems. *Bioscience*, **39**, 378–386.

Aber, J.D., Ollinger, S.V., Driscoll, C.T., *et al.* (2002) Inorganic nitrogen losses from a forested ecosystem in response to physical, chemical, biotic, and climatic perturbations. *Ecosystems*, **5**, 648–658.

Adamson, J.K., Scott, W.A., Rowland, A.P. & Beard, G.R. (2001) Ionic concentrations in a blanket peat bog in northern England and correlations with deposition and climate variables. *European Journal of Soil Science*, **52**, 69–79.

Aherne, J., Futter, M.N. & Dillon, P.J. (2008a) The impacts of future climate change and sulphur emission reductions on acidification recovery at Plastic Lake, Ontario. *Hydrology and Earth System Sciences*, **12**, 383–392.

Aherne, J., Posch, M., Forsius, M., *et al.* (2008b) Modelling the hydro-geochemistry of acid-sensitive catchments in Finland under atmospheric deposition and biomass harvesting scenarios. *Biogeochemistry*, **88**, 233–256.

Barlaup, B.T., Hindar, A., Kleiven, E. & Høgberget, R. (1998) Incomplete mixing of limed water and acidic runoff restricts recruitment of lake spawning brown trout in Hovvatn, southern Norway. *Environment Biology of Fishes*, **53**, 47–63.

Borgstrøm, R. (2001) Relationship between spring snow depth and growth of brown trout *Salmo trutta* in an alpine lake: Predicting consequences of climate change. *Arctic Antarctic Alpine Research*, **33**, 476–480.

Bull, K.R., Achermann, B., Bashkin, V., *et al.* (2001) Coordinated effects monitoring and modelling for developing and supporting international air pollution control agreements. *Water Air and Soil Pollution*, **130**, 119–130.

Clair, T.A., Dillon, P.J., Ion, J., Jeffries, D.S., Papineau, M. & Vet, R. (1995) Regional precipitation and surface water chemistry trends in southeastern Canada (1983–1991). *Canadian Journal of Fisheries and Aquatic Sciences*, **52**, 197–212.

Cosby, B.J., Hornberger, G.M., Galloway, J.N. & Wright, R.F. (1985a) Modeling the effects of acid deposition: Assessment of a lumped parameter model of soil water and streamwater chemistry. *Water Resources Research*, **21**, 51–63.

Cosby, B.J., Wright, R.F., Hornberger, G.M. & Galloway, J.N. (1985b) Modeling the effects of acid deposition: Estimation of long-term water quality responses in a small forested catchment. *Water Resources Research*, **21**, 1591–1601.

Cosby, B.J., Ferrier, R.C., Jenkins, A. & Wright, R.F. (2001) Modelling the effects of acid deposition: Refinements, adjustments and inclusion of nitrogen dynamics in the MAGIC model. *Hydrology and Earth Systems Sciences*, **5**, 499–517.

Dillon, P.J., Molot, L.A. & Futter, M. (1997) The effect of El Nino-related drought on the recovery of acidified lakes. *Environmental Monitoring and Assessment*, **46**, 105–111.

Dise, N.B. & Wright, R.F. (1995) Nitrogen leaching from European forests in relation to nitrogen deposition. *Forest Ecology and Management*, **71**, 153–162.

Driscoll, C.T., Lawrence, G.B., Bulger, A.J., *et al.* (2001) Acidic deposition in the northeastern United States: Sources and inputs, ecosystem effects, and management strategies. *Bioscience*, **51**, 180–198.

Eimers, M.C., Watmough, S.A., Buttle, J.M. & Dillin, P.J. (2008) Examination of the potential relationship between droughts, sulphate and dissolved organic carbon at a wetland-draining stream. *Global Change Biology*, **14**, 938–948.

Evans, C.D. (2005) Modeling the effects of climate change on an acidic upland stream. *Biogeochemistry*, **74**, 21–46.

Evans, A.J., Zelazny, L.W. & Zipper, C.E. (1988) Solution parameters influencing dissolved organic carbon levels in three forest soils. *Soil Science Society of America Journal*, **52**, 1789–1792.

Evans, C.D., Cullen, J., Alewell, C., *et al.* (2001) Recovery from acidification in European surface waters. *Hydrology and Earth System Sciences*, **5**, 283–298.

Evans, C.D., Reynolds, B., Hinton, C., *et al.* (2008) Effects of decreasing acid deposition and climate change on acid extremes in an upland stream. *Hydrology and Earth System Sciences*, **12**, 337–351.

Futter, M.N., Butterfield, D., Cosby, B.J., Dillon, P.J., Wade, A.J. & Whitehead, P.G. (2007) Modelling the mechanisms that control in-stream dissolved organic carbon dynamics in upland and forested catchments. *Water Resources Research*, **43**, W02434.

Futter, M., Starr, M., Forsius, M. & Holmberg, M. (2008) Modelling the effects of climate on long-term patterns of dissolved organic carbon concentrations in the surface waters of a boreal catchment. *Hydrology and Earth System Sciences*, **12**, 437–447.

Gennings, C., Molot, L.A. & Dillon, P.J. (2001) Enhanced photochemical loss of organic carbon in acidic waters. *Biogeochemistry*, **52**, 339–354.

Gilbert, B., Dillon, P.J., Somers, K.M., Reid, R.A. & Scott, L.D. (2008) Response of benthic macroinvertebrate communities to El Nino related drought events in six upland streams in south-central Ontario. *Canadian Journal of Fisheries and Aquatic Sciences*, **65**, 890–905.

Gunn, J.M. & Keller, W. (1984) Spawning site water chemistry and lake trout (*Salvelinus namaycush*) sac fry survival during spring snowmelt. *Canadian Journal of Fisheries and Aquatic Sciences*, **41**, 319–329.

Hardekopf, D., Horecky, J., Kopacek, J. & Stuchlik, E. (2008) Predicting long-term recovery of a strongly acidified stream using MAGIC and climate models (Litavka, Czech Republic). *Hydrology and Earth System Sciences*, **12**, 479–490.

Hesthagen, T., Sevaldrud, I.H. & Berger, H.M. (1999) Assessment of damage to fish populations in Norwegian lakes due to acidification. *Ambio*, **28**, 112–117.

Hindar, A., Henriksen, A., Tørseth, K. & Semb, A. (1994) Acid water and fish death. *Nature*, **372**, 327–328.

Hole, L. & Enghardt, M. (2008) Climate change impact on atmospheric nitrogen deposition in northwestern Europe: A model study. *Ambio*, **37**, 9–17.

Hole, L., de Wit, H. & Aas, W. (2008) Influence of summer and winter climate variability on nitrogen wet deposition in Norway. *Hydrology and Earth System Sciences*, **12**, 405–414.

Hruška, J., Krám, P., McDowell, W.H. & Oulehle, F. (2009) Increased dissolved organic carbon (DOC) in Central European streams is driven by reductions in ionic strength rather than climate change or decreasing acidity. *Environmental Science and Technology*, **43**, 4320–4326.

Hughes, S., Reynolds, B., Hudson, J. & Freeman, C. (1997) Effects of summer drought on peat soil solution chemistry in an acid gully mire. *Hydrology and Earth System Sciences*, **1**, 661–669.

IPCC (Intergovernmental Panel on Climate Change) (2007) Summary for policymakers. l. In: *Climate Change 2007: The Physical Science Basis. Contribution of Working Group I to the Fourth Assessment Report of the Intergovernmental Panel on Climate Change* (eds S. Solomon, D. Qin, M. Manning, Z. Chen, M. Marquis, K.B. Averyt, M. Tignor & H.L. Miller). Cambridge University Press, Cambridge and New York.

Jeffries, D.S., Clair, T.A., Couture, S., *et al.* (2003) Assessing the recovery of lakes in southeastern Canada from the effects of acid deposition. *Ambio*, **32**, 176–182.

Kaste, Ø., de Wit, H.A., Skjelkvåle, B.L. & Høgåsen, T. (2007) Nitrogen runoff at ICP Waters sites 1990–2005: Increasing importance of confounding factors? In: *Trends in Surface Water Chemistry and Biota; The Importance of Confounding Factors* (eds H.A. de Wit & B.L. Skjelkvåle), pp. 29–38. Norwegian Institute for Water Research, ICP Waters Report 87/2007, Oslo.

Kaste, Ø., Austnes, K., Vestgarden, L. & Wright, R.F. (2008) Manipulation of snow in small headwater catchments at Storgama, Norway: Effects on leaching of inorganic nitrogen. *Ambio*, **37**, 29–37.

Kopáček, J., Turek, J., Hejzlar, J., Kaňa, J. & Porcal, P. (2006a) Element fluxes in watershed-lake ecosystems recovering from acidification: Čertovo Lake, the Bohemian Forest, 2001–2005. *Biologia*, **61** (Suppl. 20), S413–S426.

Kopáček, J., Turek, J., Hejzlar, J., Kaňa, J. & Porcal, P. (2006b) Element fluxes in watershed-lake ecosystems recovering from acidification: Plešné Lake, the Bohemian Forest, 2001–2005. *Biologia*, **61** (Suppl. 20), S427–S440.

Kowalik, R.A., Cooper, D.M., Evans, C.D. & Ormerod, S.J. (2007) Acidic episodes retard the biological recovery of upland British streams from chronic acidification. *Global Change Biology*, **13**, 2439–2452.

Kroglund, F., Rosseland, B.O., Teigen, H.C., Salbu, B., Kristensen, T. & Finstad, B. (2008) Water quality limits for Atlantic salmon (*Salmo salar* L.) exposed to short-term reductions in pH and increased aluminium simulating episodes. *Hydrology and Earth System Sciences*, **12**, 491–507.

Laudon, H. (2008) Recovery from episodic acidification delayed by drought and high sea salt deposition. *Hydrology and Earth System Sciences*, **12**, 363–370.

Laudon, H. & Bishop, K.H. (1999) Quantifying sources of acid neutralisation capacity depression during spring flood episodes in northern Sweden. *Environmental Pollution*, **105**, 427–435.

Laudon, H. & Bishop, K.H. (2002a) Episodic stream water pH decline during autumn storms following a summer drought in northern Sweden. *Hydrological Processes*, **16**, 1725–1733.

Laudon, H. & Bishop, K.H. (2002b) The rapid and extensive recovery from episodic acidification in northern Sweden due to declines in SO_4^{2-} deposition. *Geophysical Research Letters*, **29**, 1594.

Laudon, H. & Buffam, I. (2008) Impact of changing DOC concentrations on the potential distribution of acid sensitive biota in a boreal stream network. *Hydrology and Earth System Sciences*, **12**, 425–435.

Laudon, H., Hruška, J., Köhler, S. & Krám, P. (2005) Retrospective analyses and future predictions of snowmelt-induce acidification: Example from a heavily impacted stream in the Czech Republic. *Environmental Science and Technology*, **39**, 3197–3202.

Majer, V., Cosby, B.J., Kopáček, J. & Veselý, J. (2003) Modelling reversibility of Central European mountain lakes from acidification: Part I – The Bohemian Forest. *Hydrology and Earth System Sciences*, **7**, 494–509.

Majer, V., Krám, P. & Shanley, J.B. (2005) Rapid regional recovery from sulfate and nitrate pollution in streams of the western Czech Republic – Comparison to other recovering areas. *Environmental Pollution*, **135**, 17–28.

Mayerhofer, P., Alcamo, J., Posch, M. & Van Minnen, J.G. (2001) Regional air pollution and climate change in Europe: An integrated assessment (AIR-CLIM). *Water Air and Soil Pollution*, **130**, 1151–1156.

Monteith, D.T., Hildrew, A.G., Flower, R.J., *et al.* (2005) Biological responses to the chemical recovery of acidified fresh waters in the UK. *Environmental Pollution*, **137**, 83–101.

Monteith, D.T., Stoddard, J.L., Evans, C.D., *et al.* (2007) Dissolved organic carbon trends resulting from changes in atmospheric deposition chemistry. *Nature*, **450**, 537–540.

Oulehle, F., McDowell, W.H., Aitkenhead-Peterson, J.A., *et al.* (2008) Long-term trends in stream nitrate concentrations and losses across watersheds undergoing recovery from acidification in the Czech Republic. *Ecosystems*, **11**, 410–425.

Overrein, L., Seip, H.M. & Tollan, A. (1980) *Acid Precipitation – Effects on Forest and Fish*. Final report of the SNSF-project 1972–1980. FR 19-80, SNSF project, Ås, Norway.

Posch, M., Aherne, J., Forsius, M., Fronzek, S. & Veijalainen, N. (2008) Modelling the impacts of European emission and climate change scenarios on acid-sensitive catchments in Finland. *Hydrology and Earth System Sciences*, **12**, 449–463.

Psenner, R. (1999) Living in a dusty world: Airborne dust as a key factor for alpine lakes. *Water Air and Soil Pollution*, **112**, 217–227.

Rodà, F., Bellot, J., Avila, A., Escarré, A., Piñol, J. & Terradas, J. (1993) Saharan dust and the atmospheric inputs of elements and alkalinity to Mediterranean ecosystems. *Water Air and Soil Pollution*, **66**, 277–288.

Rogora, M. (2007) Synchronous tends in N-NO$_3$ export from N-saturated river catchments in relation to climate. *Biogeochemistry*, **86**, 251–268.

Rogora, M. & Mosello, R. (2007) Climate as a confounding factor in the response of surface water to nitrogen deposition in an area south of the Alps. *Applied Geochemistry*, **22**, 1122–1128.

Rogora, M., Mosello, R. & Arisci, S. (2003) The effect of climate warming on the hydrochemistry of alpine lakes. *Water Air and Soil Pollution*, **148**, 347–361.

Rogora, M., Mosello, R. & Marchetto, A. (2004) Long-term trends in the chemistry of atmospheric deposition in Northwestern Italy: The role of increasing Saharan dust deposition. *Tellus*, **56B**, 426–434.

Schindler, D.W. (1988) Effects of acid rain on freshwater ecosystems. *Science*, **239**, 149–157.

Schöpp, W., Posch, M., Mylona, S. & Johansson, M. (2003) Long-term development of acid deposition (1880–2030) in sensitive freshwater regions in Europe. *Hydrology and Earth System Sciences*, **7**, 436–446.

Sjøeng, A.M.S., Kaste, Ø. & Wright, R.F. (2009a) Modelling seasonal nitrate concentrations in runoff of a heathland catchment in SW Norway, using the MAGIC model I. Calibration and specification of nitrogen processes. *Hydrology Research*, **40**, 198–216.

Sjøeng, A.M.S., Kaste, Ø. & Wright, R.F. (2009b) Modelling future NO$_3$ leaching from an upland headwater catchment in SW Norway using the MAGIC model II. Simulation of future nitrate leaching given the scenarios of climate change. *Hydrology Research*, **40**, 217–233.

Skjelkvåle, B.L. & Wright, R.F. (1990) *Overview of Areas Sensitive to Acidification in Europe*. Acid Rain Research Report 20/1990, Norwegian Institute for Water Research, Oslo.

Sommaruga-Wögrath, S., Koinig, K.A., Sommaruga, R., Tessadri, R. & Psenner, R. (1997) Temperature effects on the acidity of remote alpine lakes. *Nature*, **387**, 64–67.

Stoddard, J.L., Jeffries, D.S., Lükewille, A., *et al.* (1999) Regional trends in aquatic recovery from acidification in North America and Europe 1980–95. *Nature*, **401**, 575–578.

Tammi, J., Appelberg, M., Beier, U., Hesthagen, T., Lappalainen, A. & Rask, M. (2003) Fish status survey of Nordic lakes: Effects of acidification, eutrophication and stocking activity on present fish species composition. *Ambio*, **32**, 98–105.

Tipping, E. & Hurley, M.A. (1988) A model of solid-solution interactions in acid organic soils, based on complexation properties of humic substances. *Journal of Soil Science*, **39**, 505–519.

UNECE (1999) *The 1999 Protocol to Abate Acidification, Eutrophication and Ground-Level Ozone.* Document ECE/EB.AIR, United Nations Economic Commission for Europe, New York and Geneva.

Vesely, J., Majer, V., Kopacek, J. & Norton, S.A. (2003) Increasing temperature decreases aluminum concentrations in Central European lakes recovering from acidification. *Limnology and Oceanography*, **48**, 2346–2354.

Vuorenmaa, J., Forsius, M. & Mannio, J. (2006) Increasing trends of total organic carbon concentrations in small forest lakes in Finland from 1987 to 2003. *Science of the Total Environment*, **365**, 47–65.

de Wit, H.A. & Wright, R.F. (2008) Projected stream water fluxes of NO_3 and total organic carbon from the Storgama headwater catchment, Norway, under climate change and reduced acid deposition. *Ambio*, **37**, 56–63.

de Wit, H.A., Hindar, A. & Hole, L. (2008) Winter climate affects long-term trends in stream water nitrate in acid-sensitive catchments in southern Norway. *Hydrology and Earth System Sciences*, **12**, 393–403.

Wright, R.F. (1998) Effect of increased CO_2 and temperature on runoff chemistry at a forested catchment in southern Norway (CLIMEX project). *Ecosystems*, **1**, 216–225.

Wright, R.F. (2008) The decreasing importance of acidification episodes with recovery from acidification: An analysis of the 30-year record from Birkenes, Norway. *Hydrology and Earth System Sciences*, **12**, 353–362.

Wright, R.F. & Henriksen, A. (1978) Chemistry of small Norwegian lakes with special reference to acid precipitation. *Limnology and Oceanography*, **23**, 487–498.

Wright, R.F., Alewell, C., Cullen, J., et al. (2001) Trends in nitrogen deposition and leaching in acid-sensitive streams in Europe. *Hydrology and Earth System Sciences*, **5**, 299–310.

Wright, R.F., Larssen, T., Camarero, L., et al. (2005) Recovery of acidified European surface waters. *Environmental Science and Technology*, **39**, 64A–72A.

Wright, R.F., Aherne, J., Bishop, K., et al. (2006) Modelling the effect of climate change on recovery of acidified freshwaters: Relative sensitivity of individual processes in the MAGIC model. *Science of the Total Environment*, **365**, 154–166.

8

Distribution of Persistent Organic Pollutants and Mercury in Freshwater Ecosystems Under Changing Climate Conditions

Joan O. Grimalt, Jordi Catalan, Pilar Fernandez, Benjami Piña and John Munthe

Introduction

Persistent organic pollutants (POPs) and some trace metals (e.g. mercury, cadmium and lead) are toxic substances released into the atmosphere by a range of urban, industrial and agricultural processes. Once emitted into the atmosphere, these pollutants are dispersed widely and are deposited onto waterbodies. They can enter the food web where they bioaccumulate and become toxic to aquatic and terrestrial organisms. Many of the diverse aspects of climate change (e.g. temperature increase, variations in rainfall, wind patterns and dust deposition) affect the distribution and mobility of toxic substances in freshwater systems.

Although many toxic substances introduced into the environment by human activity have been banned or restricted in use, many persist, especially in soils and sediments, and they either remain in contact with food webs or can be eventually re-mobilized and taken up by aquatic biota (Catalan *et al.* 2004; Vives *et al.* 2005). The high levels of metals (e.g. Hg) and POPs (polychlorobiphenyls [PCBs], DDE) in the tissue of freshwater fish in arctic and alpine lakes (Grimalt *et al.* 2001; Vives *et al.* 2004a) attest to the mobility and transport of these substances in the atmosphere (Carrera *et al.* 2002; Fernandez *et al.* 2002, 2003; van Drooge *et al.* 2004) and their concentration in cold regions (Fernandez & Grimalt 2003). The possible consequences of these factors on water quality are described below.

Organohalogen compounds

POPs are characterized by high chemical stability, which stems from their halogen substituents: primarily chlorine, constituting the so-called organochlorine compounds (OCs). OCs first appeared in the environment 70 years ago, but due

Climate Change Impacts on Freshwater Ecosystems. First edition. Edited by M. Kernan, R. Battarbee and B. Moss. © 2010 Blackwell Publishing Ltd.

to their chemical stability, they have not disappeared. Some of these compounds, such as PCBs, are so stable that they are not known to be destroyed completely by any environmental process. As such, they are continuously recycled between the environment and organisms, either through death or metabolic processes. The high stability of POPs enables them to be transported over large distances and to survive the oxidative and photolytic processes that atmospheric compounds are subject to, especially in the upper layers of the troposphere.

The majority of these compounds were synthesized as pesticides. These include the insecticides DDT, lindane (γ-hexachlorocyclohexane or γ-HCH), aldrin, toxaphenes, chlordane, mirex, dieldrin and endrin. Hexachlorobenzene (HCB) was used as a fungicide and is still produced today as a by-product in the manufacturing of various chlorinated organic solvents. In contrast, PCBs were synthesized for use as dielectrics in transformers, fire retardants, high thermal stability oils and other applications. Whilst some of these compounds were synthesized as pure products, they were often produced and used as mixtures, as in the case of PCBs, HCHs and toxaphenes. Hence, numerous compounds were introduced into the environment. In some cases, these products transformed into other contaminants (e.g. DDT transforms into DDE and DDD), further increasing the number of organic pollutants in ecosystems. Current concentrations of these compounds in remote European water bodies are in the ranges of $1-10$ pg l^{-1} for HCB, $50-3000$ pg l^{-1} for HCHs, $60-500$ pg l^{-1} for endosulfan (Table 8.1), $7-14$ pg l^{-1} for DDTs and $50-120$ pg l^{-1} for PCBs (Table 8.2).

Dioxins and dibenzofurans should also be mentioned. These products are not manufactured; rather, they are generated through processes such as the combustion of organic materials containing chlorine atoms (virtually all organic matter contains chlorine in at least small amounts) or through industrial processes such as certain types of paper pulp bleaching. Polychlorostyrenes (CSs) are also by-products of industrial processes in electrolytic plants. In addition, the 1990s witnessed the introduction of a new generation of organohalogen contaminants into the environment: polybromodiphenyl ethers (PBDEs), designed as flame retardants, as well as other brominated and fluorinated compounds.

POPs are lipid soluble, semi-volatile and toxic. Hence, the majority of them are now banned from use. In 2001, EU member states signed the Stockholm Convention, aimed at both reducing levels of chlorinated POPs (primarily through eliminating their use) and stimulating research into the implications of OCs for the environment and for human health. Fewer than 60 years after the development of these compounds, fully restrictive measures had to be taken to eliminate them from production.

Pollution from OCs has been observed in zones where there are focal sources, such as the town of Flix (Catalonia, Spain) where a chlor-alkali plant is located at the shore of the Ebre River. This factory emitted large amounts of HCB to the atmosphere and OCs and mercury to the river (Grimalt *et al.* 1994). This exposure has had various effects on human health, primarily related to thyroid disfunction and cancer (Grimalt *et al.* 1994; Sala *et al.* 2001). Nevertheless, the impact of these compounds on human health is not specific to a particular town or region;

Table 8.1 Concentrations of OCs in waters from remote sites (mean ± standard deviation; pg l^{-1}). (From Vilanova et al. 2001a; Fernandez et al. 2005.)

Location	HCB	α-HCH	γ-HCH	α-Endosulfan	β-Endosulfan	Endosulfan sulphate
Ladove Lake Sept. 2000	8.5 ± 3.2	68 ± 35	139 ± 89	nd	nd	280 ± 65
Lake Redon May 2000	6.0 ± 2.4	313 ± 120	1760 ± 606	207 ± 88	157 ± 51	1246 ± 293
Lake Redon (1996–8)	8.4 ± 11	410 ± 220	2500 ± 1090	60 ± 38	84 ± 46	1000 ± 540
Lake Gossenkölle (1996–7)	4.0 ± 1.8	64 ± 53	930 ± 850	44 ± 28	28 ± 24	92 ± 72
Øvre Neådalsvatn (1998)	6.2 ± 1.0	110 ± 52	200 ± 76	nd	nd	120 ± 16
Amituk Lake (1994, Arctic)		630	169	28		
Sea water (Antarctica)		3.4	0.7			
Arctic Ocean (1990, 1992–4)	14–18	870–4700	180–700	2.0–7.2	0.35–5.3	
Lake Malawi (1996–8)	4.4 ± 1.6	9.8 ± 6.2	14 ± 8.7	3.3 ± 6.2		
Bow Lake (2000)	21	210	130	19	23	
Kananaskis (2000)	15	160	140	15	17	

nd, not detected.

Table 8.2 Concentrations of PCBs and DDTs in waters from remote sites (mean ± standard deviation). (From Vilanova *et al.* 2001a; Fernandez *et al.* 2005.)

Location	PCBs (pg l^{-1})				DDTs (pg l^{-1})
	Total	Particulate	Dissolved	% in SPM	Total
Ladove Lake Sept. 2000	64 ± 30	33 ± 8	31 ± 31	59 ± 19	12 ± 3
Lake Redon Nov. 2000	79	53	26	67	9.6
Lake Redon May 2000	56 ± 11	28 ± 8	29 ± 8		7.1 ± 2.9
Lake Redon (1996–8)	62 ± 44				16 ± 28
Lake Gossenkölle (1996–7)	110 ± 64				14 ± 6.3
Øvre Neådalsvatn (1998)	26 ± 5.4				0.59 ± 0.40
Amituk Lake (1994, Arctic)	372				27
Arctic Ocean (1990, 1992–4)					1.0 ± 0.30*
Easthwaite Water Lake (1996–7)	680 ± 190[†]				
Lakes in southern Sweden (1997)	8–144[‡]				

* Only dissolved phase.
[†] Sum of 61 PCBs.
[‡] Sum of 49 PCBs.

rather, it is general to the entire planet. These compounds pose a new toxic threat. Humans are exposed to small doses starting in the womb and continuing throughout life. The consequences of the combined long-term effects of such exposure must be evaluated. Health problems due to exposure in the developmental phase may not appear until a much later age. For example, links between oncogene mutation and higher blood concentrations of some of these compounds in patients with colon cancer (Howsam *et al.* 2004) or exocrine pancreatic cancer have been observed (Porta *et al.* 1999). These compounds may even cause problems in the earliest stages of life. Indeed, decreases in cognitive ability have been correlated with intrauterine exposure to DDT (Ribas-Fito *et al.* 2006). Wheezing at age 6 has been reported in association with prenatal exposure to DDE (Sunyer *et al.* 2005, 2006) and increases in attention deficit and hyperactivity disorders in young children have been attributed to high exposure to HCB during the neonatal period (Ribas-Fito *et al.* 2007). An increase in global temperature will lead to greater volatilization and, consequently, to higher concentrations of these compounds in the atmosphere and in aquatic systems and ultimately in the human body.

Table 8.3　Concentrations of PAHs in waters from remote sites (mean ± standard deviation). (From Vilanova *et al.* 2001b; Fernandez *et al.* 2005.)

Location	PAH particulate (Units in ng/L)	PAH dissolved (Units in ng/L)	PAH total (Units in ng/L)
Ladove Lake Sept. 2000	8.5 ± 0.7	3.4 ± 0.4	12 ± 1.0
Lake Redon May 2001	0.18 ±0.03	0.58 ± 0.2	0.77 ± 0.20
Lake Redon (1996–8)	0.41 ± 0.13	0.27 ± 0.19	0.70 ± 0.21
Lake Gossenkölle (1996–7)	0.57 ± 0.34	0.35 ± 0.19	0.86 ± 0.44
Øvre Neådalsvatn (1998)	0.50 ± 0.08	0.56 ± 0.06	1.1 ± 0.1
Esthwaite Water Lake		92 ± 32	
Raritan Bay (New Jersey)	7.0–7.1	3.2–7.4	10–15
Hamilton Harbour (Lake Ontario)			45 ± 4
Niagara River			17 ± 5
Danube Estuary	0.13–1.25	0.18–0.21	
Northern Chesapeake Bay			8.7–14
Southern Chesapeake Bay			
Hampton (urban)	2.9	3.2	
York River (semiurban)	5.2	5.2	
Elizabeth River (industrial)	23	43	
Baltic Sea	0.07–0.33	0.57–0.74	0.64–1.08

Polycyclic aromatic hydrocarbons

Polycyclic aromatic hydrocarbons (PAHs) do not contain chlorine atoms; they are formed by fused aromatic (benzene-type) rings. Due to this fusion, PAHs are very stable and, therefore, exhibit properties of POPs. However, they are more sensitive to photolysis and environmental oxidation. These compounds are primarily generated during combustion (e.g. from cars, thermal plants and forest fires) and are also present in petroleum. Hence, PAHs, in contrast to OCs, have always been present in nature: ever since the earth has had an oxygenated atmosphere they have been introduced into the environment by forest fires and geochemical processes.

Organisms have been exposed to PAHs over the course of evolution. Current concentrations in remote European water bodies range between 0.1 and 10 ng l^{-1} (Table 8.3). This continued exposure has resulted in the development of metabolic mechanisms for their efficient elimination. Therefore, PAHs do not bioaccumulate in higher organisms. However, this does not mean that PAHs do not negatively affect human health; indeed, some of these compounds are well known to be highly carcinogenic (IARC 1983). The main difference between PAHs and OCs is that the former exert their effects more strongly through direct exposure, whereas the latter tend to affect health over the long term. Nonetheless, as a consequence of the extensive use of combustible fossil fuels as energy sources starting in the mid 19th century, the levels of these compounds in ecosystems have risen by several orders of magnitude (Fernandez *et al.* 2000). Furthermore, humans live in areas where PAHs are constantly being emitted, such as cities.

Mercury

Mercury (particularly as methyl Hg) can both bioaccumulate and biomagnify in the food chains and may undergo long-range transport when released to the atmosphere. Contamination of fish by mercury has been observed in many regions. Methylmercury is the most toxic form of mercury in the environment and bioaccumulates in aquatic food chains. Many lakes have been found to have levels of Hg in fish exceeding the dietary recommendations of the World Health Organization of $0.5\,\mu g\,g^{-1}$ wet weight (ww). The United States Environmental Protection Agency is even more restrictive, and recommends $0.18\,\mu g\,g^{-1}$ ww as a Federal Advisory limit for consumption of mercury-contaminated fish (US-EPA 2003). Methylmercury is deleterious for the development of the central nervous system of unborn children (Mergler *et al.* 2007). High body burdens of mercury may also be a risk factor for acute coronary events (coronary heart disease and infarction) in middle-aged men (Virtanen *et al.* 2005).

In Scandinavia, fish in thousands of lakes have mercury levels above the health guideline, making them unsuitable for human consumption. However, ecosystem characteristics are important (Munthe *et al.* 2007a). Hg is bioaccumulated in the form of methylmercury. Thus, the specific ecosystem capacities for the production and transport of methylmercury determine the final bioaccumulation rates (Munthe *et al.* 2007b). Examples from Arctic and mountain regions include observations of high Hg levels in the piscivorous Arctic charr population of Arresjøen, Svalbard, where the highest concentration was $0.44\,\mu g\,g^{-1}$ ww, despite concentrations of this metal in the sediment and lake water being very low (Rognerud *et al.* 2002). In 1993, five brown trout from Lochnagar were analysed for Hg in muscle and the concentrations were found to range from 0.04 to $0.08\,\mu g\,g^{-1}$ ww (Rosseland *et al.* 1997). In 2001, the Hg levels were higher still, ranging from 0.035 to $0.23\,\mu g\,g^{-1}$ ww.

The effects of temperature increases

Temperature influences the majority of physico-chemical properties and processes that determine the environmental behaviour of chemical compounds, affecting thermodynamic aspects (e.g. equilibrium constants, partition constants, absorption isotherms, vapour pressure and solubility) as well as kinetic aspects (e.g. transport, and reaction rates). Hence, variations in temperature can affect the dynamics, transport and fate of contaminants in the environment, especially in the aquatic environment.

Temperature also determines the relative proportion of water phases in each ecosystem. Warmer climate leads to higher air humidity and less snow and ice in cold areas. Longer ice-free periods can increase catchment soil erosion elevating the release of previously deposited contaminants bound to soils, but modelling and quantitative evaluation of these processes is difficult due to the complexity of the interactions and to the uncertainties in many of the parameters implied.

Organohalogen compounds and temperature

Many POPs have a vapour pressure that enables a certain level of volatilization. Hence, they can enter the atmosphere in warm or temperate zones of the planet, and then condense or accumulate as solids or liquids in cold areas. This phenomenon, known as the *global distillation effect*, primarily depends on temperature (Wania & Mackay 1995).

Variations in temperature due to climatic change can have considerable influence on the dynamics of these pollutants, affecting their long-range transport, bioaccumulation, bioavailability, biodegradation and, above all, environmental persistence and incorporation into trophic chains. For example, the total quantity of PCBs produced on Earth is about 1.3 million tons, 97% of which have been produced in the northern hemisphere. The majority of these compounds have remained trapped in the soil of the zones in which they were produced or used (i.e. temperate latitudes) (Meijer *et al.* 2003). Nonetheless, a fraction entered the atmosphere and subsequently accumulated in cold zones (e.g. Arctic zones). However, it has recently been observed that this process of accumulation via condensation affects not only distant zones, but also high mountain regions (Grimalt *et al.* 2001; Vives *et al.* 2004a). In other words, the industrialized countries have not only exported part of their pollution abroad, but they have also transferred part of it to what were previously the best-preserved areas of the industrialized world.

Figure 8.1 shows the distribution of various OCs in fish from European lakes. It reveals that the most contaminated lakes are those which are furthest from the sources of the pollution (i.e. cities and factories). This observation might seem like a paradox: traditionally, dumping of waste to distant zones has been considered to be associated with dilution, which diminishes the effects on the environment. In this case, what is observed is a net transfer of pollutants to ecosystems in distant zones owing to a process of evaporation (i.e. dilution) and subsequent condensation (i.e. concentration). Hence, the effects of POPs on ecosystems are not diluted; they simply move from one location (warm) to another (cold).

The same phenomenon is observed for compounds with very different applications. DDE is present in the environment because it is the main metabolite of DDT insecticide. PCBs, including the least volatile congeners of the mixtures (e.g. No. 101, No. 118, No. 153, No. 138 and No. 180), exhibit the same trend as DDE, and behave differently from the most volatile ones (e.g. No. 28 and No. 52) (Fig. 8.1). The determining factor in their environmental behaviour is volatility. Compounds with vapour pressures lower than $10^{-2.5}$ Pa accumulate in high-mountain zones. Compounds with vapour pressures above this value are not retained in the range of the annual average temperatures represented by the mountain lake series but rather in the high latitude areas (Grimalt *et al.* 2001).

Temperature is also a key factor in the environmental redistribution of the recently manufactured or used organobromine compounds. This is true for PBDEs, which were first used upon prohibition of PCBs in the late 1980s. The distribution of these two types of compounds was studied in a transect of high-mountain Pyrenean and Tatra lakes, encompassing altitudinal differences of about 1000 and 550 m, respectively. Figure 8.2 reveals that for the Pyrenean

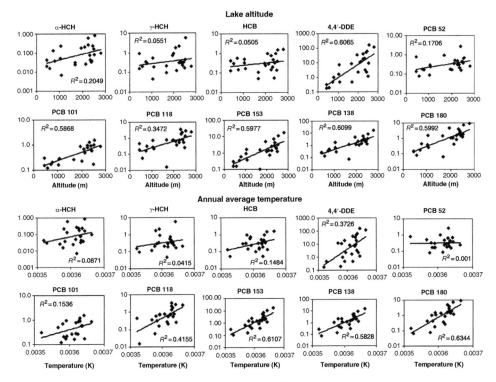

Figure 8.1 The concentrations of various OCs in fish (ng g^{-1}) from high-mountain European lakes depend on altitude and temperature (these two variables are roughly related in the series). Each point is the mean for the fish analysed in each lake. As observed, there is a correlation between the concentrations of high molecular weight compounds (4,4′-DDE, and PCBs 118, 153, 138 and 180) and altitude or temperature. This dependence, which implies greater contamination in the highest and most remote zones, is not observed for the most volatile compounds. (Based on Vives *et al.* 2004a.)

region, where PBDEs were used earlier, the high-mountain distribution is consistent with temperature dependence – which is also observed for PCBs (Gallego *et al.* 2007). In contrast, for the Tatra region, where PBDEs were not used until later, these contaminants have not yet reached a steady-state distribution in the high mountains.

This trend becomes further marked upon joint comparison of the distributions of PBDEs and PCBs in the two mountainous zones (Fig. 8.3). In the Pyrenean lakes, parallel distributions of concentrations are observed, despite the difference in time (more than 40 years) in the respective introduction of each type of compound into the environment (Gallego *et al.* 2007). However, this correlation is not observed in the Tatras.

Research on the food webs of these mountain lakes provides further clues on the physico-chemical processes leading to the accumulation of these compounds in organisms. The OC composition in water, chironomids, terrestrial insects, cladocerans, molluscs, cyanobacteria and fish (brown trout) has been investigated

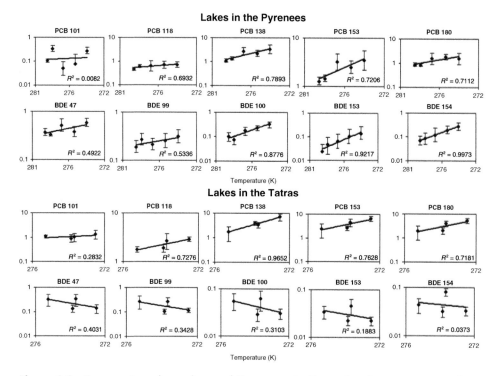

Figure 8.2   Temperature dependence of the concentrations of various congeners of PCBs and PBDEs in fish. There is a correlation between the accumulation of high molecular weight PCBs and temperature in both cases but only in the Pyrenees for PBDEs. (Based on Gallego et al. 2007.)

in Lake Redon (Pyrenees). OCs with an octanol–water partition coefficient (K_{ow}) higher than 10^6 showed lower concentrations in food than expected from theoretical octanol–water partition (Fig. 8.4), indicating that the distribution of these compounds does not reach equilibrium within the life span of the food web organisms (c.1 year). On the other hand, the degree of biomagnification in fish increased with K_{ow}, except in the case of the largest compound analysed, PCB 180, which contains seven chlorine substituents.

Examination of OC exchange at fish gill and gut using a fugacity model (expressing transfer of compounds between compartments) based on the water, food and fish concentrations showed a net gill loss and a net gut uptake for all compounds (Fig. 8.5). A pseudo-stationary state was only achieved for compounds with $\log(K_{ow}) < 6$. Calculation of average residence times for OCs in fish in apparent steady state gave values of days to a few weeks for HCHs, 1 year for HCB and 4,4′-DDE and 2–3 years for 4,4′-DDT, PCB 28 and PCB 52. Residence times longer than one decade were found for the more chlorinated PCB (Catalan et al. 2004).

The OC concentrations in aquatic insects change depending on the life-history stage. Increasing concentrations of OCs and PBDEs from larvae to pupae, independent of physico-chemical properties, has been observed. These concentration increases may result from the weight loss of pupae during metamorphosis as a

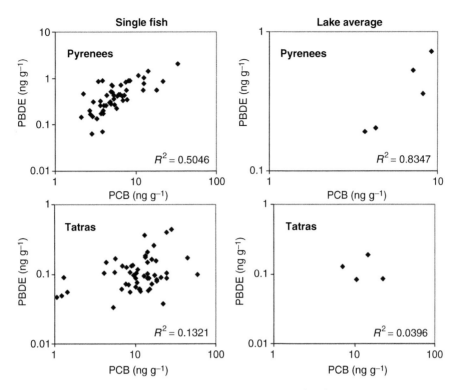

Figure 8.3   Concentrations of various PCBs and PBDEs in fish from high-mountain Pyrenean and Tatran lakes. (Based on Gallego *et al.* 2007.)

consequence of mainly protein carbon respiration and lack of feeding. Despite the lack of change in the total amount, the concentration increases from larvae to pupae are very relevant for the pollutant ingestion of higher predators. The concentration differences between larvae and pupae are apparently stronger for PBDEs than for OCs (Fig. 8.6), which suggest a higher retention of the former, consistent with the higher molecular volume of the brominated compounds. The intakes of OCs and PBDEs by trout (or any other predator) are between two- and fivefold higher per calorie gained when feeding on pupae than on larvae (Bartrons *et al.* 2007).

Temperature dependence is also observed for other environmental components, including lacustrine sediments (Europe: Grimalt *et al.* 2001), snow (Europe: Carrera *et al.* 2001; Canada: Blais *et al.* 1998), soils (Tenerife Island, Spain: Ribes *et al.* 2002) and mosses (Andes: Grimalt *et al.* 2004). PCBs were found in all snow samples collected in European sites (Carrera *et al.* 2001). HCHs were also very abundant, whereas DDTs were found at lower concentrations. Some of the major compounds found in snow samples, e.g. HCH and PCB, occur predominantly in the gas phase. Their high abundance in the snowpack reflects the occurrence of effective transfer mechanisms from gas to snow flakes. A temperature dependence of the snow concentrations of both volatile and non-volatile OCs is observed. Thus, higher levels of pollutants are found at lower temperatures. However, these dependences are more clearly defined for the less volatile compounds.

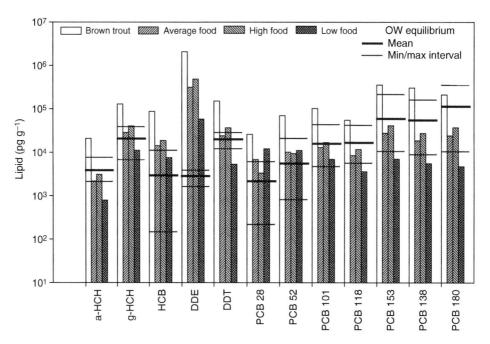

Figure 8.4 Concentrations of OCs in fish and their prey standardized by lipid content. Horizontal bars indicate the expected values according to the concentrations in water and K_{ow} (values are indicated for high (maximum) and low (minimum) food periods and mean diet). (Based on Catalan *et al.* 2004.)

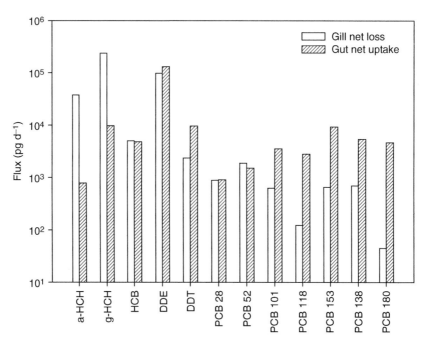

Figure 8.5 Calculated net gill loss and net gut uptake of OCs in a fish (mass 204 g) in Lake Redon compared with the concentrations in water, food and fish. (Based on Catalan *et al.* 2004.)

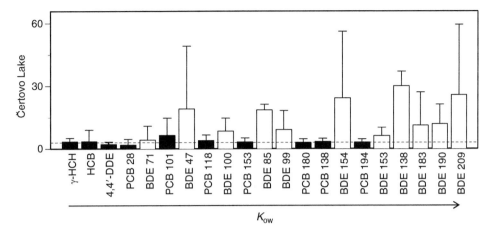

Figure 8.6 Increment ratio from larvae to pupae of the concentrations of OCs and PBDEs. The compounds are ranked by increasing K_{ow}. The slashed line indicates the average concentration increase for OCs (threefold) between the two metamorphic stages. (Based on Bartrons *et al.* 2007.)

Once these compounds are introduced into the environment, temperature becomes a determining factor in their distribution. Climatic changes imply a change in temperature, and increases in temperature have been particularly high in currently cold and mountainous regions. Hence, temperature variations should result in a redistribution of contaminants accumulated in these areas. High-mountain zones are inherently important: they represent the most remote ecosystems present in Europe. Nonetheless, these zones are primary areas of production of water resources for human use, constituting the headwaters of river systems. Maintaining these zones in environmentally healthy conditions is required for ensuring low levels of water contamination.

Studies on the atmospheric deposition of these compounds are important for gaining knowledge of the processes associated with temperature dependence. These have been performed in Izaña in Tenerife (van Drooge *et al.* 2001), Redon Lake in the Pyrenees, Gossenköllesee in the Alps and Øvre Neadålsvatn in Norway (Carrera *et al.* 2002). For all of these locations, greater deposition of these organic compounds was observed during warm periods, which reflects a greater rate of volatilization from the different environmental compartments in which they were stored (Carrera *et al.* 2002). However, the retention capacity of the compounds is greater in cooler lakes, and the least volatile compounds are the most retained (Fig. 8.7).

Another major factor is the general concentration of these compounds in the atmosphere where temperature changes associated with climate change first appear. Their seasonal concentration has been studied at Redon Lake, Skalnate Pleso in the Tatra Mountains (van Drooge *et al.* 2004) and Izaña (van Drooge *et al.* 2002). In general, the same POP distributions are found in all sites, which confirms the capacity of these compounds for volatilization and their global impact. The atmosphere acts as a store of these pollutants that are transferred to continental freshwater ecosystems by precipitation. Warmer periods correlate with greater volatilization (Fig. 8.8). This trend is clearest among the least volatile

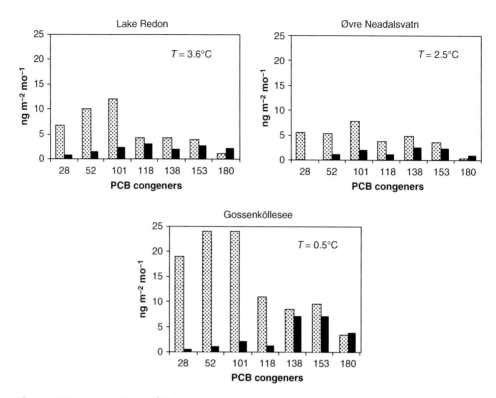

Figure 8.7 Comparison of the atmospheric deposition (dotted bars) and sedimentation flows of PCB congeners in high-mountain lakes (solid bars). Greater flows of atmospheric precipitation are observed for the lower molecular weight (i.e. most volatile) compounds, whereas sedimentation primarily affects the higher molecular weight (i.e. least volatile) compounds. This phenomenon is most pronounced in the coldest lake (Gossenköllesee). (Based on Carrera *et al.* 2002.)

compounds and at the site of lowest temperature (Skalnate Pleso). Such seasonal differences in concentration are consistent with the observations on atmospheric deposition. Accordingly, one of the direct consequences of climatic change will likely be an increase in atmospheric concentration of these compounds and, consequently, greater contamination of organisms, including humans.

The altitudinal accumulations of these compounds have toxic effects on the organisms living in these remote sites. Some of these effects have been identified by examination of cytochrome P450 1A (Cyp1A). This enzyme is an established biomarker of oxidative exposure to different environmental pollutants in many animal species, including fish (van der Oost *et al.* 2003). Expression of Cyp1A increases upon exposure to the so-called dioxin-like compounds, which include a variety of recognized pollutants, such as dioxins, coplanar PCBs and PAH, among others (Fent 2003). The effect on Cyp1A expression can be detected by measuring the levels of the corresponding RNA messenger (mRNA) (Piña *et al.* 2007).

The variability of hepatic Cyp1A gene expression levels in brown trout (*Salmo trutta*) populations from 11 European mountain lakes (101 individuals analysed) is

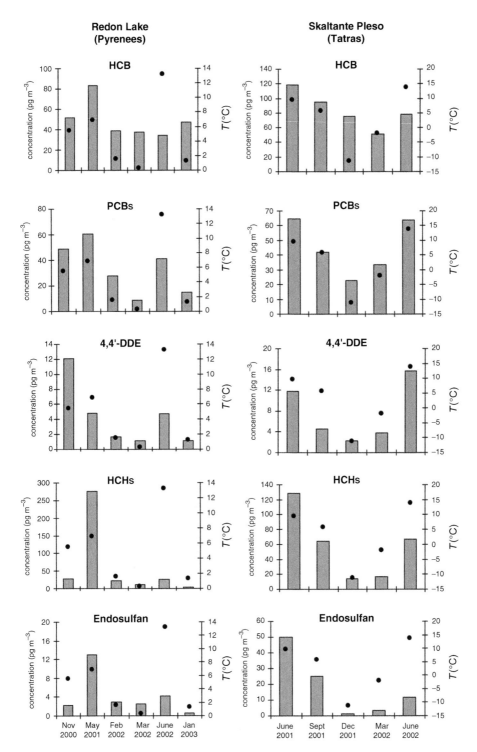

Figure 8.8 Geometric mean concentrations of various OCs in gas phase (bars) compared to mean ambient temperature (points). The warmest periods generally correlate with higher concentrations of the compounds in the gas phase. This trend is clearest for the least volatile compounds and at the site of lowest temperatures (Skalnate Pleso). (Based on van Drooge *et al.* 2004.)

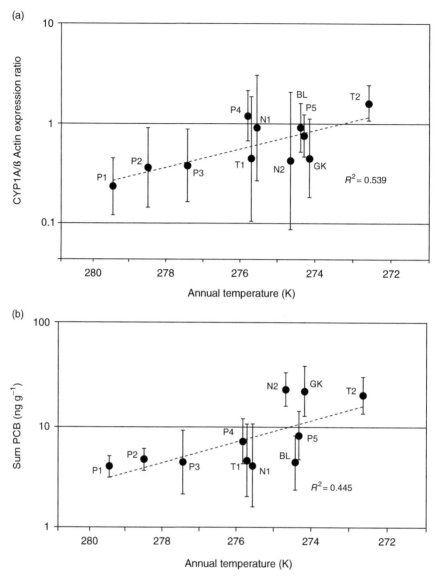

Figure 8.9 Correlation between (a) Cyp1A expression values and (b) total PCB content is *Salmo trutta* and the reciprocal of absolute year average temperatures for different European lakes. The *X*-axis represents temperature values, instead of their reciprocals, to facilitate reading. Dots represent average values for all fish from each lake; bars represent standard deviations. In both graphs, regression lines and correlation coefficient values were calculated between logarithmic transformants of gene expression or PCB contents, depending on the graph, and the reciprocal of absolute temperatures (discontinuous lines and R^2 values on the graphs). Different lakes are labelled according their geographic situation: P1–P5, lakes from the Pyrenees; T1 and T2, lakes from the Tatras; N1 and N2, lakes from central Norway; GK, lake from Austrian Alps; BL, lake from the Rila Mountains.

shown in Fig. 8.9 where it is plotted against the temperature gradient and compared with the total PCB content in muscle of the same animals. The graphs show a linear correlation between the logarithmic transformations of both series of data and the reciprocal of absolute temperatures. These trends suggest that Cyp1A induction in mountain lake trout is related to PCB exposure, more specifically to penta- and hexa-chlorobiphenyls, according to their physico-chemical properties. Likely candidates are coplanar PCB congeners, such as PCB126, although other long-transported pollutants could also be relevant for this effect. The strong dependence of hepatic Cyp1A expression in trout liver along temperature gradients suggests that POP redistribution as predicted by global warming projections may have significant effects on the physiology of the exposed fish populations even at remote sites.

Polycyclic aromatic hydrocarbons and temperature

A study of the concentrations of PAHs in the livers of high-mountain lake fish did not reveal any dependence on temperature (Vives *et al.* 2004b). This can be explained by the lack of bioaccumulation of PAHs in fish and because the global transport mechanisms of PAHs differ from those of organochlorine pollutants. The distribution of PAHs in remote (Fernandez *et al.* 2002) and urban (Aceves & Grimalt 1993) zones occurs in association with particles, especially soot. Thus, the concentrations of these compounds are more reflective of local sources than OCs.

Most organisms exhibit PAH distributions largely dominated by phenanthrene, in accordance with its predominance in atmospheric deposition, water and suspended particles (Vives *et al.* 2005). Total PAH levels are higher in organisms from littoral habitats than from the deep sediments or the pelagic water column. However, organisms from deep sediments exhibit higher proportions of higher molecular weight PAH than those in other lake areas. Different organisms exhibit species-specific features in their relative PAH composition that point to different capacities for uptake and metabolic degradation. Brown trout show an increased capacity for metabolic degradation since they have lower PAH concentrations than in their food and they are strongly enriched in lower molecular weight compounds (Fig. 8.10). The PAH levels in trout depend greatly on prey living in the littoral areas.

The atmospheric concentrations of PAHs tend to be higher in winter, when more fuel is consumed, than in summer (Aceves & Grimalt 1993; Fernandez *et al.* 2002). Increasing use of air conditioners may lead to a reversal of this trend in the future because increasing energy consumption means higher PAH emission. This would imply a higher proportion of PAHs in the gas phase than in association with particles compared with the present phase distribution, particularly in cities (Aceves & Grimalt 1993).

In any case, based on the gradual rise in the use of energy – 85% of which comes from burning of fossil fuels – increased concentrations of these compounds can be predicted for the future. However, the concentrations of PAHs in high-mountain lakes decreased by 30% from the 1970s to the present (Fernandez *et al.* 2000). This decrease is due to improvements in combustion processes, primarily at power stations, leading to a lower emission of pollutants. The emission of these carcinogenic hydrocarbons can be further limited through

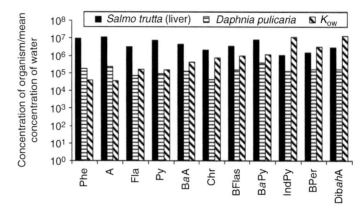

Figure 8.10 Comparison between ratios of lipid normalized PAHs in *Salmo trutta* (liver), *Daphnia pulicaria* and water (dissolved phase) and the expected values from the octanol–water partition coefficients of the compounds. Water concentrations in Vilanova *et al.* (2001b) and Fernandez *et al.* (2005). Phe, phenanthrene; A, anthracene; Fla, fluoranthene; Py, pyrene; BaA, benz[a]anthracene; Chr, chrysene; BFlas, benzo[b+k] fluoranthenes; BaPy, benzo[a]pyrene; IndPy, indeno[1,2,3-cd]pyrene; BPer, benzo[ghi] perylene; DibahA, dibenz[a,h]anthracene. (Based on Vives *et al.* 2005.)

additional improvements in combustion processes and through greater use of solar and wind power. The future global environmental occurrence of these compounds will depend on the balance between fossil fuel energy consumption and improvement of combustion methods and the degree of replacement of fossil fuel use by renewable sources. This area remains plagued with uncertainties. The effects of future improvements in combustion technology on the total amount of PAH generated may be offset by increased energy consumption.

Mercury and temperature

While mercury has many characteristics similar to the organic compounds discussed above (e.g. semi volatility, persistence, toxicity), it differs in the importance of the different forms of mercury existing in the environment. In air, elemental mercury vapour is predominant while oxidized divalent compounds are the most common in water, soils and sediments. In these media, methylmercury generally accounts for a very small fraction ($<1\%$) but since this species is the most toxic form and one that is capable of bioaccumulation, it is the most important from environmental and health perspectives.

Increased temperatures will change the environmental cycling of mercury because the rates of transformation between chemical species (e.g. oxidation, reduction, methylation) and the rates of transport between compartments (e.g. exchange between air and surfaces or water and sediments) depend on this variable. Very little is known about the overall impacts of temperature on mercury cycling. In general, increased oxidation rates of atmospheric mercury would be expected, which would involve increased deposition on land and water. This may be counterbalanced by increased volatilization of elemental mercury from land and water. Increased temperatures would most likely increase rates of methylation of inorganic mercury, but again, the rate of demethylation may also be enhanced. Increased temperatures in Arctic and Northern environments, where cold conditions currently limit methylation and mercury mobility, are of concern.

The effects of changes in precipitation

The IPCC (2007) predicts that precipitation will increase in Northern Europe and decrease in the Mediterranean zones by 10%–20% due to climatic change, and will become more unpredictable (Christensen *et al.* 2007). This translates to an increase in the frequency of extreme hydrological phenomena such as major droughts and flash floods. These changes in precipitation patterns will influence the transport and distribution of pollutants as well as their impact on aquatic environments.

For OCs, greater deposition is observed for volatile contaminants (primarily HCHs, HCB and the most volatile congeners of PCBs, DDEs and DDTs) at higher levels of wet deposition as exemplified at Teide in Tenerife (Fig. 8.11) (van Drooge *et al.* 2001). These results are consistent with the differences in atmospheric deposition observed in other sites such as Redon Lake, Øvre Neadålsvatn and Gossenköllesee (Fig. 8.7) (Carrera *et al.* 2002). Hence, any future decrease in precipitation would imply a drop in the amounts of these contaminants that enter water bodies.

In contrast, PAH deposition is primarily controlled by particle settling, followed by wet precipitation, and lastly, air temperature. The two first aspects are fundamental for high molecular weight hydrocarbons, whereas temperature is most important for the low molecular weight ones (Fernandez *et al.* 2003).

Considering that precipitation is a determinant factor in river flow, changes in this flow can be expected to influence the concentration of dissolved chemical species – including those of anthropogenic origin. As a general principle, decreases in stream flow may imply concentration increases if the pollution load is not changed. However, there are other issues to consider: major flash floods cause, among other effects, major resuspension and transport of sediments. Floods can therefore entrain contaminated sediments, causing them to enter into circulation in streams and rivers.

Figure 8.12 shows an example of this phenomenon in the Ebre River. The Flix reservoir (Catalonia, Spain) houses 500,000 tons of contaminated sediments containing OCs, mercury, cadmium and other metals, plus radioactive material, all of which stemmed from over 60 years of operation of a local chloro-alkali factory. From several studies performed in the past few years, during periods of high water in the Ebre River there has clearly been mobilization of sediments and, consequently, release of these contaminants, namely HCB, DDT and PCBs.

Such mobilization will probably have its greatest general effect once permanent snow in the mountains disappears due to climatic change, because erosion will be greater and POPs and metals will be more easily mobilized from the soil. Re-mobilization of metals as a result of increased soil erosion has been invoked to explain the continued high concentrations of Hg and Pb in Scottish lake sediments (Yang *et al.* 2002).

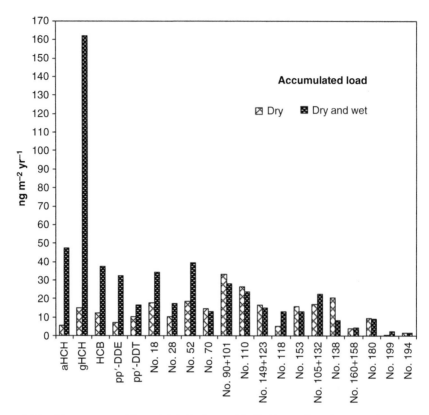

Figure 8.11 Mean composition of OCs in dry and bulk (dry and wet) atmospheric deposition over the course of 1 year in Teide. The intervals represent the standard deviation. As observed, the compounds that are the most volatile and also associated with wet precipitation have the greatest deposition (based on van Drooge *et al.* 2001). Note that these results are consistent with those shown in Fig. 8.7, which were obtained in other locations.

In the boreal forest zone of Northern Europe, climate change may lead to methylmercury concentration increases in fish. Thousands of lakes in Scandinavia already have mercury levels in fish that exceed the health guideline of 0.5 mg kg^{-1}, making them unsuitable for human consumption. In this region, climate models predict winter precipitation increases which may lead to higher groundwater levels and more water to flow through organic-rich soil horizons where a large fraction of the soil-bound mercury is accumulated, potentially causing direct mobilization of mercury and methylmercury. Changing redox conditions and higher release of DOC and nutrients may further enhance the concentrations of methylmercury in aquatic ecosystems.

Manipulation of precipitation and hydrology at Gårdsjön, an experimental lake site in south-west Sweden, has shown increases in both total mercury (i.e. sum of all forms of mercury) and methylmercury in run-off water. The transport of total mercury was found to be proportional to the increases in run-off amounts whereas

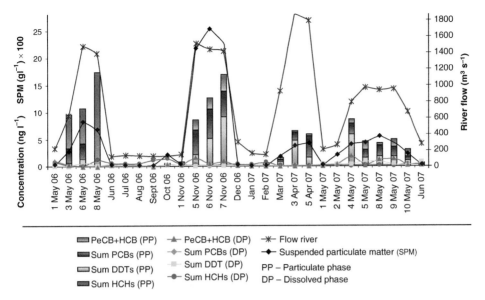

Figure 8.12 Concentrations of suspended particulate matter and pentachlorobenzene (PeCB), HCB, PCBs, HCHs and DDTs plotted against river flow for the Ebre River. As observed, the concentrations of these compounds increase with increasing flow and with higher levels of suspended particles. This trend reflects the mobilization of contaminated material located in the Flix reservoir due to increased river flow. DP, dissolved phase; PP, particulate phase. SPM: black thick line. River flow: blue thick line.

the relative increase in methylmercury was larger than the relative increase in water transport. Increased soil wetness therefore enhanced methylmercury formation. Concurrent increases of dissolved organic carbon and sulphide in run-off water suggested that soil was saturated with water and that anaerobic conditions had been established, favouring mercury methylation via sulphate-reducing micro-organisms. The results of these experiments are consistent with the scenario mentioned above, indicating that future climate change may lead to increased loadings of methylmercury to aquatic ecosystems in areas where increased precipitation is expected.

Anthropogenic effects associated with water management

Apart from the effects that are most directly linked to climate change, there is a second group of factors that can influence the quality of water resources. These can be defined as negative collateral effects derived from new water management strategies used by humans to address scarce water supplies. As these causes are indirect, their influence is much more difficult to ascertain. For example, climate change can lead to changes in land use, forcing farmers to adapt agricultural practices (e.g. changing tilling methods, extending production seasons, modifying irrigation plans and confronting new pests). These new practices can have major consequences on the types of pesticides used, which in turn can ultimately influence the quality of water that enters rivers or phreatic waters. Minimizing the impact of contaminants in water will require a major effort to control pollution by pesticides and other toxic pollutants.

Conclusions

Volatile heavy metals and POPs present much more cryptic problems than those of habitat change, acidification or eutrophication. Their effects on communities are largely unknown and difficult to trace, because they are generally sub-lethal. Health implications for fish and their human consumers are nonetheless clear and the volatility and condensation effects make it very likely that a warmer climate in particular will increase their transfer to polar and mountain regions. One of the more poignant observations of recent years has been that of the widespread contamination even of Antarctic ecosystems with these substances (Bargagli 2008).

However, in the areas closer to emission sites, warmer temperatures may result in higher concentrations of these compounds in air than at present. This change will increase the dispersion capacity of hydrophobic organic pollutants and mercury, leading to higher rates of toxification of humans and higher organisms through respiration. This effect will also be important for PAHs although in this case the ultimate environmental impact will depend on the balance between improved combustion sources and future energy demands, which may increase following population and wealth growth.

The implications of increased temperatures will be similar for mercury. Transformation into methylmercury may be enhanced at higher temperatures due to increased microbial activity. Higher temperatures in Arctic and northern environments, where cold conditions currently limit methylation and mercury mobility, are of concern. In Scandinavia, thousands of lakes have fish with mercury levels above the health guidelines. The projected decreases in atmospheric precipitation, e.g. rain and snow may further increase these levels due to higher proportions of groundwater percolating through mercury-rich soils.

References

Aceves, M. & Grimalt, J.O. (1993) Seasonally dependent size distributions of aliphatic and polycyclic aromatic hydrocarbons in urban aerosols from densely populated areas. *Environmental Science and Technology*, **27**, 2896–2908.

Bargagli, R. (2008) Environmental contamination in Antarctic ecosystems. *The Science of the Total Environment*, **400**, 212–226.

Bartrons, M., Grimalt, J.O. & Catalan, J. (2007) Concentration changes of organochlorine compounds and polybromodiphenyl ethers during metamorphosis of aquatic insects. *Environmental Science and Technology*, **41**, 6137–6141.

Blais, J.M., Schindler, D.W., Muir, D.C.G., Kimpe, L.E., Donald, D.B. & Rosenberg, B. (1998) Accumulation of persistent organochlorine compounds in mountains of Western Canada. *Nature*, **395**, 585–588.

Carrera, G., Fernández, P., Vilanova, R.M. & Grimalt, J.O. (2001) Persistent organic pollutants in snow from European high mountain areas. *Atmospheric Environment*, **35**, 245–254.

Carrera, G., Fernandez, P., Grimalt, J.O., *et al.* (2002) Atmospheric deposition of organochlorine compounds to remote high mountain lakes of Europe. *Environmental Science and Technology*, **36**, 2581–2588.

Catalan, J., Ventura, M., Vives, I. & Grimalt, J.O. (2004) The roles of food and water in the bioaccumulation of organochlorine compounds in high mountain lake fish. *Environmental Science and Technology*, 38, 4269–4275.

Christensen, J.H., Hewitson, B., Busuioc, A., *et al.* (2007) Regional climate projections. In: *Climate Change 2007: The Physical Science Basis. Contribution of Working Group I to the Fourth Assessment Report of the Intergovernmental Panel on Climate Change* (eds S. Solomon, D. Qin, M. Manning, *et al.*). Cambridge University Press, Cambridge and New York.

van Drooge, B.L., Grimalt, J.O., Torres-García, C.J. & Cuevas, E. (2001) Deposition of semi-volatile organochlorine compounds in the free troposphere of the Eastern North Atlantic Ocean. *Marine Pollution Bulletin*, 42, 628–634.

van Drooge, B.L., Grimalt, J.O., Torres-García, C.J. & Cuevas, E. (2002) Semivolatile organochlorine compounds in the free troposphere of the northeastern Atlantic. *Environmental Science and Technology*, 36, 1155–1161.

van Drooge, B.L., Grimalt, J.O., Camarero, L., Catalan, J., Stuchlik, E. & Torres-Garcia, C.J. (2004) Atmospheric semivolatile organochlorine compounds in European high-mountain areas (Central Pyrenees and High Tatras). *Environmental Science and Technology*, 38, 3525–3532.

Fent, K. (2003). Ecotoxicological problems associated with contaminated sites. *Toxicology Letters*, 140–141, 353–365.

Fernandez, P. & Grimalt, J.O. (2003) On the global distribution of persistent organic pollutants. *Chimia*, 57, 514–521.

Fernandez, P., Vilanova, R.M., Martínez, C., Appleby, P. & Grimalt, J.O. (2000) The historical record of atmospheric pyrolytic pollution over Europe registered in the sedimentary PAH from remote mountain lakes. *Environmental Science and Technology*, 34, 1906–1913.

Fernandez, P., Grimalt, J.O. & Vilanova, R.M. (2002) Atmospheric gas-particle partitioning of polycyclic aromatic hydrocarbons in high mountain regions of Europe. *Environmental Science and Technology*, 36, 1162–1168.

Fernandez, P., Carrera, G., Grimalt, J.O., *et al.* (2003) Factors governing the atmospheric deposition of polycyclic aromatic hydrocarbons to remote areas. *Environmental Science and Technology*, 37, 3261–3267.

Fernandez, P., Carrera, G. & Grimalt, J.O. (2005) Persistent organic pollutants in remote freshwater ecosystems. *Aquatic Sciences*, 67, 263–273.

Gallego, E., Grimalt, J.O., Bartrons, M., *et al.* (2007) Altitudinal gradients of PBDEs and PCBs in fish from European high mountain lakes. *Environmental Science and Technology*, 41, 2196–2202.

Grimalt, J.O., Sunyer, J., Moreno, V., *et al.* (1994) Risk excess of soft-tissue sarcoma and thyroid cancer in a community exposed to airborne organochlorinated compound mixtures with a high hexachlorobenzene content. *International Journal of Cancer*, 56, 200–203.

Grimalt, J.O., Fernandez, P., Berdié, L., *et al.* (2001) Selective trapping of organochlorine compounds in mountain lakes of temperate areas. *Environmental Science and Technology*, 35, 2690–2697.

Grimalt, J.O., Borghini, F., Sanchez-Hernandez, J.C., Barra, R., Torres-Garcia, C. & Focardi, S. (2004) Temperature dependence of the distribution of organochlorine compounds in the mosses of the Andean mountains. *Environmental Science and Technology*, 38, 5386–5392.

Howsam, M., Grimalt, J.O., Guino, E., *et al.* (2004) Organochlorine exposure and colorectal cancer risk. *Environmental Health Perspectives*, 112, 1460–1466.

IARC (1983) WHO, Vol. 32, Lyons, France.

IPCC (Intergovernmental Panel on Climate Change) (2007) Summary for policymakers. l. In: *Climate Change 2007: The Physical Science Basis. Contribution of Working Group I to the Fourth Assessment Report of the Intergovernmental Panel on Climate Change* (eds S. Solomon, D. Qin, M. Manning, *et al.*). Cambridge University Press, Cambridge and New York.

Meijer, S.N., Ockenden, W.A., Seetman, A., Breivik, K., Grimalt, J.O. & Jones, K.C. (2003) Global distribution and budget of PCBs and HCB in background surface soils: Implications for sources and environmental processes. *Environmental Science and Technology*, 37, 667–672.

Mergler, D., Anderson, H.A., Chan, L.H.M., *et al.* (2007) Methylmercury exposure and health effects in humans: A worldwide concern. *Ambio*, 36, 3–11.

Munthe, J., Bodaly, R.A., Branfireun, B.A., *et al.* (2007a) Recovery of mercury-contaminated fisheries. *Ambio*, 36, 33–44.

Munthe, J., Wängberg, I., Rognerud, S., *et al.* (2007b). Mercury in Nordic ecosystems. IVL Report B1761. Available at http://www.ivl.se.

van der Oost, R., Beyer, J. & Vermeulen, N. (2003) Fish bioaccumulation and biomarkers in environmental risk assessment: A review. *Environmental Toxicology and Pharmacology*, **13**, 57–149.

Piña, B., Casado, M. & Quirós, L. (2007) Analysis of gene expression as a new tool in ecotoxicology and environmental monitoring. *TrAC – Trends in Analytical Chemistry*, **26**, 1145–1154.

Porta, M., Malats, N., Jariod, M., *et al.* (1999) Serum levels of organochlorine compounds and K-*ras* mutations in exocrine pancreatic cancer. *The Lancet*, **354**, 2125–2129.

Ribas-Fito, N., Torrent, M., Carrizo, D., *et al.* (2006) In utero exposure to background concentrations of DDT and cognitive functioning among preschoolers. *American Journal of Epidemiology*, **164**, 955–962.

Ribas-Fitó, N., Torrent, M., Carrizo, D., Júlvez, J., Grimalt, J.O. & Sunyer, J. (2007) Exposure to hexachlorobenzene during pregnancy and children's social behavior at 4 years of age. *Environmental Health Perspectives*, **115**, 447–450.

Ribes, A., Grimalt, J.O., Torres-Garcia, C.J. & Cuevas, E. (2002) Temperature and organic matter dependence of the distribution of organochlorine compounds in mountain soils from the subtropical Atlantic (Teide, Tenerife Island). *Environmental Science and Technology*, **36**, 1879–1885.

Rognerud S., Grimalt, J.O., Rosseland, B.O., *et al.* (2002) Mercury, and organochlorine contamination in brown trout (*Salmo trutta*) and Arctic charr (*Salvelinus alpinus*) from high mountain lakes in Europe and the Svalbard archipelago. *Water Air and Soil Pollution Focus*, **2**, 209–232.

Rosseland B.O., Grimalt J.O., Lien L., *et al.* (1997) Population structure and concentrations of heavy metals and organic micropollutants. In: *AL:PE – Acidification of Mountain Lakes: Palaeolimnology and Ecology. Part 2 – Remote Mountain Lakes As Indicators of Air Pollution and Climate Change* (eds B. Wathne, S. Patrick & N. Cameron), pp. 4-1–4-73. NIVA Report 3638-97.

Sala, M., Sunyer, J., Herrero, C., To-Figueras, J. & Grimalt, J.O. (2001) Association between serum concentration of hexachlorobenzene and polychlorobiphenyls with thyroid hormone and liver enzymes in a sample of the general population. *Occupational and Environmental Medicine*, **58**, 172–177.

Sunyer, J., Torrent, M., Muñoz-Ortiz, L., *et al.* (2005) Prenatal dichlorodiphenyldichloroethylene (DDE) and asthma in children. *Environmental Health Perspectives*, **113**, 1787–1790.

Sunyer, J., Torrent, M., Garcia-Esteban, R., *et al.* (2006) Early exposure to dichlorodiphenyldichloroethylene, breastfeeding and asthma at age six. *Clinical and Experimental Allergy*, **36**, 1236–1241.

US-EPA (2003) Chapter 2.5 Consumption of fish and shellfish. *Draft Report on the Environment 2003*. Technical Document.

Vilanova, R., Fernández, P., Martinez, C. & Grimalt, J.O. (2001a) Organochlorine pollutants in remote mountain lake waters. *Journal of Environmental Quality*, **30**, 1286–1295.

Vilanova, R.M., Fernandez, P., Martinez, C. & Grimalt, J.O. (2001b) Polycyclic aromatic hydrocarbons in remote mountain lake waters. *Water Research*, **35**, 3916–3926.

Virtanen, J.K., Voutilainen, S., Rissanen, T.H. *et al.* (2005) Mercury, fish oils, and risk of acute coronary events and cardiovascular disease, coronary heart disease, and all-cause mortality in men in eastern Finland. *Arteriosclerosis, Thrombosis, and Vascular Biology*, **25**, 228–233.

Vives, I., Grimalt, J.O., Catalan, J., Rosseland, B.O. & Battarbee, R.W. (2004a) Influence of altitude and age in the accumulation of organochlorine compounds in fish from high mountain lakes. *Environmental Science and Technology*, **38**, 690–698.

Vives, I., Grimalt, J.O., Fernandez, P. & Rosseland, B.O. (2004b) Polycyclic aromatic hydrocarbons in fish from remote and high mountain lakes in Europe and Greenland. *Science of the Total Environment*, **324**, 67–77.

Vives, I., Grimalt, J.O., Ventura, M. & Catalan, J. (2005) Distribution of polycyclic aromatic hydrocarbons in the food web of a high mountain lake, Pyrenees, Catalonia, Spain. *Environmental Toxicology and Chemistry*, **24**, 1344–1352.

Wania, F. & Mackay, D. (1995) A global distribution model for persistent organic chemicals. *Science of the Total Environment*, **160/161**, 211–232.

Yang, H., Rose, N.L., Battarbee, R.W. and Boyle, J.F. (2002) Mercury and lead budgets for Lochnagar, a Scottish mountain lake and its catchment. *Environmental Science and Technology*, **36**, 1383–1388.

9

Climate Change: Defining Reference Conditions and Restoring Freshwater Ecosystems

Richard K. Johnson, Richard W. Battarbee, Helen Bennion, Daniel Hering, Merel B. Soons and Jos T.A. Verhoeven

Introduction

Natural ecosystems, already under considerable stress from land-use change and pollution, now face additional pressures from climate change (e.g. Mann *et al.* 1998), both directly and indirectly through interactions with other drivers. At the same time, the world's biodiversity is also being reduced at an unprecedented rate, which has recently manifested in a growing concern that when species are lost, ecosystem services may also become impaired. Such concerns have resulted in a growing interest in restoring both the biodiversity and functioning of ecosystems to a more natural or reference state. Restoration usually entails removing or mitigating the activities causing the degradation, but may also require more active management such as physical reconstruction of freshwater ecosystems (e.g. by re-meandering streams) or by altering internal habitat conditions (e.g. by replacing large woody debris in streams, either directly, or indirectly by promoting the growth of riparian forests).

To determine if restoration has been successful requires knowledge of the condition that is expected to occur in the absence of human-induced stress. Specifically, reference conditions are needed: (i) to understand how the current condition differs from the ecological target/reference condition; (ii) to determine what factors have been degraded and by how much; (iii) to identify the drivers of the observed change; and (iv) to decide what steps are needed to restore the ecosystem to the desired condition. Ideally, restoration studies, and indeed all studies of disturbance and recovery, should be based on deviation from an undisturbed condition (e.g. Downes *et al.* 2002). However, finding adequate controls at the scales that are usually relevant for ecosystem studies, such as contemporary landscapes/ecosystems where disturbances have not occurred, is often problematic. Consequently, the most common approach is to establish the reference condition

Climate Change Impacts on Freshwater Ecosystems. First edition. Edited by M. Kernan, R. Battarbee and B. Moss. © 2010 Blackwell Publishing Ltd.

from putatively comparable contemporary landscapes or by using historical or palaeoecological methods to reconstruct a past undisturbed condition.

In this chapter, we focus on the use of different methods to establish reference conditions, as well as on how climate change might affect contemporary baselines or restoration targets. In particular, we are interested in comparing different methods for establishing reference states. Our first main objective is to understand the inherent variability that might be associated with the use of different methods to establish reference conditions. Secondly, with the growing awareness that climate change is currently affecting ecosystems, we are interested in determining how climate change might affect how we view contemporary reference conditions or baselines, and hence influence interpretation of restoration. Finally, we consider how freshwater ecosystems can best be restored in a world of ongoing climate change.

Detecting the effects that human activities are having on freshwater ecosystems is complex, as ecosystems themselves are complex and often diffuse entities. A challenge therefore for ecologists is to understand the importance of linkages between the structure and function of ecosystems as well as to understand ecosystem properties such as resistance and resilience to natural and human-induced stress. Often restoration endeavours have focused on a single system of interest (i.e. a single lake or stream), usually ignoring that freshwater systems are intricately linked with their surroundings (e.g. lakes may be perceived as islands nested in a terrestrial environment). However, there is now an increasing recognition that ecosystems are strongly interconnected, and this understanding needs to reach land-use managers.

Establishing reference conditions

Effective management of aquatic resources requires knowledge of when a water body differs from the natural condition (i.e. absence of human disturbance) and what has caused the deviation from the expected unimpaired condition. Reference conditions (*sensu* Bailey *et al.* 2004) are being increasingly used to gauge the effects and magnitude of human intervention. A number of problems emerge, however, in the use of reference conditions. Although seemingly trivial, the definition of what constitutes a reference condition often results in misunderstanding and contention. Definitions can be based on narrative or empirical data and range from the natural condition where human influence is lacking or minimal, to the best attainable condition which recognizes that humans are an inherent part of the ecosystem (e.g. Nowicki 2003).

According to the European Water Framework Directive (European Commission 2000; Annex 5, section 1.2), high ecological status (i.e. reference conditions) for biological attributes is defined as having 'no, or only very minor' evidence of distortion. Consequently, the Directive does not sanction the use of the best available sites within a region, unless these sites can be shown to reflect a natural state with no or only minor human influence. In an attempt to add clarity to the use of reference condition terminology, Stoddard *et al.* (2006) proposed that the term reference should be reserved solely for 'naturalness' or 'biological integrity', and when referring to reference conditions that deviate from naturalness or integrity, the authors propose the use of four terms: (i) minimally disturbed condition, to describe the condition in the absence of significant human disturbance; (ii) historical condition, to refer to a condition at some previous time; (iii) least disturbed condition is reserved for sites in the landscape having the best available physico-chemical and biological conditions; and (iv) best attainable condition is equivalent to the expected ecological or the least disturbed condition if the best possible management practice were in use for some time.

Many approaches are used to establish reference conditions (e.g. Stevenson *et al.* 2004). The most common can be grouped into four categories: (i) spatial approaches such as survey; (ii) temporal approaches such as contemporary time series, historical data and palaeo-reconstruction; (iii) modelling approaches such as hindcasting; and (iv) expert judgement. In areas where land use has not drastically altered the landscape, the identification of reference conditions is rather straightforward and *spatial methods* are frequently used. In such areas, use of survey data is common since the approach either explicitly (sites are sampled to include among-year variability) or implicitly (space-for-time substitution) includes natural variability. Another reason for the popularity of using survey data is transparency – the definition of what constitutes a reference condition is established *a priori*.

In many areas of Europe and elsewhere, humans have, however, extensively altered the landscape over long periods and hence reference conditions cannot adequately be determined using space-for-time approaches. When contemporary reference sites are lacking, models and hindcasting are commonly used to establish a reference condition. For instance, relationships between response and predictor variables can be used to predict the expected reference condition (e.g. community composition or palaeo-reconstruction of water chemistry and biota) in the absence of stress (e.g. Wright 1995).

Recognition that even relatively pristine systems may change a great deal, periodically cycling through different successional states and exhibiting erratic or unexpected changes in system behaviour (Holling 1992), has been valuable in promoting new approaches to resource management and for establishing reference conditions and ecological targets. The use of palaeo-reconstruction and contemporary time series data are two *temporal* methods that may capture the dynamic nature of aquatic ecosystems. For lakes, reconstructing past conditions is often done either directly using the remains of taxa stored in the sediment to reconstruct an assemblage, or indirectly using taxon information to infer past water chemistry. For example, weighted-averaging transfer functions have allowed for quantitative inferences of nutrient enrichment, hypolimnetic oxygen, pH and temperature (Bennion & Battarbee 2007). By contrast, the use of sediment remains to reconstruct community assemblages directly is less frequent due to high costs and spatial limitations (reconstructions are site-specific). In contrast to the use of time-series data, use of historical data is often a static measure of the reference condition; another weakness is that data availability is often limited (e.g. only qualitative data are available), although some interesting, novel approaches, such as the use of journals of pirates and century old cookbooks are being used in marine reconstruction (Schrope 2006).

A problem with many of these approaches is that they are essentially reductionist, often providing information on a single population or assemblage of organisms, and are seldom used to reconstruct the elements of whole communities or ecosystems. *Expert judgement* is one way of amalgamating many different types of information, such as empirical data, opinion and present-day concepts,

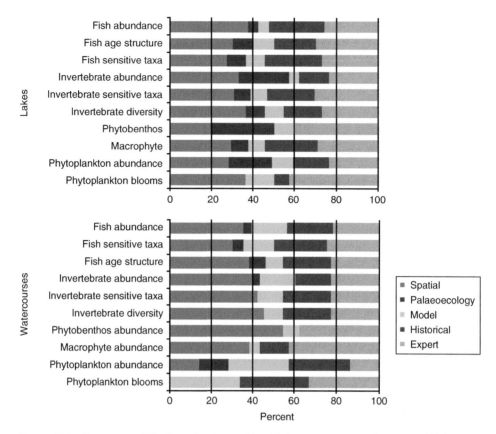

Figure 9.1   Frequency (%) of methods used by 12 European countries to establish reference conditions. (Data are taken from the REFCOND, http://www-nrciws.slu.se/REFCOND/doc_rew2.html.)

to define a more holistic reference condition that includes ecosystem-level attributes. However, one caveat with expert judgement is that it often consists of a narrative articulation of a perceived reference condition, and consequently this approach may introduce subjectivity (e.g. the common perception that it was always better in the past) and bias (e.g. experts may disregard sites having naturally low diversity) and can be extremely inaccurate. Another weakness is that the measure obtained is often static and therefore neglects the dynamic and inherent variability often associated with natural ecosystems. Consequently, expert judgement is seldom used as a single method to establish reference conditions, but may be a strong complement when combined with other methods.

   Methods used to establish reference conditions of European inland surface waters were reviewed by the EU-funded project REFCOND (Development of a protocol for identification of reference conditions, and boundaries between high, good and moderate status in lakes and watercourses) (Wallin *et al.* 2003). Approaches were found to vary between lakes and watercourses and with biological quality elements (Fig. 9.1). Spatial approaches were the most common methods used by the 12 countries that participated in the survey (mean = 31%

for lakes and 34% for watercourses), followed by expert judgement (30% lakes, 27% watercourses), historical data (18% lakes, 21% watercourses), predictive models (9% lakes, 15% watercourses) and palaeo-reconstruction (12% lakes, 3% watercourses). The selection of methods seemed to reflect the difficulty associated with establishing reference conditions for the various quality elements and habitat types. For example, spatial approaches were commonly used to establish reference conditions for phytobenthos in watercourses (54%), whereas expert judgement was more commonly used for macrophytes (43%). Not surprisingly, the use of palaeo-reconstruction was almost twice as common in lakes (e.g. 30% for phytobenthos) compared with watercourses (14% for phytoplankton). Although a number of methods are currently being used for establishing reference conditions, surprisingly little is known of their inherent error, in particular how different levels of uncertainty may affect determinations of disturbance and recovery.

Comparison of different approaches

Typology- and model-based approaches

Ecologists have for some time recognized the importance of biogeographical drivers for species distribution patterns, and the use of spatial typologies for partitioning natural variability (e.g. Hawkins *et al.* 2000; Johnson *et al.* 2007). Establishing reference conditions and ecological targets for restoration is a direct application of this knowledge. For example, cognizant of the importance of regional drivers for aquatic biodiversity, the European Water Framework Directive (European Commission 2000) specifies two alternative spatial approaches for partitioning biological variability and assessing ecological quality. System-A consists of four categories (e.g. ecoregion, altitude, catchment area and geology for streams), whereas system-B consists of a mixture of obligatory as well as optional factors. The value of typology-based approaches for explaining and partitioning biological variability has, however, been much debated and findings are hitherto equivocal. For instance, a number of studies have shown that aquatic assemblages are related to large-scale patterns in vegetation and climate (e.g. Feminella 2000; Rabeni & Doisy 2000; Verdonschot & Nijboer 2004), whilst others have questioned the efficacy of classifications based solely on landscape patterns in vegetation and climate, generally arguing that regional classifications need to be augmented with other factors such as altitude, size and catchment characteristics to discriminate communities effectively (e.g. Sandin & Johnson 2000a; Van Sickle & Hughes 2000).

Although many studies have focused on the importance of regional and local scale variables for predicting the composition of aquatic habitats, few have tested the efficacy of *a priori* classification systems (like system-A variables) for predicting biotic assemblages and fewer still have quantified the uncertainties associated with these predictions. In agreement with findings from earlier studies, R.K. Johnson (unpublished) found that latitude, altitude and catchment size were important predictors of invertebrate assemblages in boreal streams (e.g. Hawkins *et al.* 2000; Johnson *et al.* 2004). However, use of system-A variables alone resulted in many sites being misclassified and model uncertainty

Table 9.1   Standard deviations of the observed/expected ratios of biotic indices for
RIVPACS-type model reference sites based on null models, WFD system-A models and
the RIVPACS-type models used in Great Britain, Sweden and the Czech Republic. (Results
are taken from Davy-Bowker *et al.* 2006.)

Index	Country	Season	Null model*	System-A	RIVPACS-type model
Number of taxa	Great Britain	Autumn	0.263	0.251 (−5)	0.218 (−17)
	Sweden	Autumn	0.355	0.345 (−3)	0.312 (−12)
	Czech Republic	Autumn	0.383	0.317 (−17)	0.276 (−28)
ASPT	Great Britain	Autumn	0.132	0.113 (−14)	0.086 (−35)
	Sweden	Autumn	0.139	0.122 (−12)	0.112 (−19)
	Czech Republic	Autumn	0.091	0.089 (−2)	0.085 (−7)

The per cent reduction in SD(O/E) compared to null models is shown in parentheses. ASPT (average
score per taxon of BMWP families) (National Water Council, 1981).
*The null model, the predicted reference condition index value for a site, was calculated as the
average observed value of the index for all reference sites (Van Sickle *et al.* 2005).

was relatively high (c. 57.3% prediction error). Taking this one step further,
Davy-Bowker *et al.* (2006) compared the efficacy of typology- (system-A) and
model-based (RIVPACS-type models) approaches for establishing reference
conditions. Comparisons were made using predictive models calibrated for
Great Britain (RIVPACS, e.g. Clarke *et al.* 2003), Sweden (SWEPAC$_{SRI}$, R.K.
Johnson, unpublished) and the Czech Republic (PERLA, Kokes *et al.* 2006).
The effectiveness of the system-A typology was compared directly with the
RIVPACS-type model predictions of expected index values. The null model,
the predicted reference condition index value for a site, was calculated as the
average observed value of the index for all reference sites (Van Sickle *et al.*
2005), and prediction accuracy was measured as the standard deviation (SD) of
the ratios of the observed (O) to expected (E) values of each biotic index for
the reference sites. Davy-Bowker *et al.* (2006) found that typology-based
approaches were inferior to model predictions. For all four indices studied
(only the results for two indices are shown here) and for all seasons and season
combinations (only results from autumn shown), SD(O/E) ratios were
consistently highest for the null models, indicating relatively high uncertainty,
and lowest for the RIVPACS-type models (Table 9.1).

On average, RIVPACS-type models resulted in an 11% decrease in uncertainty
compared with system-A. Although the environmental variables used by the three
RIVPACS-type predictive models differed, several variables, such as latitude,
longitude, altitude and substratum type, were used in either two or all three of
the models. In all three country models, the variables explained a larger proportion
of the variation in invertebrate communities than the WFD system-A variables.
The authors argued that predictive models were more effective due to the use of
continuous rather than categorical predictor variables and because the models

were not constrained by the use of a limited number of *a priori* selected variables. Realization that not only environmental factors but also biological communities are best described as gradients and not discrete entities has also been proposed by Hawkins and Vinson (2000). These authors argued that since assemblages generally vary continuously along environmental gradients, the classification methods that seek to place sites into discrete categories are fundamentally limited compared with approaches that recognize biological continua.

In conclusion, there are now indications that classifications based solely on large-scale (e.g. landscape-level) predictors are not sufficient to capture the fine-scale variability of aquatic communities. In other words, fixed typologies, such as system-A classification, may provide a useful framework for setting ecological targets, but need to be augmented by the use of more site-specific predictors such as in-stream water chemistry and substratum. A predictive approach using continuous environmental and taxonomic data is a robust method for establishing reference conditions and one that should be considered. Lastly, regardless of the method used to determine the reference condition, assessments of ecological quality are of limited value without knowledge of their precision and confidence and these need to be quantified.

Historical information and palaeo-records

For some aquatic system types, it is now difficult to find good examples of reference sites and this is especially true of European shallow lowland lakes, many of which have been long disturbed (e.g. Bradshaw & Rasmussen 2004; Leira *et al.* 2006) and particularly during the 20th century (e.g. Bennion *et al.* 2004). In such cases, it is necessary to use historical information to determine the reference conditions. Unfortunately, long-term data sets are rare for most ecosystems, and where they do exist, monitoring programmes tend to have begun after disturbance has occurred. However, for lakes, palaeoecological techniques can be employed, whereby the remains of aquatic organisms preserved in sediment cores are examined to provide historical information and hence to define reference conditions (Bennion & Battarbee 2007).

In a few rare instances, where historical data sets are of sufficient length, contemporary time series and palaeolimnological methods can be used in combination to describe the past state with increased precision. One such example is provided by Groby Pool, a small (0.12 km^2) shallow (mean depth < 1.1 m) lake in England that has experienced nutrient enrichment over at least the last two centuries (Sayer *et al.* 1999; Davidson *et al.* 2005). This lake has attracted numerous botanists since the mid-1700s, and an extensive botanical record has been collated from several sources including four county floras (1850, 1886, 1933 and 1988), UK museum herbaria, journal articles, notebook entries, unpublished manuscripts/sketches, academic theses and various reports of conservation bodies and local interest groups. A second example is Loch Leven, a key study site in Euro-limpacs, which is a significantly larger (13.3 km^2) but relatively shallow (mean depth 3.9 m) lowland lake in Scotland. The site underwent early nutrient enrichment in the mid-1800s associated with the lowering of the lake, followed by a second eutrophication phase during the mid-1900s when phosphorus-rich effluent from a large woollen mill discharged to the loch (Carvalho & Kirika 2003). The lake has an extensive historical macrophyte record available both from literature sources (e.g. Hooker 1821; Balfour & Sadler 1863; West 1910; Spence 1964; Jupp & Spence 1977a, b) and contemporary monitoring data (e.g. Robson 1986, 1990; Murphy & Milligan 1993; Griffin & Milligan 1999; Carvalho *et al*. 2004). Plant macrofossils have been analysed in sediment cores from Groby Pool (Davidson *et al*. 2005) and Loch Leven (Salgado 2006; Salgado *et al*. 2009) and the records compared with the old plant survey data. Hence, the data sets for these two lakes provide a rare opportunity for a comparative study of the botanical record with the aquatic plant history as revealed by the sediment record.

At both sites, there was good agreement in the timing and nature of changes in the selected aquatic taxa between the two data sources, with both data sets suggesting a clear succession in the aquatic flora over approximately the last 150 years (Figs. 9.2 and 9.3). However, both methods are affected by inherent bias. Old plant records are generally patchy and in the Groby Pool dataset, a gap in the historical record between 1747 and 1835 gives a false impression of species absence (Fig. 9.2), whilst at Loch Leven there is a paucity of survey data for the period 1910–66 (Fig. 9.3). Furthermore, the botanical surveys are likely to be skewed towards rarities which the sediment record is less likely to capture (e.g. Zhao *et al*. 2006). Macrofossils generally underestimate past species richness, with approximately 40% of the historically recorded aquatic taxa in Groby Pool represented by macro-remains, and pondweeds being particularly poorly represented (Davidson *et al*. 2005). Nevertheless, when comparing the macrofossil remains against the historical species described for the eastern side of St Serf Island, the coring location in Loch Leven, the fossil record captured 79% of the historically recorded species. Conversely, several taxa were absent from the old plant surveys but were present in the macrofossil data including, at Groby Pool, *Utricularia vulgaris* agg., *Myriophyllum alterniflorum* (Fig. 9.2) and Characeae (Fig. 9.4) and, at Loch Leven, *Isoetes lacustris*, *Lobelia dortmanna* and *Elatine hexandra* (Fig. 9.3).

The Groby Pool example also highlights the complementarity of the pollen record which adds information on species that leave few macro-remains such as *Littorella uniflora* and can provide more accurate information on the occurrence of *Potamogeton* spp. than seeds and leaves (Fig. 9.2). Given that historical records are lacking for most sites, these two case studies illustrate that macrofossils can be used on their own to provide a reliable record of the dominant components of the plant community. Additionally, these studies demonstrate the value of using a combination of methods (here historical plant records and macrofossils) to describe the ecological reference condition (albeit not the reference condition *sensu* the European Water Framework Directive definition of 'no, or only very minor' evidence of distortion (European Commission 2000), and indeed the use of multiple groups of organisms to establish reference conditions is gaining in popularity.

Multiple (palaeo) indicators

The sediment record contains the remains of a range of biotic components which can be examined individually or in combination (multi-proxy study) to assess changes in biological structure. Although much early work focused on

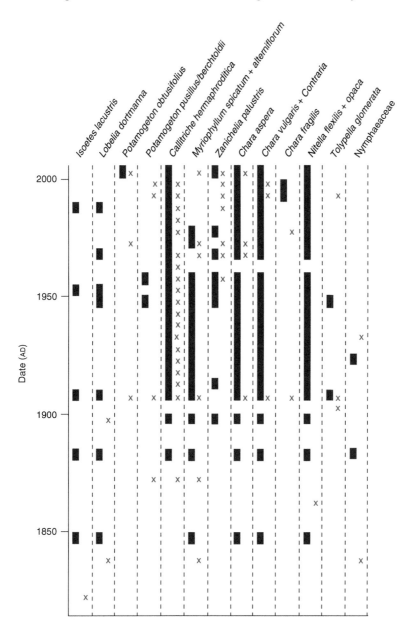

Figure 9.2 Comparison of the historical, macrofossil and pollen records for several taxa selected on the basis of their presence as sedimentary remains in a core from Groby Pool. (From Davidson *et al.* 2005.)

the response of single groups of organisms, recently the multi-indicator approach has increased as the need for a more complete understanding of the ecological history of aquatic systems has grown (e.g. Birks & Birks 2006). The use of multiple organism groups is also a particularly powerful tool for determining

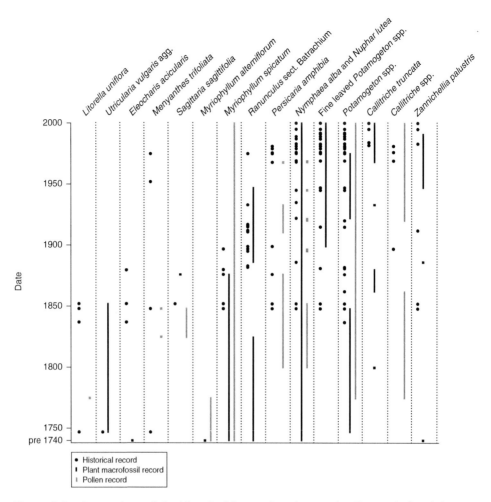

Figure 9.3 Comparison of the historical (crosses) and macrofossil records (bars) for several taxa selected on the basis of their presence as sedimentary remains in a core from Loch Leven (Salgado *et al.* 2009).

site-specific reference conditions (e.g. Guilizzoni *et al.* 2006; Taylor *et al.* 2006; Bennion & Battarbee 2007).

The approach is illustrated in Fig. 9.4 by a comparison of the remains of a range of biota (diatoms, plant macrofossils and zooplankton ephippia) in 20 year increments from the surface (~AD 1980–2000) and reference samples (~AD 1700–20) of dated sediment cores from Groby Pool (see above). The ecological reference condition is one of benthic diatom dominance (*Fragilaria* spp.), a relatively diverse submerged aquatic plant community (e.g. Characeae including *Nitella* and *Chara*, *Myriophyllum* spp. (both *spicatum* and *alterniflorum*), *Utricularia vulgaris* agg.) and plant-associated Chydoridae. In contrast, the surface sample is dominated by planktonic forms of diatoms (e.g. *Stephanodiscus parvus*) and zooplankton (*Bosmina* and *Daphnia* spp.) and elodeid macrophytes

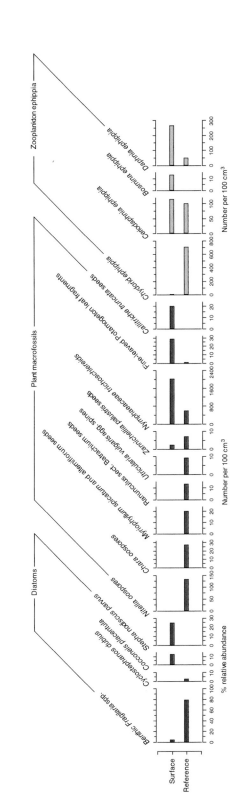

Figure 9.4 Summary diagram of multi-proxy palaeoecological analysis of 20 year time slices from the surface (~AD 1980–2000) and reference (~AD 1700–20) samples from Groby Pool, England. Note the variable scaling on the x-axis. (Reproduced from Bennion & Battarbee 2007 with permission from Springer.)

associated with higher nutrient concentrations (e.g. fine-leaved *Potamogeton* spp. including *P. pectinatus* L. and *P. pusillus* L., *Callitriche truncata* Guss. and Nymphaeaceae).

This approach is especially valuable in shallow, lowland lakes as there are few undisturbed examples of this lake type remaining in the current population of European water bodies and, therefore, spatial approaches are often inappropriate. The direct ecological information recorded by the sediments provides a sounder basis on which to establish reference conditions and formulate management decisions such as the definition of restoration targets. In the future it would be interesting to quantify the uncertainties associated with different sediments – macrofossil approaches to better understand how variability in establishing the reference condition affects inferences of anthropogenic-induced change.

Reference conditions and uncertainty

Although palaeolimnological records offer an excellent method for establishing reference conditions for lakes, as discussed above, the methods used in reconstructing past chemical and biological conditions have inherent uncertainty – as exemplified above in the studies comparing discrepancies between historical plant records and use of macrofossils. Although there is now a growing interest in using multiple methods, both direct (e.g. multiple groups of organisms) and indirect (e.g. different forms of inference), few studies have compared and contrasted the use of different methods to better understand their intrinsic uncertainties. One notable exception is a recent study by Battarbee *et al.* (2005), which assessed the relative accuracy of different approaches in reconstructing the past pH of a Scottish loch.

Battarbee *et al.* (2005) used contemporary monitoring data (pH and diatoms from 1991 to the present day) from the Round Loch of Glenhead, an acidified Scottish loch (cf. Flower & Battarbee 1983), to evaluate the performance of three pH inference models. Each of the pH inference models, based on different calibration training sets, was used to reconstruct pH from the diatom assemblages of three cores taken in different years, resulting in nine reconstructions of pH (Fig. 9.5). Taking the year AD 1800 as the reference date, the pH values for this date, derived using the transfer functions, were compared with the pH inferred using a diatom–cladocera modern analogue technique (Simpson *et al.* 2005) and with hindcast values using the MAGIC model (Model of Acidification of Groundwaters in Catchments, Cosby *et al.* 1985).

Comparison of the different methods showed that the inferred reference pH for the Scottish loch in AD 1800 varied by less than an order of magnitude (between 5.5 and 6.1). Relatively good agreement, with respect to both between-core and between-training set differences, was found between the diatom-pH transfer functions. Diatom-pH transfer functions indicated a reference pH of between 5.5 and 5.7, and the weighted average pH using the modern analogue method gave a similar value (5.8). By contrast, the MAGIC-model hindcast estimate for 1850 was somewhat higher (6.1). Although there is no way to ascertain which estimate is the most accurate, the discrepancy between the MAGIC hindcast and the diatom-based values is the most striking. According to Battarbee *et al.* (2005), one explanation is that the MAGIC model does not allow for DOC concentrations

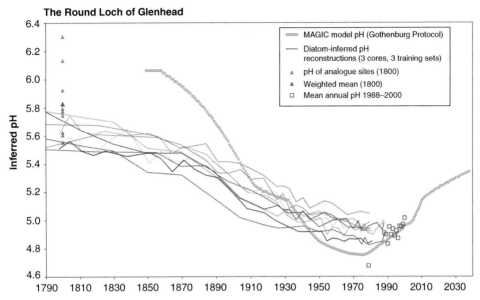

Figure 9.5 Comparison of pH reconstruction outputs and annual measured pH. Chronology of diatom inferred pH according to SWAP (The Surface Water Acidification Programme), UK and EDDI models (fine lines) for 210Pb dated samples from three sediment cores (RLGH 81, RLGH 3 and K05). The RMSEP of the SWAP, UK and EDDI training sets are 0.38, 0.31 and 0.25 pH units respectively. Modern annual pH of nine lakes providing the strongest biological analogues for a pre-acidification (c. 1800) sediment sample (open triangles) and the weighted average of these (filled triangle). MAGIC model pH reconstruction (open circles) and mean annual average pH for the period 1988e2000 and the year 1979 (open squares). (From Battarbee *et al.* 2005.)

to have been higher in the past when sulphur deposition was lower. Although not enough is known about the relationship between sulphur deposition and soil water DOC concentration for DOC to be incorporated into the MAGIC model, evidence is growing that the relationship is important (e.g. Monteith *et al.* 2007).

With regard to pH reconstruction using sediment records, uncertainty could still be reduced by increasing the size and representativeness of the calibration data-sets and by continued verification against observations from monitoring programmes. Overall, the finding that different methods to establish a reference condition were congruent, and, importantly, that model predictions agreed with contemporary monitoring data are encouraging. This study clearly demonstrates the usefulness of hindcasting methods for establishing the reference condition of lakes where no present-day spatial analogues exist. However, the modelling and contemporary time-series also indicate substantial variability on decadal and yearly time scales, trends that are clearly the result of human-generated changes in surface water quality. More knowledge is needed concerning how climate change, with both relatively short- (interannual) and long-term (shifts in baseline conditions) variability, affects our ability to detect degradation and recovery.

The latter is particularly important as many European countries move towards restoring the ecological quality of inland waters as mandated by the Water Framework Directive.

Detecting recovery with changing baseline conditions

A fundamental part of ecological assessment in general, and restoration in particular, is to isolate change induced by degradation/mitigation measures from natural spatial and temporal variability. Statistical power to detect change can be increased by reducing natural variability as much as possible (e.g. by stratifying sampling effort) and increasing the signal or effect size by selecting variables with strong responses (e.g. Johnson 1998; Sandin & Johnson 2000b). However, adequate controls and replication are equally crucial for determining the effectiveness of restoration.

Although at first glance, detecting human-generated disturbance, and assessing the efficacy of restoration endeavours may seem to be two sides of the same coin, there is one important difference. In determining ecological degradation, the null hypothesis is that there is no difference between the putative perturbed site and target, whereas for restoration, the criterion for success is to find no difference between the putative restored and the target state (e.g. Downes *et al.* 2002). In essence, this means that we have to test our hypothesis and not the null hypothesis. This quandary can be circumvented if we hypothesize that during the extent of the study, the putative restored site will differ from the initial or starting condition. Ideally, such a statistical design would include three types of sites: (i) restored sites; (ii) target or control sites; and (iii) sites similarly impaired as those restored but not restored. A number of factors, such as study design, the confounding effects of natural, spatial and temporal variability and a poor understanding of the mechanisms driving recovery affect our ability to determine the success of restoration.

Changing baselines confound interpretation of recovery from acidification

International agreements and actions to protect and restore natural resources threatened by acidification have resulted in marked reductions in the emission and deposition of acidifying compounds (Stoddard *et al.* 1999) and concomitant increases in surface-water pH (Stoddard *et al.* 1999; Lynch *et al.* 2000; Skjelkvåle *et al.* 2000, 2003). However, despite putative recovery of surface water chemistry, records of biological recovery are scarce and results are equivocal (Skjelkvåle *et al.* 2000; Alewell *et al.* 2001; Stendera & Johnson 2008). Consequently, acidification is still considered as one of the foremost problems affecting the biodiversity of inland surface waters in Northern Europe (Johnson *et al.* 2003; see also Chapter 7) and elsewhere (e.g. Kowalik *et al.* 2007; Burns *et al.* 2008).

Acidification often affects aquatic biota in a predictable way (Økland & Økland 1986), even though the processes and mechanisms governing degradation are not always clearly understood (Hildrew & Ormerod 1995; Strong & Robinson

2004). Likewise, the mechanisms important for biological recovery are not that well understood, and many factors such as habitat connectivity, dispersal abilities, food availability and species interactions may singly or in concert be important (see below), as may the influence of other drivers such as climate change or nutrient enrichment. In addition, detection of recovery or restoration success may also depend on the response variable selected, such as the choice of indicator type (chemical, biological), choice of habitat (stream, lake; pelagic, benthic) (Wright 2002; Johnson *et al*. 2006a, b; Stendera & Johnson 2008) and choice of metric (e.g. diversity, taxonomic composition) (Johnson & Hering 2009), as well as many site-specific factors.

Differences in site-specific characteristics, such as latitude, altitude, catchment characteristics, can result in lag responses that differ among sites, resulting in high levels of uncertainty and confounding interpretation of the efficacy of restoration. For instance, Wright (2002) postulated that lags in recovery from acidification might increase with trophic level and vary with water-retention times. To test the first assumption, Stendera and Johnson (2008) evaluated the recovery rates of different biological components and habitats over 16 years in 10 boreal lakes recovering from acidification. Expectations were that chemical recovery would be quicker than biological recovery and that biological responses would vary among organism groups and trophic levels. Recovery times were expected to be shorter for phytoplankton than for littoral invertebrates because of shorter generation times and higher recolonization and dispersal rates of phytoplankton. Littoral invertebrate assemblages were expected to respond more quickly than sublittoral or profundal assemblages. Finally, sublittoral and profundal assemblages, although not directly affected by surface-water chemistry, were expected to respond more to changes in food (bottom-up) and/or predation (top-down), resulting in relatively long lag times.

Stendera and Johnson (2008) showed that measuring biological recovery from acidification often requires the use of holistic and multiple assessment approaches. Several measures showed significant, positive trends over time that supported expectations of biological recovery (Table 9.2). For instance, taxonomic diversity of phytoplankton assemblages, but not taxonomic composition, showed early signs of recovery related to pH increase. Significant trends were, however, also noted for non-acidified reference lakes. Recovery trends of benthic invertebrate assemblages were more equivocal. Littoral invertebrate taxon richness and diversity increased in both acidified and reference lakes, whereas sublittoral and profundal invertebrate assemblages of both lake groups showed clear negative trends. The authors suggested that sublittoral and profundal assemblages might be influenced by factors other than lake acidity, such as coincident changes in habitat quality (e.g. ambient O_2 and temperature).

Interestingly, it was anticipated that communities in reference boreal lakes would be fluctuating around a long-term mean (Fig. 9.6), but they did not, implying the influence of some other driver. A gradual shift in baseline conditions brought about by the effects of global warming might be one explanation for the long-term trends noted in both the reference and acidified lakes. For example, a shift in spring temperatures caused by warmer winters in the early years of the study might have caused changes in the timing or duration of summer stratification.

Table 9.2   Mean slopes with respect to time (± 1 standard deviation) of selected indicators from four acidified and six reference lakes sampled from 1988 until 2003. (Taken from Stendera and Johnson 2008.)

Variable	pH	Taxon richness	Diversity
Acidified	0.0218 ± 0.009		
Phytoplankton		0.138 ± 0.042	0.043 ± 0.017
Benthic invertebrates			
Littoral		0.845 ± 0.307	0.488 ± 0.120
Sublittoral		0.038 ± 0.238	−0.055 ± 0.136
Profundal		−0.043 ± 0.107	−0.045 ± 0.084
Reference	0.006 ± 0.016		
Phytoplankton		0.072 ± 0.034	0.0367 ± 0.010
Benthic invertebrates			
Littoral		0.980 ± 0.64	0.422 ± 0.401
Sublittoral		−0.665 ± 0.496	−0.400 ± 0.307
Profundal		−0.082 ± 0.272	−0.055 ± 0.146

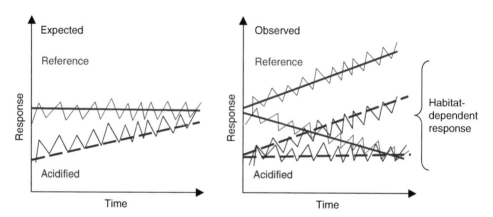

Figure 9.6   Conceptual diagram illustrating a comparison of expected and observed responses of invertebrates (expressed as taxon richness) in acidified and reference lakes in southern Sweden to a reduction in sulphur deposition.

Prolonged summer stratification periods are predicted to increase hypolimnetic anoxia (Magnuson *et al.* 1997), and if sublittoral assemblages are periodically situated under the summer thermocline, the lower O_2 concentrations or lower ambient temperatures might negatively affect growth and survival. However, since both sublittoral and profundal assemblages showed similar trends, a more plausible explanation is that the deeper benthic habitats might be negatively affected by increased concentrations of organic matter. In nearly all lakes, water colour and total organic carbon (TOC) increased during the study period, a trend that has been recorded elsewhere in the Nordic countries (Skjelkvåle *et al.* 2005) and in the United States (Stoddard *et al.* 2003) over the last decade (see also Chapter 7). Increased inputs of DOC might have resulted in a decrease in ambient

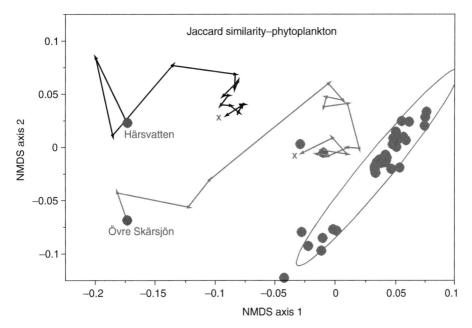

Figure 9.7 Non-metric multidimensional scaling (NMDS) of phytoplankton assemblages using Jaccard similarity of two acidified (Härsvatten and Övre Skärsjön) and four reference lakes sampled between 1992 and 2007. Reference lakes are shown as dark dots and with a 95% CI (ellipse). Between-year changes in acidified lakes are shown with grey arrows. The start of the time series is marked with a dot and the end with an X.

O_2 and the loss of O_2-sensitive taxa. This conjecture is also supported by the increase in populations of *Chaoborus flavicans* (Meigen), a taxon that is relatively mobile and can avoid hypoxia. The findings of this study are alarming, in that they indicate that reference or baseline conditions may be steadily changing and that such changes can confound the interpretation of recovery success.

Clearly, long-term studies of both natural and disturbed systems are of interest for determining the effects of large-scale regional drivers (e.g. climate) on recovery pathways and the underlying mechanisms. For example, phytoplankton assemblages in acidified lakes have been shown to respond rapidly (high between-year change) and positively (progress towards the reference condition) to decreased acidity (R.K. Johnson & D.G. Angeler, in press) (Fig. 9.7). That phytoplankton assemblages are responding quickly to declines in lake acidity is consistent with other studies (Wright 2002; Findlay 2003; Stendera & Johnson 2008), and supports the conjecture of Wright (2002) who argued that phytoplankton assemblages often showed indications of recovery within 1–4 years after chemical improvement. However, the environmental factors responsible for the recovery of phytoplankton assemblages are less clear. As discussed above, changes in phytoplankton taxon richness were related not only to increased pH, but also positively correlated with other environmental variables such as increased water colour (a proxy for dissolved organic carbon, DOC) and nutrient concentrations.

Although increased pH should allow for the colonization of acid-sensitive species, the simultaneous increases in nutrient and DOC concentrations could also result in shifts in species composition, such as an increase of mixotrophic algae (Wetzel 1995) that is not solely related to declines in acidity.

It is important to understand how climate change may affect recovery pathways and the underlying mechanisms. The North Atlantic Oscillation (NAO) is an important regional-level driver of climate in southern Sweden (Weyhenmeyer 2004), with positive NAO values signifying higher precipitation and thus more variable hydraulic and chemical conditions. Stendera and Johnson (2008) found that interannual shifts in phytoplankton assemblages were positively correlated with regional patterns that increased water colour, thereby lending further support to the importance of large-scale, climatic drivers for changes in biodiversity. Several studies have shown how interannual variability in NAO affects lake plankton assemblages by affecting temperature, ice cover and timing of spring algal blooms (Straile 2000; Weyhenmeyer *et al.* 1999). In the boreal lakes studied by Stendera and Johnson (2008), the effect of NAO_{winter} on phytoplankton assemblages may be manifested through increased DOC (water colour) and nutrients, resulting in shifts to more mixotrophic taxa, although the increase in DOC may also be related to a reduction in acid deposition (Monteith *et al.* 2007; see also Chapter 7).

In combination, these studies show that long-term data sets are needed to better understand not only the drivers of regional changes in chemistry and biology of aquatic systems, but also that knowledge of the direction and magnitude of changing baseline conditions is needed to make informed decisions regarding the efficacy of restoration efforts. In addition, these and other studies (e.g. Johnson *et al.* 2007) support the contention that the scale of perturbation (e.g. local- vs region-scale drivers of change) and the scale of restoration (e.g. individual habitats within a stream or streams within a catchment or landscape) need to be considered to design robust and cost-effective restoration programmes.

Global change and restoration

The presence or absence of organisms may depend on rare or large-scale dispersal and colonization, while local abundance is more a function of continual, local biotic interactions and habitat heterogeneity (e.g. Ricklefs 1987). Understanding the processes and mechanisms that integrate patterns and scale is a growing theme in restoration ecology (e.g. Bruinderink *et al.* 2003; Hanski 2005). For aquatic ecosystems, position in the landscape along with catchment geology and land use have been known for nearly a century to be good predictors of the chemistry and biology of lakes and streams (e.g. Thienemann 1925; Naumann 1932; Vannote *et al.* 1980). Many studies have shown the importance of local factors such as riparian vegetation, water chemistry and substratum as predictors of lake and stream invertebrate assemblages (e.g. Ormerod *et al.* 1993; Johnson & Goedkoop 2002; Stendera & Johnson 2005), while other studies have stressed the importance of regional or landscape factors to be more important predictors of communities (e.g. Allan *et al.* 1997; Heino 2002; Townsend *et al.* 2003).

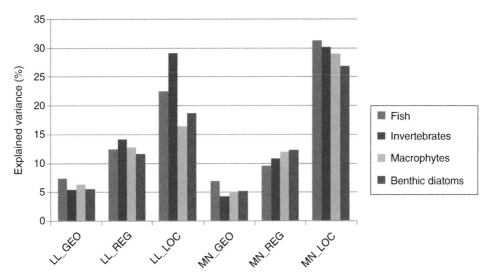

Figure 9.8 Among-stream variance (%) in fish (blue bars), invertebrate (red), macrophyte (green) and benthic diatom (purple) assemblages explained by geographic (GEO), regional (REG) and local (LOC) factors. LL, lowland; MN, mountain streams. (Results are taken from Johnson *et al.* 2007.)

Ecological relationships between communities and spatial scale

Recognition of how aquatic systems are linked with their surrounding landscapes and how linkages affect resistance and resilience to change is important to underpin management decisions. Johnson *et al.* (2007), analysing relations between benthic diatoms and fish and different aspects of spatial scale, postulated that organism-response would vary according to life histories. They hypothesized that fish, being long-lived, mobile organisms, would be more related to large-scale, regional variability, while small organisms, like benthic diatoms and invertebrates, would be more related to local factors. They found that local scale variables such as water chemistry and substratum explained, on average, more than four times the amount of variance than was accounted for by geographic position (latitude and longitude) and more than twice the variance that was accounted for by regional-scale variables such as ecoregion and catchment land use/cover (Fig. 9.8). However, contrary to expectations, no support was found for the conjecture that large organisms (fish) are responding more to large-scale (regional) factors, while small organisms (e.g. benthic diatoms) are responding more to small scale (habitat) factors. The finding that local scale variables were important predictors of stream assemblages agrees with several previous studies (e.g. Hawkins *et al.* 2000; Paavola *et al.* 2006; Hering *et al.* 2006), and has direct implications for designing programmes for stream restoration (see the use of woody debris in stream restoration below).

Furthermore, even though in-stream variables were better predictors of taxonomic composition, this does not imply that large-scale factors such as catchment land use should be ignored in the planning and implementation of conservation and management programmes. Indeed, since many disturbances are the result of socio-economic processes, conservation and restoration strategies will be more effective if treated as part of the landscape development rather than as isolated entities (Holling & Meffe 1996; Meir *et al.* 2004). One example of this involves the interaction of catchment forest and stream structure as illustrated in the case study of the next section.

Use of large woody debris in stream restoration

The term 'stream (or river) restoration' is used for a wide variety of project objectives, ranging from conventional bio-engineering to the restoration of natural processes aiming to generate natural in-stream structures and a natural channel pattern (Kondolf 1996). In the United States, stream restoration is defined as 'the return of an ecosystem to a close approximation of its condition prior to disturbance' (National Research Council USA 1992). However, in many parts of the world, irreversible changes of the natural setting have occurred and hence the re-creation of a previous historical state is impossible (Kauffman *et al.* 1997; Brown 2002). This is especially true for a densely populated cultural landscape in Europe and other parts of the Old World, where streams have been altered by human activity since the Mesolithic period. Therefore, the equilibrium state of a stream, which potentially would develop under the present natural setting without further human intrusion, is frequently used in Central Europe as a reference and target condition in stream restoration projects (Deutscher Verband für Wasserwirtschaft und Kulturbau 1996). The present natural setting may differ from the historical natural setting as a result of irreversible changes of the natural setting, for example as a result of sediment deposition in floodplains since the Late Holocene caused by anthropogenic deforestation and increased erosion (late Holocene alluvium), mineralization of organic soils, large-scale excavation (surface mining) and landfills, subsidence caused by mining and recent climate change.

The restoration of hydromorphologically degraded rivers has become a widely accepted social objective and scientific interest in stream restoration has been steadily increasing over the last two decades (Shields *et al.* 2003; Bernhardt *et al.* 2005). In densely populated areas such as Central Europe, a large proportion of rivers are heavily degraded; thus, there is a strong demand for simple and cost-effective restoration measures. Large wood (defined as logs with a diameter > 0.1 m and a length > 1 m according to Gregory *et al.* 2003) is an important component of stream ecosystems in temperate forested ecoregions. It influences stream hydrology, hydraulics, sediment budget, morphology and biota (e.g. Gregory *et al.* 2003). Considering these beneficial effects, even in a densely populated region like Central Europe, up to one third of the streams could potentially be improved by restoration with wood (Kail & Hering 2005). After evaluating the use of large wood in 50 stream restoration projects, Kail *et al.* (2007) concluded that (i) potential effects of wood placement must be evaluated within a catchment- and reach-scale context; (ii) wood measures are most successful if they mimic natural wood; (iii) effects of wood structures on stream morphology are strongly dependent on conditions such as stream size and hydrology; (iv) wood placement has positive effects on several fish species; and (v) most projects revealed a rapid improvement of their hydromorphological status.

The high natural variability of large wood standing stock and ongoing climate change makes it difficult to define reference and target conditions for streams. Therefore, the use of 'passive restoration' methods (restoring the process of wood recruitment on larger scales) instead of 'active restoration' (placement of wood structures on a reach scale) is desirable for a number of reasons. First, local, active restoration measures are at risk of neglecting the more general processes at the catchment scale causing degradation, and often treat symptoms rather than causes of stream degradation, thus being prone to failure in the long run (e.g. Kauffman *et al.* 1997). Along these lines, it has been proposed that restoration projects are more likely to be successful if they are undertaken in the context of entire catchments (Wohl *et al.* 2005). The findings of Kail & Hering (2005) lend further support to the conjecture that restoration activities using large wood must consider other factors such as catchment land use. Secondly, active restoration measures may create conditions that do not correspond to the potential natural state. It is doubtful that the target condition for stream restoration (the potential natural state) can be accurately described for stream reaches if controls like discharge and sediment supply have been altered by human activity and, therefore, historic conditions cannot be used as reference or target conditions.

Moreover, climate change will probably have a strong impact on stream hydraulics, morphology, biota and hence the potential natural state. In contrast, the establishment of natural riparian vegetation and natural wood input will initiate natural channel dynamics, ultimately resulting in natural channel dimensions which correspond to the potential natural state. Thirdly, even in a densely populated region like Central Europe the process of natural wood input can be restored. A conservative estimate of Kail and Hering (2005) showed that about 7% of the streams in Central Europe can potentially be restored by large wood recruitment from native or non-native riparian forests. In many parts of Central Europe, land-use pressure may decrease in the future, due to decreasing population size, alternative agricultural techniques and increasing flood risks following climate change. In the long term, this may increase chances for the establishment of riparian forests and passive restoration, particularly in remote parts of mountainous areas.

Ecosystem connectivity and species' dispersal ability

For the restoration of freshwater ecosystems to be considered successful, the reference conditions need to be attained or at least closely approximated as discussed above. According to the European Water Framework Directive (European Commission 2000) this involves more than just the restoration of abiotic reference conditions such as water nutrient status and pH. It also involves the restoration of reference communities, i.e. the assemblages of the fauna and flora typical of unperturbed conditions. Whereas, as a first step, abiotic reference conditions can be restored by applying external measures (such as reduced nutrient loading or addition of large woody debris), the restoration of reference communities is in most cases expected to follow a natural recovery. This often requires that species that have been absent from the restored site re-colonize from outside source populations. However, even when abiotic conditions have been successfully restored, the re-colonization of reference species may be disappointing if dispersal is a limiting factor (e.g. Donath *et al.* 2003; Jähnig 2007).

Dispersal of organisms to restored sites is dependent on (i) the connection of the site to source populations or the 'connectivity' of the site; and (ii) the dispersal abilities of the organisms themselves. The connectivity of a site is dependent not only on the distance between the site and potential source populations, but also on the pathways through which organisms may reach the site. Sites located at different positions in the landscape may be connected to each other, dependent on position and type of connections (e.g. Soons *et al.* 2005; Soons 2006). Generally, organisms with source populations nearby are the first to colonize restored sites (e.g. Donath *et al.* 2003), although the dispersal ability of freshwater organisms varies widely. A main categorization is often made between organisms that can actively move substantial distances, such as macroinvertebrates with an aerial life-stage and fish, and those that are predominantly passively dispersed over larger distances, such as phytoplankton, macrophytes and most invertebrate fauna lacking an aerial life-stage. Dispersal of active dispersed species is dependent on species movement ecology and landscape structure and spatial configuration. In addition, the dispersal of individual species can be enhanced by the construction of corridors or the removal of barriers (e.g. fish passages in rivers, Schilt 2007). Passively dispersed species in freshwater ecosystems can be transported by three main dispersal vectors (namely, water, wind and water birds – each with their own dispersal potential) either in their adult life stages or as dispersal propagules. Dispersal by wind is generally confined to relatively short- to landscape-scale distances (e.g. Nathan *et al.* 2002; Soons *et al.* 2004a, b), whereas dispersal by water and water birds can occur over much longer distances, from landscape to regional and even continental scales (e.g. Clausen *et al.* 2002; Figuerola & Green 2002; Charalambidou *et al.* 2003; Boedeltje *et al.* 2004; Soons *et al.* 2008). While dispersal distances are highly species-specific, in general it can be expected that freshwater species that produce many small and drying-tolerant dispersal propagules have the greatest dispersal abilities.

When restoring sites, connections to potential source populations and the dispersal abilities of reference organisms need consideration. The often much delayed re-colonization of restored sites, because of spatial isolation and/or low dispersal abilities of colonizers, may result in failure to reach the species composition characteristic of reference conditions. Therefore, it has been argued that restoration should occur in a landscape context, where connections in the landscape are taken into account (e.g. Verhoeven *et al.* 2008 and see below). A useful approach for this is the operation landscape unit (OLU) approach (described below), which considers how disrupted connections in the landscape may impede restoration success, and provides a tool to include crucial landscape connections in a restoration plan. This is especially important in fragmented landscapes where natural connections between freshwater ecosystems have been disrupted by diversion and canalization of water flows and extraction of water for human purposes (e.g. Gordon *et al.* 2008). Reference communities derived from less fragmented or natural landscapes may exist due to connections between populations throughout the landscape and have natural spatial species dynamics (such as metapopulation dynamics or other spatial dynamics; Hanski 1999; Freckleton & Watkinson 2002) that cannot be attained in restored sites in fragmented or otherwise heavily modified landscapes (Hanski 1994). This is an issue that requires careful consideration in the planning of restoration projects, and requires further investigation for general guidelines.

In addition, it is important to consider connections and dispersal abilities in the context of climate change. Climate change may not only affect reference conditions (as discussed above) but may also alter connections between sites and dispersal abilities of species. Important mechanisms through which climate change is anticipated to alter the connectivity of freshwater ecosystems are (i) changed hydrodynamics of streams and rivers due to more extreme rain events (see Chapter 4) which will alter the connections between sites by surface water flows. For example, changes in river flooding events will change connections between sites via seed dispersal by water, because seed deposition during flooding events will be affected (Goodson *et al.* 2003); (ii) faster drying of ponds and shallow lakes, due to warmer and drier summers, which will reduce 'stepping stone' connections between wetlands. For example, the disappearance of ponds increases the isolation of pond habitats and reduces zooplankton dispersal between ponds (Allen 2007). Wind-dispersed organisms will be dispersed over longer distances when storms increase in intensity and frequency, but this effect is predicted to be relatively small (Soons *et al.* 2004b). In summary, measures of the effects of climate change on the connectivity among freshwater ecosystems are needed to predict future impacts on the success of restoration measures. However, general guidelines are difficult to provide because effects are likely to be highly specific per site and landscape, and will require landscape-scale predictions of climate change and detailed knowledge of the spatial configuration and connections in the landscape. A landscape-scale approach to restoration that addresses these spatial issues, such as the OLU approach, in combination with climate change predictions at the same working scale, will result in more successful, realistic and 'climate change–proof' restoration activities.

Operational landscape units: an approach for enhancing success of wetland restoration

Nature conservation and restoration has focused traditionally on protecting individual sites. However, in parts of the world where the natural landscape has been severely altered for agricultural or urban use, individual patches are often too small and isolated to ensure effective nature protection. Dispersal of organisms and the balance between local extinction and recolonization are impeded, resulting in populations, even as metapopulations, being less stable and more subject to species loss. Another effect of landscape fragmentation is that natural linkages in the landscape, such as water flows, are severed due to modifications in geomorphology and hydrology.

Lateral connections in landscapes maintain biodiversity and ecological functioning at a much higher level than is possible in separate ecosystems without these spatial interactions. Lateral connections include corridors between landscape patches, such as riparian forests connecting extensive floodplain forests, but also connections by water flow or wind. Hence, conservation efforts should always look beyond the boundaries of the ecosystem to protect its connections to other systems in the landscape. The need for this approach to nature conservation arises from the increasing degree of landscape fragmentation (e.g. Soons *et al.* 2005). Indeed, fragmentation into small patches has dramatic consequences for biodiversity and for environmental quality in catchments that originally had a high degree of landscape connectivity. A small patch size leads to high extinction and the high degree of isolation to low colonization, resulting in a severe loss of species richness of patches and of the landscape as a whole (Hanski 2005).

A second consequence of landscape fragmentation (e.g. from agricultural intensification) is a drastic change in hydrology. Drainage of agricultural land has resulted in lower groundwater tables, loss of groundwater discharge in (semi-) natural landscape patches, straightening of low-order streams, dampening of stream water level fluctuations, loss of floodplain habitat, loss of meandering as a natural stream habitat process and a deterioration of stream water quality (Brierley & Stankoviansky 2002). These changes have led to drastic changes in diversity of habitats across the landscape, even in protected reserves with appropriate internal management. It is evident that these landscape-scale modifications have resulted in additional losses of biodiversity, which have probably been just as severe as those caused by fragmentation *per se.*

Identification of OLUs, which are defined as 'combinations of landscape patches with their biotic and hydrogeological connections', has recently been proposed as a tool to facilitate wetland restoration in catchments with a high degree of fragmentation and strongly altered hydrology (Verhoeven *et al.* 2008). The combined consideration of biotic (i.e. dispersal, transports of organisms) and hydrological (flooding events, groundwater flowpath connections across the landscapes) factors is novel in this context. To synthesize the OLU for a specific restoration project, the following three-step approach can be used. In the first step, focus is on defining the restoration targets. Which plant or animal species or community needs to be restored? Alternatively, which ecosystem function is to be restored (e.g. water quality enhancement or floodwater detention)? What are the reference conditions for the restoration that can be used to set the restoration target? The second step identifies the features of hydrology and dispersal that are crucial for the species, community or ecosystem to be restored. It may be that such spatial mechanisms are not relevant for the species or system to be restored, in which case conservation or restoration of individual patches is sufficient and the OLU approach may not be appropriate. But in most cases, linkages play an important role. The third step is to define the extent of the OLU by identifying the landscape components which are necessary as building blocks for restoration of (i) hydrological functioning and water regime and (ii) dispersal of plant and/or animal species.

The OLU approach will often involve the identification of 'source areas' (i.e. intact, species-rich nature reserves which need to be conserved), as well as 'receptor areas' which would be suitable for restoration and encompass the restoration target area and connecting pathways such as streams or other hydrological flow paths. To identify the receptor areas and connecting pathways, historical maps and other records of present and past hydrological functioning are useful. A necessary further step for the identification of the OLU is the creation of a map that represents the combination of landscape elements that are required for regional survival of a species or a self-sustaining ecosystem function or for creating target conditions for plant communities or ecosystem functions.

The usefulness of the OLU concept in land management is best exemplified through a case study. Restoration of stream valleys (e.g. Antheunisse *et al.* 2006) provides examples of this approach in the Ottershagen area of the Dinkel catchment in N.E. Twente, the Netherlands. A plan to restore the Ottershagen area arose from a meeting attended by representatives from the water authorities, nature protection agencies and farmers, who have contributed to the discussion of options for land and water use. Restoration of the Ottershagen floodplain area specifically requires an analysis of the hydrology and the current as well as historic land use of the Hollandsegraven sub-catchment, which is part of the Dinkel catchment (Fig. 9.9a).

The Hollandsegraven sub-catchment is currently mostly in intensive agricultural use for dairy farming. In the southern part there is a relatively small wetland reserve, the Agelerbroek, with wet species-rich grasslands and brook forest with many waterfowl and amphibian species. The Ottershagen area in the north, now an extensive flat area with intensively fertilized and grazed meadows, was once a large, species-rich wetland which was drained and reclaimed for agriculture in the 1950s. Flooding by the Hollandsegraven and Dinkel streams has only very exceptionally occurred in the Ottershagen or in the Agelerbroek since then because of restrictions posed by canalization and bank reconstructions.

The restoration targets are twofold: (i) the restoration of the large floodplain area Ottershagen to increase regional wetland biodiversity and (ii) the creation of a floodwater storage basin to cope with high river water discharges which are expected to increase in frequency due to climate change. Hence, floodwater detention measures need to be identified that would result in regular inundations of the Ottershagen area with water from the Hollandsegraven stream. At the same time, the goal is to create opportunities for vegetation succession in the current agricultural pastures towards floodplain plant communities such as *Caricion gracilis, Caricion elatae, Calthion, Phragmition* and *Alnion glutinosae.*

Figure 9.9b shows how the Operational Landscape Unit in the Hollandsegraven sub-catchment has been delineated. The boundaries were drawn according to a combination of the extent of the wetland areas indicated on the historical map (1900) and the extent of the areas that were flooded during a short, extremely wet period in 1998. The historical map shows that the Agelerbroek wetland area was more than twice as large in 1900 and that Otterhagen was another large floodplain wetland in the northern half of the Hollandsegraven catchment. The OLU now encompasses the Agelerbroek as a source area for plant and animal diaspores, an extensive wetland area immediately to the north and east of it, the Tilligterbeek, Hamburgerbeek and Hollandsegraven streams (actually one continuous major stream in the centre of the OLU, see Fig. 9.9) and the northern Ottershagen floodplain area. The delineated OLU shows that the area suitable for floodplain restoration is not only the former floodplain at Ottershagen, but also the former wetlands adjacent to the present Agelerbroek. Both former wetland areas have the potential to be restored to regularly flooded marshes if the hydrological connections between the Agelerbroek and the Tilligterbeek are optimized and the stream discharge characteristics changed through hydrological measures.

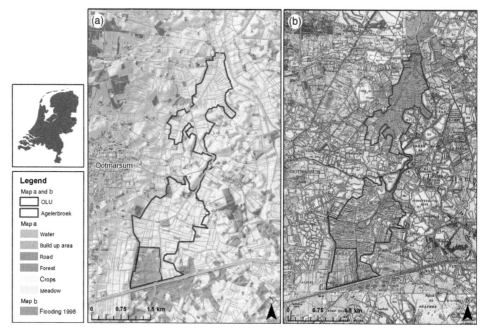

Figure 9.9 Topographic maps of the study area in N.E. Twente, the Netherlands, with the boundaries of the OLU. (a) Recent topographic map, 1 : 10,000 (source: Topografische Dienst Kadaster, Emmen 2003). The topography is indicated through light grey shading. Most of the area within the OLU is agricultural pasture, with some arable land and forest. The Agelerbroek is a wetland nature reserve with forest and meadows. The main stream system (Tilligterbeek–Hamburgerbeek–Hollandsegraven, actually one stream) connects the Agerlerbroek area with the Ottershagen area (northern section of the OLU). Both areas are low-lying plains and the connecting stream system flows through higher terrain. (b) The boundaries of the OLU depicted on a historical map from 1900, 1 : 25,000 (source: Topografische Dienst Kadaster, Emmen 1900). The purple areas are the extent of flooding after extremely high rainfall in 1998. The delineation of the OLU follows the original outlines of wetlands on the historical map and the extent of the 1998 flooding. (Reproduced with permission from *Journal of Applied Ecology* (2008) Vol. 45, pp. 1496–1503.)

The plant communities in the Agelerbroek are typical for stream valleys with regular flooding and are suitable as a source of species for the restoration areas.

The hydrological and spatial mechanisms to be restored are the natural flooding frequency of the Ottershagen area and the hydrological connection of the Agelerbroek reserve with the Hollandsegraven stream network. The Ottershagen area can be turned into a floodplain again by a different regulation of stream discharges in the Hollandsegraven sub-catchment, resulting in flooding events with highest frequency and duration in the winter season. The Agelerbroek nature reserve, which currently is only partly and rather indirectly connected to the Tilligterbeek-Hollandsegraven stream system, should be fully connected again, so that there will be a direct connection between the Agelerbroek and the Ottershagen areas during periods of flooding. There is an elevation gradient sloping downward from the Agelerbroek northwards, so that floodwater from that area would flow into the Ottershagen area. By allowing flooding also to take place in the Agelerbroek, diaspores could be carried northward in the flood-water. These restoration measures will create potential for development of floodplain areas typical for medium-sized (stream order 3–4) streams with their characteristic gradients in sediment deposition and create opportunities for water resource managers to discharge water in these areas to prevent flooding in economically vital areas. Hence this case study demonstrates how the OLU approach can increase the awareness of resource managers to the importance of spatial processes and connectivity in landscapes and, if properly applied, can lead to more long-term successful restoration projects.

Conclusions

Studies comparing different methods to establish reference conditions, such as spatial typology, predictive models, palaeo-reconstruction and historical and contemporary data are encouraging. More comprehensive studies are, however, needed to quantify better the uncertainties associated with different approaches. Somewhat disconcerting is the finding that spatial typologies (probably the most common approach used by European member states in implementing the WFD) are significantly poorer at partitioning biological variance than model-based approaches. Furthermore, there is now a growing awareness that the reference condition is not static or as previously presumed, oscillating around long-term means, but showing monotonic climate-driven trends. Hence approaches that explicitly include interannual variability in weather and/or long-term trends in climate, such as modelling approaches that include predictor variables like number of degree-days and hydrological variables should be given greater consideration. Along these lines, studies focussed on trying to disentangle the effects of short-term climatic events such as the influence of interannual NAO oscillations from long-term climate-driven trends would be of interest.

Predicting the effects of global change on freshwater ecosystems is complicated, and there is now an increased understanding that complexity increases many-fold as ecosystems already affected by anthropogenic stress now experience and respond to the effects of climate change. How ecosystems respond to the effects of multiple stressors is currently under debate. Nevertheless, it is clear that managers need to consider the importance of how multiple stressors may affect ecosystem structure and function of aquatic ecosystems if management endeavours are to be successful. Restoration measures such as the use of woody debris are considered a cost-effective approach to increase habitat complexity and improve hydromorphology in streams. However, more robust statistical designs, including multiple restored, unrestored and reference sites and several years of study both before and after intervention are needed to better understand the short- and long-term effects of this and other types of restoration.

Importantly, connections between restored sites and their surroundings, both in terms of dispersal of organisms and flows of water and solutes, deserve more consideration in future restoration projects. Furthermore, the scale of restoration plans needs to be placed in the context of the scale of the perturbation. The use of conceptual models, like OLU, that consider the connectivity between landscape patches, are useful for disseminating ideas and creating a platform for discussion. For example, these approaches can be excellent tools to aid managers in communicating with others in trying to simplify the complexity of interactions such as the importance of connectivity, between terrestrial and aquatic systems and among freshwater ecosystems, for resilience and resistance to change. They also provide a platform for testing and developing ideas to understand and mitigate future effects of global change on ecosystem biodiversity and function.

References

Alewell, C., Armbruster, M., Bittersohl, J., *et al.* (2001) Are there signs of acidification reversal in freshwaters of the low mountain ranges in Germany? *Hydrology and Earth System Sciences*, 5, 367–378.

Allan, J.D., Erickson, D.L. & Fay, J. (1997) The influence of catchment land use on stream integrity across multiple spatial scales. *Freshwater Biology*, 37, 149–161.

Allen, M.R. (2007) Measuring and modeling dispersal of adult zooplankton. *Oecologia*, 153, 135–143.

Antheunisse, A.M., Loeb, R., Lamers, L.P.M. & Verhoeven, J.T.A. (2006) Regional differences in nutrient limitations in floodplains of selected European rivers: Implications for rehabilitation of characteristic floodplain vegetation. *River Research and Applications*, 22, 1039–1055.

Bailey, R.C., Norris, R.H. & Reynoldson, T.B. (2004) *Bioassessment of Freshwater Ecosystems: Using the Reference Condition Approach*. Kluwer Academic Publishers, New York.

Balfour, J.H. & Sadler, J. (1863) *Flora of Edinburgh*. Black, Edinburgh.

Battarbee, R.W., Monteith, D.T., Juggins, S., Evans, C.D., Jenkins, A. & Simpson, G.L. (2005) Reconstructing pre-acidification pH for an acidified Scottish loch: A comparison of palaeolimnological and modelling approaches. *Environmental Pollution*, 137, 135–149.

Bennion, H. & Battarbee, R. (2007) The European Union Water Framework Directive: Opportunities for palaeolimnology. *Journal of Paleolimnology*, 38, 285–295.

Bennion, H., Fluin, J. & Simpson, G.L. (2004) Assessing eutrophication and reference conditions for Scottish freshwater lochs using subfossil diatoms. *Journal of Applied Ecology*, 41, 124–138.

Bernhardt, E.S., Palmer, M.A., Allan, J.D., *et al.* (2005) Synthesizing U.S. river restoration efforts. *Science*, 308, 636–637.

Birks, H.H. & Birks, H.J.B. (2006) Multi-proxy studies in palaeolimnology. *Vegetation History and Archaeobotany*, 15, 235–251.

Boedeltje, G., Bakker, J.P., Ten Brinke, A., Van Groenendael, J.M. & Soesbergen, M. (2004) Dispersal phenology of hydrochorous plants in relation to discharge, seed release time and buoyancy of seeds: The flood pulse concept supported. *Journal of Ecology*, 92, 786–796.

Bradshaw, E.G. & Rasmussen, P. (2004) Using the geological record to assess the changing status of Danish lakes. *Geological Survey of Denmark and Greenland Bulletin*, 4, 37–40.

Brierley, G. & Stankoviansky, M. (2002) Geomorphic responses to land use change: Lessons from different landscape settings. *Earth Surface Processes and Landforms*, 27, 339–341.

Brown, A.G. (2002) Learning from the past: Palaeohydrology and palaeoecology. *Freshwater Biology*, 47, 817–829.

Bruinderink, G.G., Van Der Sluis, T., Lammertsma, D., Opdam, P. & Pouwels, R. (2003) Designing a coherent ecological network for large mammals in northwestern Europe. *Conservation Biology*, 17, 549–557.

Burns, D.A., Riva-Murray, K., Bode, R.W. & Passy, S. (2008) Changes in stream chemistry and biology in response to reduced levels of acid deposition during 1987–2003 in the Neversink River Bain, Catskill Mountains. *Ecological Indicators*, 8, 191–203.

Carvalho, L. & Kirika, A. (2003) Changes in shallow lake functioning: Response to climate change and nutrient reduction. *Hydrobiologia*, 506 (1–3), 789–796.

Carvalho, L., Kirika, A. L. & Gunn, I. (2004) *Long-Term Patterns of Change in Physical, Chemical and Biological Aspects of Water Quality at Loch Leven*. Centre for Ecology and Hydrology, Edinburgh.

Charalambidou, I., Santamaria, L. & Figuerola, J. (2003) How far can the freshwater bryozoan *Cristatella mucedo* disperse in duck guts? *Archiv fur Hydrobiologie*, **157**, 547–554.

Clarke, R.T., Wright, J.F. & Furse, M.T. (2003) RIVPACS models for predicting the expected macroinvertebrate fauna and assessing the ecological quality of rivers. *Ecological Modelling*, **160**, 219–233.

Clausen, P., Nolet, B.A., Fox, A.D. & Klaassen, M. (2002) Long-distance endozoochorous dispersal of submerged macrophyte seeds by migratory waterbirds in northern Europe – A critical review of possibilities and limitations. *Acta Oecologica-International Journal of Ecology*, **23**, 191–203.

Cosby, B.J., Hornberger, G.M., Galloway, J.N. & Wright, R.F. (1985) Time scales of catchment acidification. *Environmental Science and Technology*, **19**, 1144–1149.

Davidson, T., Sayer, C., Bennion, H., David, C., Rose, N. & Wade, M. (2005) A 250 year comparison of historical, macrofossil and pollen records of aquatic plants in a shallow lake. *Freshwater Biology*, **50**, 1671–1686.

Davy-Bowker, J., Clarke, R.T., Johnson, R.K., Kokes, J., Murphy, J.F. & Zahrádková, S. (2006) A comparison of the European Water Framework Directive physical typology and RIVPACS-type models as alternative methods of establishing reference conditions for benthic macroinvertebrates. *Hydrobiologia*, **566**, 91–105.

Deutscher Verband für Wasserwirtschaft und Kulturbau (DVWK) (1996) *Fluß und Landschaft – ökologische Entwicklungskonzepte*. DVWK-Merkblätter zur Wasserwirtschaft 240, 285.

Donath, T.W., Holzel, N. & Otte, A. (2003) The impact of site conditions and seed dispersal on restoration success in alluvial meadows. *Applied Vegetation Science*, **6**, 13–22.

Downes, B.J., Barmuta, L.A., Fairweather, P.G., *et al.* (2002) *Monitoring Ecological Impacts: Concepts and Practice in Flowing Waters*. Cambridge University Press, Cambridge.

European Commission (2000) *Directive 2000/60/EC of the European Parliament and of the Council – Establishing a Framework for Community Action in the Field of Water Policy*. Brussels, Belgium, 23 October 2000.

Feminella, J.W. (2000) Correspondence between stream macroinvertebrate assemblages and 4 ecoregions of the southeastern USA. *Journal of North American Benthological Society*, **19**, 442–461.

Figuerola, J. & Green, A.J. (2002) Dispersal of aquatic organisms by waterbirds: A review of past research and priorities for future studies. *Freshwater Biology*, **47**, 483–494.

Findlay, D.L. (2003) Response of phytoplankton communities to acidification and recovery in Killarney Park and the Experimental Lakes Area, Ontario. *Ambio*, **32**, 190–195.

Flower, R.J. & Battarbee, R.W. (1983) Diatom evidence for recent acidification of two Scottish Lochs. *Nature*, **305**, 130–133.

Freckleton, R.P. & Watkinson, A.R. (2002) Large-scale spatial dynamics of plants: Metapopulations, regional ensembles and patchy populations. *Journal of Ecology*, **90**, 419–434.

Goodson, J.M., Gurnell, A.M., Angold, P.G. & Morrissey, I.P. (2003) Evidence for hydrochory and the deposition of viable seeds within winter flow-deposited sediments: The River Dove, Derbyshire, UK. *River Research and Applications*, **19**, 317–334.

Gordon, L.J., Peterson, G.D. & Bennett, E.M. (2008) Agricultural modifications of hydrological flows create ecological surprises. *Trends in Ecology & Evolution*, **23**, 211–219.

Gregory, S.V., Boyer, K.L. & Gurnell, A.M. (2003) *The Ecology and Management of Wood in World Rivers*. American Fisheries Society, Symposium 37, Bethesda, MD, 431.

Griffin, L.R., & Milligan, A.L. (1999) *Submerged Macrophytes of Loch Leven, Kinross*. Division of Evolutionary and Environmental Biology, Institute of Biomedical and Life Sciences, University of Glasgow, Glasgow.

Guilizzoni, P., Lami, A., Manca, M., Musazzi, S. & Marchetto, A. (2006) Palaeoenvironmental changes inferred from biological remains in short lake sediment cores from the Central Alps and Dolomites. *Hydrobiologia*, **562**, 167–191.

Hanski, I. (1994) Patch-occupancy dynamics in fragmented landscapes. *Trends in Ecology & Evolution*, **9**, 131–135.

Hanski, I. (1999) *Metapopulation Ecology*. Oxford University Press, Oxford.

Hanski, I. (2005) Landscape fragmentation, biodiversity loss and the societal response – The longterm consequences of our use of natural resources may be surprising and unpleasant. *EMBO Reports*, **6**, 388–392.

Hawkins, C.P. & Vinson, M.R. (2000) Weak correspondence between landscape classifications and stream macroinvertebrate assemblages: Implications for bioassessment. *Journal of North American Benthological Society*, **19**, 501–517.

Hawkins, C.P., Norris, R.H., Gerritsen, J., *et al.* (2000) Evaluation of landscape classifications for biological assessment of freshwater ecosystems: Synthesis and recommendations. *Journal of the North American Benthological Society*, **19**, 541–556.

Heino, J. (2002) Concordance of species richness patterns among multiple freshwater taxa: A regional perspective. *Biodiversity and Conservation*, **11**, 137–147.

Hering, D., Johnson, R.K., Kramm, S., Schmutz, S., Szoszkiewicz, K. & Verdonschot, P.F.M. (2006) Assessment of European streams with diatoms, macrophytes, macroinvertebrates and fish: A comparative metric based analysis of organism response to stress. *Freshwater Biology*, **51**, 1757–1785.

Hildrew, A.G. & Ormerod, S.J. (1995) Acidification – causes, consequences and solutions. In: *The Ecological Basis for River Management*, pp. 147–160. Wiley, Chichester.

Holling, C.S. (1992) Cross-scale, morphology, geometry, and dynamics of ecosystems. *Ecological Monographs*, **62**, 447–502.

Holling, C.S. & Meffe, G.K. (1996) Command and control and the pathology of natural resource management. *Conservation Biology*, **10**, 328–337.

Hooker, W.J. (1821) *Flora Scotica: Or a Description of Scottish Plants*. Archibald Constable & Co., London.

Jähnig, S.C. (2007) *Comparison between multiple-channel and single-channel stream sections – Hydromorphology and benthic macroinvertebrates*. Dissertation, Universität Duisburg-Essen.

Johnson, R.K. (1998) Spatio-temporal variability of temperate lake macroinvertebrate communities: Detection of impact. *Ecological Applications*, **8**, 61–70.

Johnson, R.K. & Angeler, D.G. (in press) Tracing recovery under changing climate: response of phytoplankton and invertebrate assemblages to decreased acidification. *Journal of North American Benthological Society*.

Johnson, R.K. & Goedkoop, W. (2002) Littoral macroinvertebrate communities: Spatial scale and ecological relationships. *Freshwater Biology*, **47**, 1840–1854.

Johnson, R.K. & Hering, D. (2009) Response taxonomic groups in streams to gradients in resource and habitat characteristics. *Journal of Applied Ecology*, **46**, 175–186.

Johnson, R.K., Bell, S., Davies, L., *et al.* (2003) Wetlands. In: *Conflicts Between Human Activities and the Conservation of Biodiversity in Agricultural Landscapes, Grasslands, Forests, Wetlands, and Uplands in Europe* (eds J. Young, P. Nowicki, D. Alard, *et al.*), pp. 98–122. A report of the BIOFORUM project. Centre for Ecology and Hydrology, Banchory, Scotland.

Johnson, R.K., Goedkoop, W. & Sandin, L. (2004) Spatial scale and ecological relationships between the macroinvertebrate communities of stony habitats of streams and lakes. *Freshwater Biology*, **49**, 1179–1194.

Johnson, R.K., Hering, D., Furse, M.T. & Verdonschot, P.F.M. (2006a) Indicators of ecological change: Comparison of the early response of four organism groups to stress gradients. *Hydrobiologia*, **566**, 139–152.

Johnson, R.K., Hering, D., Furse, M.T. & Clarke, R.T. (2006b) Detection of ecological change using multiple organism groups: Metrics and uncertainty. *Hydrobiologia*, **566**, 115–137.

Johnson, R.K., Furse, M.T., Hering, D. & Sandin, L. (2007) Ecological relationships between stream communities and spatial scale: Implications for designing catchment-level monitoring programs. *Freshwater Biology*, **52**, 939–958.

Jupp, B.P. & Spence, D.H.N. (1977a) Limitations of macrophytes in a eutrophic lake, Loch Leven. I. Effects of phytoplankton. *Journal of Ecology*, **65**, 175–186.

Jupp, B.P. & Spence, D.H.N. (1977b) Limitations of macrophytes in a eutrophic lake, Loch Leven. II. Wave action, sediments and waterfowl grazing. *Journal of Ecology*, **65**, 431–446.

Kail, J. & Hering, D. (2005) Using large wood to restore streams in Central Europe: Potential use and likely effects. *Landscape Ecology*, **20**, 755–772.

Kail, J., Hering, D., Gerhard, M., Muhar, S. & Preis, S. (2007) The use of large wood in stream restoration: Experiences from 50 projects in Germany and Austria. *Journal of Applied Ecology*, **44**, 1145–1155.

Kauffman, J.B., Beschta, R.L., Otting, N. & Lytjen, D. (1997) An ecological perspective of riparian and stream restoration in the Western United States. *Fisheries*, **22**, 12–24.

Kokeš, J., Zahrádková, S., Němejcová, D., Hodovský, J., Jarkovský, J. & Soldán, T. (2006) The PERLA system in the Czech Republic: A multivariate approach for assessing the ecological status of running waters. *Hydrobiologia*, **566**(1), 343–354.

Kondolf, G.M. (1996) A cross section of stream channel restoration. *Journal of Soil and Water Conservation*, **51**, 119–125.

Kowalik, R.A., Cooper, D.M., Evans, C.D. & Ormerod, S.J. (2007) Acidic episodes retard the biological recovery of upland British streams from chronic acidification. *Global Change Biology*, **13**, 2439–2452.

Leira, M., Jordan, P., Taylor, D., *et al.* (2006) Assessing the ecological status of candidate reference lakes in Ireland using palaeolimnology. *Journal of Applied Ecology*, **43**, 816–827.

Lynch, J.A., Bowersox, V.C. & Grimm, J.W. (2000) Acid rain reduced in eastern United States. *Environmental Science and Technology*, **34**, 940–949.

Magnuson, J. J., Webster, K.E., Assel, R.A., *et al.* (1997) Potential effects of climate changes on aquatic systems: Laurentian Great Lakes and Precambrian Shield Region. *Hydrological Processes*, **11**, 825–871.

Mann, M.E., Bradley, R.S. & Hughes, M.K. (1998) Global-scale temperature patterns and climate forcing over the past six centuries. *Nature*, **392**, 779–787.

Meir, E., Andelman S. & Possingham, H.P. (2004) Does conservation planning matter in a dynamic and uncertain world? *Ecology Letters*, **7**, 615–622.

Monteith, D.T., Stoddard, J.L., Evans, C.D., *et al.* (2007) Dissolved organic carbon trends resulting from changes in atmospheric deposition chemistry. *Nature*, **450**, 537–540.

Murphy, K.J. & Milligan, A. (1993) *Submerged Macrophytes of Loch Leven, Kinross*. The Centre for Research in Environmental Science and Technology, University of Glasgow, Glasgow.

Nathan, R., Katul, G.G., Horn, H.S., *et al.* (2002) Mechanisms of long-distance dispersal of seeds by wind. *Nature*, **418**, 409–413.

National Research Council USA (1992) *Restoration of Aquatic Ecosystems*. National Academy Press, Washington, DC.

National Water Council (1981) *River Quality: The 1980 Survey and Future Outlook*. National Water Council, London.

Naumann, E. (1932) Grundzüge der regionalen Limnologie. *Die Binnengewässer*, **11**, 176.

Nowicki, P. (2003) The ecosystem approach. In: *Conflicts Between Human Activities and the Conservation of Biodiversity in Agricultural Landscapes, Grasslands, Forests, Wetlands and Uplands in Europe* (eds J. Young, P. Nowicki, D. Alard, *et al.*), pp. 13–24. A report of the BIOFORUM project. Centre for Ecology and Hydrology, Banchory, Scotland.

Økland, J. & Økland, K.A. (1986) The effects of acid deposition on benthic animals in lakes and streams. *Experentia*, **42**, 471–486.

Ormerod, S.J., Rundle, S.D., Lloyd, E.C. & Douglas, A.A. (1993) The influence of riparian management on the habitat structure and macroinvertebrate communities of upland streams draining plantation forests. *Journal of Applied Ecology*, **30**, 13–24.

Paavola, R., Muotka, T., Virtanen, R., Heino, J., Jackson, D. & Mäki-Petäys, A. (2006) Spatial scale affects community concordance among fishes, benthic macroinvertebrates and bryophytes in streams. *Ecological Applications*, **16**, 368–379.

Rabeni, C.F. & Doisy, K.E. (2000) Correspondence of stream benthic invertebrate assemblages to regional classification schemes in Missouri. *Journal of North American Benthological Society*, **19**, 419–428.

Ricklefs, R.E. (1987) Community diversity: Relative roles of local and regional processes. *Science*, **235**, 167–171.

Robson, T.O. (1986) *Final Report on the Survey of the Submerged Macrophyte Population of Loch Leven, Kinross*. Nature Conservancy Council.

Robson, T.O. (1990) *Final Report on the Distribution, Species Diversity and Abundance of Macrophytes in Loch Leven, Kinross*. Nature Conservancy Council.

Salgado, J. (2006) *Assessing macrophyte responses to eutrophication and recovery using both contemporary survey and palaeolimnological data, Loch Leven, Kinross, Scotland*. MSc dissertation, University College London.

Salgado, J., Sayer, C.D., Carvalho, L., Davidson, T.A. & Gunn, I. (2009) Assessing aquatic macrophyte community change through the integration of palaeolimnological and historical data at Loch Leven, Scotland. *Journal of Paleolimnology*, **43** (1), 191–204.

Sandin, L. & Johnson, R.K. (2000a) Ecoregions and benthic macroinvertebrate assemblages of Swedish streams. *Journal of North American Benthological Society*, **19**, 462–474.

Sandin, L. & Johnson, R.K. (2000b) Statistical power of selected indicator metrics using macroinvertebrates for assessing acidification and eutrophication of running waters. *Hydrobiologia*, **422/423**, 233–243.

Sayer, C., Roberts, N., Sadler, J., David, C. & Wade, P.M. (1999) Biodiversity changes in a shallow lake ecosystem: A multi-proxy palaeolimnological analysis. *Journal of Biogeography*, **26**, 97–114.

Schilt, C.R. (2007) Developing fish passage and protection at hydropower dams. *Applied Animal Behaviour Science*, **104**, 295–325.

Schrope, M. (2006) The real sea change. *Nature*, **443**, 622–624.

Shields, F.D., Cooper, C.M., Knight, S.S. & Moore, M.T. (2003) Stream corridor restoration research: A long and winding road. *Ecological Engineering*, **20**, 441–454.

Simpson, G.L., Shilland, E.M., Winterbottom, J.M. & Keay, J. (2005) Defining reference conditions for acidified waters using a modern analogue approach. *Environmental Pollution*, **137**, 119–133.

Skjelkvåle, B.L., Andersen, T., Halvorsen, G., *et al.* (2000) *The 12-Year Report: Acidification of Surface Waters in Europe and North America: Trends, Biological Recovery, and Heavy Metals*. Norwegian Institute for Water Research, Oslo, Norway.

Skjelkvåle, B.L., Evans, C.D., Larssen, T., Hindar, A. & Raddum, G. (2003) Recovery from acidification in European surface waters: A view to the future. *Ambio*, **32**, 170–175.

Skjelkvåle, B.L., Stoddard, J.L., Jeffries, D.S., *et al.* (2005) Regional scale evidence for improvements in surface water chemistry 1990–2001. *Environmental Pollution*, **137**, 165–176.

Soons, M.B. (2006) Wind dispersal in freshwater wetlands: Knowledge for conservation and restoration. *Applied Vegetation Science*, **9**, 271–278.

Soons, M.B., Heil, G.W., Nathan, R. & Katul, G.G. (2004a) Determinants of long-distance seed dispersal by wind in grasslands. *Ecology*, **85**, 3056–3068.

Soons, M.B., Nathan, R. & Katul, G.G. (2004b) Human effects on long-distance wind dispersal and colonization by grassland plants. *Ecology*, **85**, 3069–3079.

Soons, M.B., Messelink, J.H., Jongejans, E. & Heil, G.W. (2005) Habitat fragmentation reduces grassland connectivity for both short-distance and long-distance wind-dispersed forbs. *Journal of Ecology*, **93**, 1214–1225.

Soons, M.B., Van der Vlugt, C., Van Lith, B., Heil, G.W. & Klaassen, M. (2008) Small seed size increases the potential for dispersal of wetland plants by ducks. *Journal of Ecology*, **96**, 619–627.

Spence, D.H.N. (1964) The macrophytic vegetation of freshwater lochs, swamps and associated fens. In: *The Vegetation of Scotland* (ed. J. H. Burnett). Oliver & Boyd, Edinburgh.

Stendera, S. & Johnson, R.K. (2005) Additive partitioning of macroinvertebrate species diversity across multiple scales. *Freshwater Biology*, **50**, 1360–1375.

Stendera, S. & Johnson, R.K. (2008) Tracking recovery trends of boreal lakes: Use of multiple indicators and habitats. *Journal of North American Benthological Society*, **27**, 529–540.

Stevenson, R.J., Bailey, R.C., Harrass, M., *et al.* (2004) Designing data collection for ecological assessments. In: *Ecological Assessment of Aquatic Resources: Linking Science to Decision-Making* (eds M.T. Barbour, S.B. Norton, K.W. Thornton & H.R. Preston), pp. 55–84. Society of Environmental Toxicology and Chemistry, Pensacola, FL.

Stoddard, J.L., Jeffries, D.S., Lukewille, A., *et al.* (1999) Regional trends in aquatic recovery from acidification in North America and Europe. *Nature*, **401**, 575–578.

Stoddard, J., Kahl, J.S., Deviney, F., *et al.* (2003) *Response of Surface Water Chemistry to the Clean Air Act Amendments of 1990*. EPA/620/R-03/001. US EPA, Washington, DC.

Stoddard, J.L., Larsen, P., Hawkins, C.P., Johnson, R.K. & Norris, R.H. (2006) Setting expectations for the ecological condition of running waters: The concept of reference condition. *Ecological Applications*, **16**, 1267–1276.

Straile, D. (2000) Meteorological forcing of plankton dynamics in a large and deep continental European lake. *Oecologia*, **122**, 44–50.

Strong, K.F. & Robinson, G. (2004) Odonate communities of acidic Adirondack Mountain lakes. *Journal of the North American Benthological Society*, **23**, 839–852.

Taylor, D., Dalton, C., Leira, M., *et al.* (2006) Recent histories of six productive lakes in the Irish Ecoregion based on multiproxy palaeolimnological evidence. *Hydrobiologia*, **571**, 237–259.

Thienemann, A. (1925) Die Binnengewässer Mitteleuropas. Eine limnologische Einführung. *Die Binnengewässer*, **1**, 255.

Townsend, C.R., Doledec, S., Norris, R., Peacock, K. & Arbuckle, C. (2003) The influence of scale and geography on relationships between stream community composition and landscape variables: Description and prediction. *Freshwater Biology*, **48**, 768–785.

Van Sickle, J. & Hughes, R.M. (2000) Classification strengths of ecoregions, catchments, and geographical clusters for aquatic vertebrates in Oregon. *Journal of North American Benthological Society*, **19**, 370–384.

Van Sickle, J., Hawkins, C.P., Larsen, D.P. & Herlihy, A.H. (2005) A null model for the expected macroinvertebrate assemblage in streams. *Journal of North American Benthological Society*, **24**, 178–191.

Vannote, R.L., Minshall, G.W., Cummins, K.W., Sedell, J.R. & Cushing, C.E. (1980) The river continuum concept. *Canadian Journal of Fisheries and Aquatic Sciences*, 37, 130–137.

Verdonschot, P.F.M. & Nijboer, R.C. (2004) Testing the European stream typology of the Water Framework Directive for macroinvertebrates. *Hydrobiologia*, 516, 35–54.

Verhoeven, J.T.A., Soons, M.B., Janssen, R. & Omtzigt, N. (2008) An Operational Landscape Unit approach for identifying key landscape connections in wetland restoration. *Journal of Applied Ecology*, 45, 1496–1503.

Wallin, M., Wiederholm, T. & Johnson, R.K. (2003) *Guidance on Establishing Reference Conditions and Ecological Status Class Boundaries for Inland Surface Waters*. Produced by CIS working group 2.3 – REFCOND, 93 pp.

West, G. (1910) A further contribution to a comparative study of the dominant phanerogamic and higher cryptogamic flora of aquatic habit in Scottish lakes. *Proceedings of the Royal Society Edinburgh*, 30 (Part 2, 6), 65–181.

Wetzel, R.G. (1995) Death, detritus, and energy flow in aquatic ecosystems. *Freshwater Biology*, 33, 83–89.

Weyhenmeyer, G.A. (2004) Synchrony in relationships between the North Atlantic Oscillation and water chemistry among Sweden's largest lakes. *Limnology and Oceanography*, 49, 1191–1201.

Weyhenmeyer, G.A., Blenckner, T. & Petterson, K. (1999) Changes of the plankton spring outburst related to the North Atlantic Oscillation. *Limnology and Oceanography*, 44, 1788–1792.

Wohl, E., Angermeier, P.L., Bledsoe, B., *et al.* (2005) River Restoration. *Water Resources Research*, 41, W10301.

Wright, J.F. (1995) Development and use of a system for predicting the macroinvertebrate fauna in flowing waters. *Australian Journal of Ecology*, 20, 181–197.

Wright, R.F. (2002) *Workshop on models for biological recovery from acidification in a changing climate*, 42. Norwegian Institute for Water Research, Grimstad, Norway.

Zhao, Y., Sayer, C., Birks, H., Hughes, M. & Peglar, S. (2006) Spatial representation of aquatic vegetation by macrofossils and pollen in a small and shallow lake. *Journal of Paleolimnology*, 35, 335–350.

10

Modelling Catchment-Scale Responses to Climate Change

Richard A. Skeffington, Andrew J. Wade, Paul G. Whitehead, Dan Butterfield, Øyvind Kaste, Hans Estrup Andersen, Katri Rankinen and Gaël Grenouillet

Introduction

The focus of the Euro-limpacs project was on responses of aquatic ecosystems (rivers, lakes and wetlands) to climate change, but these responses cannot be fully understood or predicted without considering the connections to other earth systems. Rivers, lakes and wetlands are connected to each other and to other water bodies such as groundwater and estuarine and coastal waters. Most of the water in these aquatic systems has passed through the terrestrial environment at some stage. A catchment-scale approach that considers these different environments is thus essential for predicting how European aquatic ecosystems might respond to climate change.

Typically, measurements of the aquatic and terrestrial environments and experimental manipulations are done in small ($<10 \, km^2$) research catchments, in the laboratory or in small *in situ* tanks which represent the larger system to be studied. Management decisions are, however, typically made at much larger scales ($>1000 \, km^2$), as in the EU Water Framework Directive, in which the River Basins are all large catchments. Furthermore, projections of future climates made by the models of atmospheric and oceanic circulation (General Circulation Models, GCMs) are produced at a coarse scale greater in size than many catchments. Models can help fill the gaps between the mismatch of scales between scientific measurement, management and climate projections. The complexity of the interactions between all these aquatic and terrestrial systems also necessitated a modelling approach: individual experiments and manipulations alone cannot consider this complexity or integrate the different processes.

Modelling catchment responses to climate change is a very demanding undertaking, requiring a number of tasks that are themselves very challenging. Firstly, in order to make predictions of the effects of climate change, it is necessary

Climate Change Impacts on Freshwater Ecosystems. First edition. Edited by M. Kernan, R. Battarbee and B. Moss. © 2010 Blackwell Publishing Ltd.

to produce climate change scenarios at the catchment scale. A credible methodology for converting the predictions of GCMs to the spatial scale required (known as downscaling) and of generating the resultant 'weather' needed by the models (e.g. daily precipitation and temperature) must be available. Secondly, models must be developed which connect catchment climates to variables that can be measured and which are features of interest in aquatic systems, such as water flows, water quality or the abundance of aquatic organisms. These models may involve detailed representations of catchment structure and function and their interactions with climate, or they may be more empirical. All models need to be tested to determine whether they represent observed data adequately, normally in an iterative cycle of testing and revision. Once a model is performing satisfactorily as judged by its ability to reproduce observations and conform to notions of how a catchment functions, a set of changed climates can be used to drive the model to produce a set of changed response variables. Thus, models can potentially provide an estimate of the effects of the changed climate on nitrate concentrations or fish biodiversity, for instance. Potential changes in catchment structure and function due to climate change must be considered during this process (e.g. the alteration of vegetation types in the catchment). Finally, the influence of changes in catchment management (e.g. novel crops or agricultural practices) can be assessed. These might, for instance, be due to changed climates, socio-economic factors or adaptive responses of catchment managers attempting to mitigate climate change effects.

This chapter outlines how this approach was used within Euro-limpacs, illustrates how the modelling process was applied in a range of case studies and describes how a consistent modelling approach for assessing flow and water quality across Europe was developed. The science of modelling was taken further by chaining models to simulate the response of flow and nitrogen at the catchment scale. Models that incorporate ecological effects have been developed for lakes, but for rivers these remain a research goal owing to the dynamic, complex nature of the river environment (Chapra 1997). The main focus of integrated modelling in the Euro-limpacs project was the development of catchment-scale models of flow and water quality. The applications described are a small sample of those undertaken. As the plethora of abbreviations and acronyms used in modelling work can rapidly become confusing, Table 10.1 is provided for explanation and reference.

The Euro-limpacs modelling strategy

Developing an integrated toolkit of models for catchment analysis and assessment has been central to the Euro-limpacs project, based on six key questions: (i) Can the impacts of climate change, land-use change and pollution be evaluated using modelling? (ii) How can models be used to assess likely effects of climate change on freshwater systems? (iii) Can models simulate the spatial/temporal variation in pollutant behaviour in freshwater systems? (iv) Can the uncertainty associated with these models be quantified? (v) Can socio-economic scenarios be incorporated into modelling assessments of climate change effects? (vi) How can models be best used to assist the management of surface waters influenced by climate

Table 10.1 List of abbreviations and acronyms

Abbreviation	Meaning (if any)	Description
AET	Actual Evapotranspiration	
CATCHMOD	Catchment Model	UK water balance model
CLUAM	Climate and Land-Use Allocation Model	
CGCM2	Canadian Global Coupled Model	GCM from the Canadian Centre for Climate Modelling and Analysis
CSIRO2	Commonwealth Scientific and Industrial Research Organisation	GCM from the CSIRO in Australia
EARWIG	Environment Agency Rainfall and Weather Impacts Generator	Model that generates weather data from downscaled GCMs in the United Kingdom
ECHAM4	European Centre Hamburg Model	GCM developed by the Max Planck Institute
GCM	General Circulation Model	Model used for understanding and predicting global-scale climate
GLUE	Generalised Likelihood Uncertainty Estimation	Technique for investigating model uncertainties
HadCM3	Hadley Centre Coupled Model	GCM from the Hadley Centre, UK Meteorological Office
HBV	Hydrologiska Byråns Vattenbalansavdelning	Scandinavian hydrological model
HER	Hydrologically Effective Rainfall	Rainfall potentially available to recharge rivers
HIRHAM	–	RCM developed for Europe by a number of meteorological institutes
INCA	Integrated Catchment Model	Suite of catchment models developed at the University of Reading for N, P, etc.
IPCC	Intergovernmental Panel on Climate Change	International Organisation for assessing Climate Change
MAGIC	Model of Acidification of Groundwater in Catchments	Acidification model, dealing mostly with soil and surface water (in spite of the title)
MIKE-11	Named after the model author	Hydrological model from the Danish Hydrological Institute
MPI	Max Planck Institute	German Research Institute
NAM	–	Rainfall-run-off model used with MIKE-11
RCM	Regional Climate Model	Model used for understanding and predicting climate at a smaller scale than a GCM
PET	Potential Evapotranspiration	
SDSM	Statistical Downscaling Model	UK model used for downscaling GCMs
SRES	Special Report on Emission Scenarios	IPCC report, which defined a number of standard greenhouse gas emission scenarios
TRANS	Transport	Hydrochemical model used with MIKE-11

Acronyms defined in the text and not used again are not covered in the table.

change? To address such questions, both new and existing techniques have been used. In this chapter, we describe these techniques before considering answers to the questions.

Downscaling

An essential first step in predicting the effects of future climates on aquatic ecosystems is to forecast what these climates are likely to be. On a global scale, future climate change is modelled using GCMs, which are mechanistic models of the climate system built on physical principles (IPCC 2007). Assumptions about greenhouse gas emissions, population growth and economic development have also to be made, and within Euro-limpacs, a standardized set of assumptions is used based on the Special Report on Emission Scenarios (SRES) of the IPCC (Nakićenović *et al.* 2000). These scenarios are explained in Chapter 3. GCMs are currently too coarse in resolution (~270 km × 270 km) for catchment-scale modelling, though finer-scale models are close to release. Methods are therefore required to 'downscale' the outputs from the GCMs to the appropriate scale for modelling effects. This is more problematic than might be imagined. There are two main approaches, variously called dynamic or model-based and statistical or empirical (Fowler *et al.* 2007). Dynamic downscaling uses regional climate models (RCMs) nested within the GCMs, which are used to provide input data and boundary conditions.

RCMs can simulate processes important on catchment scales and provide outputs on scales down to about 5 km. These are computationally expensive, however, and a more common approach is to use statistical downscaling methods. These rely on observed quantitative relationships between the small-scale climates and the large-scale climates. These relationships are then used to generate the large-scale or high-resolution climate from the GCM output, one major assumption being that the empirical relationships will remain the same in all projected climates, including those affected by enhanced greenhouse warming. Tisseuil *et al.* (2009) discuss further problems and refinements of statistical downscaling methods.

In Euro-limpacs, we standardized downscaling methods. Dynamically downscaled data across Europe were available from the EU-funded PRUDENCE (Prediction of Regional scenarios and Uncertainties for Defining EuropeaN Climate change risks and Effects, 2001–04) website (http://prudence.dmi.dk), for the periods 1961–90 and 2071–2100. The data were generated by nesting an RCM within two GCMs, but the output cell size (0.5° × 0.5°) was still too coarse for most catchment applications and required further downscaling using the Statistical Downscaling Method (SDSM; Wilby *et al.* 2002) with refinements based on 'local methods', as described by Wade *et al.* (2008). For instance, GCM and RCM temperature predictions are for the average altitude of a grid cell. To correct this to the altitude of a catchment, a lapse rate (the rate of change of temperature with altitude) based correction was proposed, preferably using a site-specific lapse rate or alternatively a 'standard' lapse rate of $-0.6\,°C$ per 100 m.

In some instances, it was appropriate to use a GCM cell different from that in which the site lies to build a relationship between GCM or RCM output and local conditions. For example, if the site is in a mountainous region, then a

GCM cell which is dominated by mountains is the most appropriate; this may be the cell which includes the site or an adjacent cell. The SDSM was apparently successful in some cases (Wilby *et al.* 2006; Whitehead *et al.* 2006), but in other applications, it failed to produce reliable reconstructions of the monthly mean rainfall totals and the seasonal patterns in rainfall for the control periods. It was therefore abandoned in favour of a standardized delta-method approach which used change factors derived from the GCMs and applied to individual catchments (Wade *et al.* 2008). For each month in a control period (1961–90), a factor consisting of the mean observed precipitation divided by the mean RCM-modelled precipitation was derived. These factors were then applied to the RCM-modelled precipitation for the period 2071–2100 to calculate catchment precipitation under a particular change scenario. For temperature, a similar procedure was applied except that the factor was additive rather than a ratio (Wade *et al.* 2008).

River Kennet case study

Table 10.2 shows an example of some results from a change factor analysis. The aim was to calculate flows in the river Kennet in southern England under a variety of climate change scenarios, as part of an attempt to model the effects of climate and socio-economic changes on the river (Skeffington 2008; see also Chapter 11). In this case, the climate scenarios were derived from the UK Climate Impacts Programme (UKCIP02, Hulme *et al.* 2002). In UKCIP02, the predictions of the HadCM3 GCM were dynamically downscaled to a 50-km grid in a double-step procedure using two regional climate models.

A selection from the SRES – the A1F1, A2, B1 and B2 scenarios (see Chapter 3) – was run for three periods, the 2020s, 2050s and 2080s, to give a number of scenario-period combinations. Due to computational limitations, only the A2–2080 combination was dynamically downscaled, the others being interpolated using pattern recognition (Hulme *et al.* 2002). The A2 and B2 scenario predictions for the Kennet catchment were used as inputs to the 'weather generator' programme EARWIG (Environment Agency Rainfall and Weather Impacts Generator: Kilsby *et al.* 2007), which generated daily values for meteorological parameters including temperature, rainfall and potential evapotranspiration. EARWIG works by fitting a sophisticated stochastic model of daily rainfall to observed data and using change factors calculated from the UKCIP02 scenarios to do the same for future climates.

Other climatic variables are calculated from rainfall using regression relationships: an approach that works well for the variables controlling river discharge (Kilsby *et al.* 2007). This calculated meteorology was then used to generate daily values for river discharge under different scenarios by feeding it through the hydrological model embedded in the INCA-N Model (Wade *et al.* 2002a). Temperature and potential evapotranspiration (PET) were used directly from EARWIG, but the INCA-N model also requires actual evapotranspiration (AET) and hydrologically effective rainfall (HER). These were calculated from EARWIG daily rainfall and PET using a simple spreadsheet model (Bernal *et al.* 2004; Durand 2004), which works by calculating a soil moisture deficit which must be satisfied before any HER occurs. It is clear from this account that even addressing relatively simple

Table 10.2 Observed and modelled meteorological and hydrological data for the river Kennet under various climate change scenarios

Variable	Units	Observed* 1961–90	Modelled 1961–90	2020s A2/B2	2050s B2	2050s A2
Annual rainfall	mm yr^{-1}	759	759	778	757	758
Days with:						
<0.2 mm	%	56	54.7	55.5	57.7	57.3
<1.0 mm	%	67	66.1	65.9	67.4	67.2
Temperature						
Mean	°C	9.2	9.2	10.2	11.0	11.3
Mean daily min	°C	5.1	5.4	6.4	7.2	7.5
Mean daily max	°C	13.0	12.9	14.0	14.8	15.1
Max daily max	°C	33	31.2	33.6	38.1	38.5
Min daily min	°C	−16	−14.8	−12.0	−12.8	−13.5
PET[†]	mm yr^{-1}		536	641	728	750
AET[‡]	mm yr^{-1}		459	481	503	512
HER[§]	mm yr^{-1}		299	298	254	247
Discharge in river Kennet						
Annual mean	m^3 s^{-1}	9.60	9.83	9.87	8.38	8.15
Minimum	m^3 s^{-1}	0.93	2.12	1.76	1.29	0.74
Maximum	m^3 s^{-1}	46.7	46.6	61.8	59.3	48.9
5th percentile	m^3 s^{-1}	3.88	3.43	3.38	2.60	2.44
1st percentile	m^3 s^{-1}	2.37	2.65	2.39	1.99	1.62

*Meteorological data are catchment means from the UK Meteorological Office at http://www.metoffice.gov.uk/climate/uk/averages/ukmapavge.html#. Hydrological data are from the UK National River Flow Archive at http://www.nwl.ac.uk/ih/nrfa/webdata/039016/g.html. INCA-predicted discharge has been adjusted to take into account drinking water abstraction from the catchment.
†PET, potential evapotranspiration (calculated without restrictions due to water availability).
‡AET, actual evapotranspiration.
§HER, hydrologically effective rainfall (rainfall potentially available to recharge rivers).

questions requires a long chain of models with associated uncertainties, suggesting that there might be some benefit in developing methods to reduce the number of models needed. Tisseuil *et al.* (2009) had reasonable success in predicting river flows in a variety of river types in the Garonne Basin by direct downscaling from GCMs using a variety of statistical models.

The results of the Kennet study (Table 10.2) show that when comparing observed and modelled data for the validation period of 1961–90, EARWIG is successful in reproducing the observed mean and distribution of temperature and rainfall. Likewise, the INCA-N model successfully uses these data to reproduce the observed mean and distribution of discharge in the river, except that the observed absolute minimum values, due to the exceptionally dry period of 1975–6, are not simulated. The success at reproducing the observed data with minimal calibration gives some confidence in the future predictions (Table 10.2).

Rainfall increases slightly in the 2020s but decreases somewhat by 2050. PET, however, increases due to rising temperatures, until in the 2050 A2 scenario it is almost equal to rainfall. Fortunately for river flows, most of the decrease in rainfall and increase in PET occur in summer: this means that the change in AET is much smaller, as the soil moisture deficit will limit evapotranspiration. As the Kennet is a groundwater-dominated river, flow will continue through the summer droughts as at present, but low flows will become more common by the 2050s.

Though these results appear credible and self-consistent, other attempts to model the same system have yielded a range of results. Arnell and Reynard (1997) predicted a decrease in run-off of 21% (range +24% to −37%) in the Lambourn, the major tributary of the Kennet, by 2050 given climate models then available and a range of hydrological assumptions. Limbrick *et al.* (2000) modelled hydrological changes in the Kennet due to climate change, using earlier versions of the Hadley Centre models (HadCM1 and HadCM2) as climate drivers and the hydrological model incorporated in INCA. The average reduction in annual flows by 2050 was 19%, similar to that described here (15%–17%), as were many of the seasonality features, such as a reduction in minimum flow of 46% (51% here). Whitehead *et al.* (2006) used INCA, and the HadCM3 model as climate driver statistically downscaled to the Kennet, to explore the implications for flow and N concentration. In contrast to this study, to Arnell and Reynard (1997) and to Limbrick *et al.* (2000), they predicted an *increase* in mean flow rates of 2%–5% by 2050. Wilby *et al.* (2006) used statistically downscaled GCM data and a more sophisticated hydrological model, CATCHMOD, coupled with INCA, to study N concentration and flow. For the HadCM3 model, this predicted a small (c. 5%) decrease in median flow in 2050 for the A2 storyline, and an even smaller increase (c. 2%) in median flow for the B2 storyline. Both Whitehead *et al.* (2006) and Wilby *et al.* (2006) used three different GCMs to generate climate drivers, with strongly contrasting results. In particular, the CGM2 model predicted large increases in mean flow by 2050 of about 35% (Whitehead *et al.* 2006) or 80% (Wilby *et al.* 2006), in contrast to the 'dry' Hadley Centre model. Clearly, the implications of these differences are large.

Statistical models

The quantification of the relationships between the distribution of biota and environmental conditions is a central theme of ecology, and there are existing methodologies available for predicting the effects of climate change on species in freshwaters. One common approach involves modelling the current distribution of an organism in relation to habitat variables, in particular temperature and flow regime, predicting changes in the habitat variables due to climate change, and using the model to predict the change in species distributions. A large number of studies of this type have been carried out (e.g. Berry *et al.* 2002; Mohseni *et al.* 2003), and the techniques being used are rapidly developing (e.g. Rushton *et al.* 2004). There are, however, problems with the approach. One is how to validate and test such models (Vaughan & Ormerod 2005). Another is that different models applied to the same data may predict significantly different impacts (e.g. Lawler *et al.* 2006). One technique for evaluating such models is to use ensemble forecasting (Araújo & New 2007) by applying several different models to the same problem and evaluating areas of agreement and disagreement. As described below, Buisson *et al.* (2008) adopted such an approach to model the future distribution of 35 fish species in French streams.

Seven different statistical techniques were used. Fish presence–absence data were extracted from the Office National de l'Eau et des Milieux Aquatiques (ONEMA) for 1110 sites distributed over the whole country. The sites chosen were judged to have minimal human disturbance, and the data were obtained by standardized electrofishing techniques. Data from fish species present in more than 25 sites were retained for study, amounting to 35 species (see Table 10.3). These data were then related to climatic and environmental variables. Three variables were used to describe climatic conditions: mean annual precipitation, mean annual air temperature and annual air temperature amplitude, derived from the difference between mean air temperature of the warmest month and mean air temperature of the coldest month. Future values for each of these three descriptors were derived for the 2080s from three GCMs: HadCM3 (Hadley Centre for Climate Prediction and Research, the United Kingdom), CGCM2 (Canadian Centre for Climate Modelling and Analysis) and CSIRO2 (Commonwealth Scientific and Industrial Research Organisation, Australia). Predictions of future climate were made for each of these using four greenhouse gas emission scenarios. These were scenarios A1, A2, B1 and B2 from the IPCC SRES (Nakićenović *et al.* 2000). Thus, 84 separate modelling runs were performed for each fish species: seven species distribution models × three GCMs × four emission scenarios.

The study also recognized that climatic variables are not the only ones to affect fish distribution: rivers and streams have a variety of habitats, and a large lowland river will clearly be different from a small headwater stream even if the climate is the same. Some of this habitat variation is correlated with information that can be obtained from databases and thus can be taken into account in the analysis. In this study, three derived variables were used to represent habitat variation: (i) elevation above sea level; (ii) a parameter representing position on the upstream–downstream gradient, derived from catchment area and distance from the source and (iii) stream velocity derived from width, depth and slope at the sampling site. As these derived variables are likely to be correlated with the climatic ones (e.g. high elevation with low annual air temperature), the deviations from the expected values at each site were used in the analysis to give six independent variables.

The species distribution models were then calibrated for each species using a randomly selected 777 river sites from the database. The remaining 333 sites were used to validate the calibration in an iterative process, and when these were satisfactory, predictions of probability of occurrence of each species at each of 1110 sites were converted into presence–absence values using a threshold maximizing the sum of two measures: sensitivity (i.e. the percentage of presence correctly predicted) and specificity (i.e. the percentage of absence correctly predicted). The calibrated models were then used to predict fish species distributions for 2080 for each of the 12 scenarios. The future probabilities of occurrence were transformed into presence–absence values by using the same threshold values as for current predictions.

Table 10.3 Analysis of agreement between projections of fish species' presence or absence in France for 2080

	Fish species				Source of variability (%)§		
Code*	Genus	Species	English name†	Consensus‡ (%)	SDM	GCM	GES
Les	Leuciscus	souffia	Varione	48	38	48	15
Lec	Leuciscus	cephalus	Chub	69	30	54	16
Ana	Anguilla	anguilla	Eel	68	56	14	30
Bar	Barbus	barbus	Barbel	62	55	35	9
Cht	Chondrostoma	toxostoma	SW European Nase	44	80	7	12
Alb	Alburnoides	bipunctatus	Spurlin	54	49	48	3
Leg	Lepomis	gibbosus	Pumpkinseed	55	85	11	4
Sas	Salmo	salar	Salmon	52	70	24	6
Lel	Leuciscus	leuciscus	Common dace	65	62	35	3
Bam	Barbus	meridionalis	Southern barbel	45	96	0	4
Rur	Rutilus	rutilus	Roach	69	48	48	4
Sce	Scardinius	erythrophthalmus	Rudd	45	58	40	2
Cyc	Cyprinus	carpio	Carp	59	89	8	3
Tht	Thymallus	thymallus	Grayling	42	41	55	4
Amm	Ameiurus	melas	Black bullhead	41	77	22	1
Gog	Gobio	gobio	Gudgeon	70	95	2	3
Blb	Blicca	bjoerkna	White bream	50	67	30	3
Tit	Tinca	tinca	Tench	58	61	36	2
Sal	Sander	lucioperca	Zander	45	88	11	1
Cac	Carassius	carassius	Crucian carp	45	93	2	5
Chn	Chondrostoma	nasus	Sneep	50	61	38	1
Rha	Rhodeus	amareus	Bitterling	58	44	55	1
Ala	Alburnus	alburnus	Bleak	70	88	11	1
Esl	Esox	lucius	Pike	65	79	20	0
Gaa	Gasterosteus	aculeatus	Three-spined stickleback	31	76	23	1

Code	Genus	species	English name				
Bab	*Barbatula*	*barbatula*	Stone loach	62	94	6	0
Lol	*Lota*	*lota*	Burbot	36	88	12	0
Abb	*Abramis*	*brama*	Common bream	63	97	2	1
Gyc	*Gymnocephalus*	*cernuus*	Ruffe	59	94	6	0
Pup	*Pungitius*	*pungitius*	Nine-spined stickleback	41	87	8	5
Php	*Phoxinus*	*phoxinus*	Minnow	56	62	35	2
Pef	*Perca*	*fluviatilis*	Perch	61	52	33	15
Lap	*Lampetra*	*planeri*	Brook lamprey	46	39	44	18
Cog	*Cottus*	*gobio*	Bullhead	41	79	5	16
Sat	*Salmo*	*trutta fario*	Brown trout	63	61	24	15

Species are arranged in order of the magnitude of change predicted, the most positive first.
*Code is the three-letter code used to identify species in Fig. 10.1.
[†]English names from Froese and Pauly (2009).
[‡]Consensus is the measure of agreement between the 84 model combinations used for the fish projections.
[§]Source of variability in the model projections for each species: SDM, species distribution model; GCM, global climate model; GES, greenhouse gas emission scenario.

To evaluate agreement between the model runs, a 'consensus' parameter was calculated as the first axis of a principal components analysis of the predictions for each species. The mean consensus between models was 57% (Table 10.3), but values for each species differed considerably. It ranged from 31% for *Gasterosteus aculeatus* (three-spined stickleback) to 70% for *Gobio gobio* (gudgeon). Generally, the rarer species had lower consensus values. Table 10.3 also shows the sources of variability in projections for each fish species. For the entire data set, 70% of the variability was due to the species distribution models, 24% to the GCMs and only 6% to the emission scenarios. However, the pattern varied somewhat between species, and for some, such as the chub (*Leuciscus cephalus*), the choice of GCM was a more important source.

Given the dominant role of the species distribution model in the fish species projections, a single GCM (HadCM3) combined with a single emission scenario (A1FI) was arbitrarily chosen to assess the potential impacts of climate change on stream fish assemblages (see Buisson *et al.* 2008). These predictions are shown in Fig. 10.1 as changes in the probability of occurrence of each species. Buisson *et al.* (2008) found that all 35 fish species would be positively or negatively affected by climate change. On average, changes in the probability of occurrence ranged from −36.6% for *Salmo trutta fario* (brown trout) to +44.6% for *Leuciscus souffia* (varione). Those gaining the most appear to be warm-water species or those with a large range of thermal tolerance. Most negatively affected were two cold-water species – *Cottus gobio* (bullhead) and *Salmo trutta fario* (brown trout). On average, however, the 35 fish species would change their probability of occurrence by +5.6% by 2080, indicating a slightly positive response to climate change. Even though a species increases overall, local extinctions in some areas might occur, compensated by colonizations of new sites. For some species, the changes in spatial distribution were calculated and are illustrated for three representative species in Fig. 10.2: *Barbus barbus* (barbel), *Esox lucius* (pike) and *Salmo trutta fario* (brown trout).

Barbel, *Barbus barbus*, a rheophilic species relatively common in French streams, was predicted to expand its range greatly under climate change. The consensual model did not predict any extinctions and forecast that it could colonize a large number of areas where it does not currently occur, for example, NW France, the Pyrenees, the Massif Central and the Jura Mountains. Pike (*Esox lucius*), a predatory species living among dense vegetation, was predicted to move to new suitable habitats mainly in eastern France and in mountainous areas, whereas it could suffer from local extinctions in the western part of France where it currently occurs in many sites. Finally, brown trout *Salmo trutta fario*, a species characteristic of cold and well-oxygenated streams, was predicted to be the most severely affected by climate change. By 2080, its distribution would be restricted to the most upstream sites of the mountainous regions and some streams of NW France. The results thus predict that cold-water species living in headwater streams would suffer the most deleterious effect of climate change as their potential distributional area reduced, while other fish species occurring currently downstream would be able to expand their range upstream.

Overall, these results were consistent with those obtained in North America, which predicted a decrease in salmonid distribution and contrasting results for

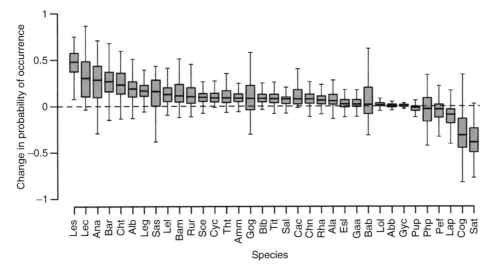

Figure 10.1 Changes in the probability of occurrence for each of 35 fish species in France predicted for 2080 using the scenario HadCM3 A1FI. Each box and whisker plot represents 84 model runs: the median is the line within the box; the edges of the box are the first and third quartiles, and the whiskers are the upper and lower extremes. See Table 10.3 for species codes.

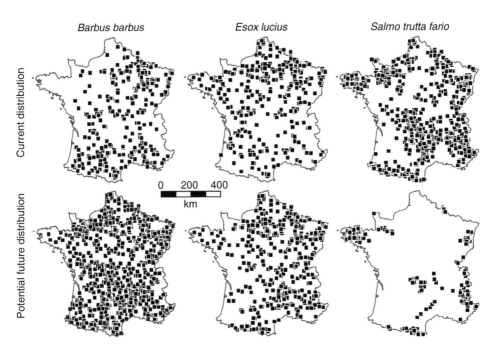

Figure 10.2 Predicted distributions in French rivers of three fish species: *Barbus barbus* (barbel), *Esox lucius* (pike) and *Salmo trutta fario* (brown trout) at the present day and in 2080.

cool- and warm-water species (e.g. Mohseni *et al.* 2003). Nevertheless, compared with other taxa for which the impacts of climate change could be very detrimental (e.g. Thomas *et al.* 2004, 2006), this assessment for French stream fish species was rather positive, as most fish were predicted to expand their distributional area rather than reduce it. This may result from the scarcity of cold-water species in French fish assemblages compared to cool- and warm-water species which have a larger range of tolerance.

Dynamic models

A fuller understanding of the potential effects of climate change on aquatic ecosystems can be obtained using dynamic, mechanistic models. Dynamic models allow model components to change with time: mechanistic models attempt to capture all the important processes as connected systems of equations. Dynamic, mechanistic models should be able to represent the transient changes in the pollutant stores in catchments and represent the response of biogeochemical cycles to altered precipitation and temperature inputs. Thus, they are eminently suited to modelling climate change effects. The disadvantages of these models are well known (e.g. Chapra 1997): they can require large amounts of data, which may not be available, they require a full knowledge of the important processes and how they are connected, they are difficult and time-consuming to write and programme and uncertainty is a major issue if a model aims for complete understanding of a system (see below). Nevertheless, they represent the best hope for accurate predictions of climate change effects, and building and improving them is a priority for research.

The most widely used dynamic models in the Euro-limpacs project have been the INCA models. The INCA (INtegrated CAtchment) models are dynamic computer models that predict aspects of water quantity and quality in rivers and catchments (Whitehead *et al.* 1998a, b; Wade *et al.* 2002a). They are designed to represent the factors and processes controlling flow and water quality dynamics in both the land and the in-stream components of river catchments, whilst minimizing data requirements and model structural complexity (Whitehead *et al.* 1998a, b). INCA can produce daily estimates of discharge, stream water concentrations and fluxes over a period of many years at any point along a river's main channel. The model provides a number of tools to aid understanding of the system, and statistics to allow comparison with observed data can be generated.

The original INCA-N model was developed and used to model nitrogen in catchments (Wade *et al.* 2002a, Wade 2006). The INCA framework has now been extended to phosphorus (INCA-P, Wade *et al.* 2002b, c), particulates (INCA-SED, Jarritt & Lawrence 2007), dissolved organic carbon (INCA-C, Futter *et al.* 2007a, 2008), mercury (INCA-Hg, Futter *et al.* 2007b) and macrophyte and epiphytic algal dynamics (Wade *et al.* 2002b; Whitehead *et al.* 2008). A notable example is the application of the INCA-N model to the Garonne river system in France (Tisseuil *et al.* 2008), at 62,700 km² the largest river basin modelled so far. The spatial and temporal dynamics in the stream water nitrate concentrations were described and related to variations in climate, land management and effluent point sources using multivariate statistics. In conjunction with the hydrological model HBV (Hydrologiska Byråns Vattenbalansavdelning), the INCA-N model was used to understand which factors and processes control the flow and N dynamics in different climate zones and to assess the relative inputs from diffuse and point sources across the catchment. The simulations suggested that, in the lowlands, seasonal patterns in the stream water nitrate concentrations were dominated by diffuse agricultural and point-source effluents with an estimated 75% of the river load in the lowlands derived from arable farming. The model proved able to simulate observed seasonal nitrate patterns at large spatial (>300 km^2) and temporal ($\geq$ monthly) scales using available national data sets. The model was equally good at simulating observations in the upper, mid and lower reaches of the Garonne. This application of the linked HBV and INCA-N models to a major European river system showed that it was possible to simulate observed behaviour in a catchment commensurate with the largest basins to be managed under the Water Framework Directive.

The success of the model in simulating observed nitrogen behaviour in catchments allows its use to project changes in nitrogen fluxes in future as a result of climate change. As described above, summer flow rates in the river Kennet in SE England are likely to fall in the future as drought periods become more extreme. Extending the modelling to simulate nitrate-N, Whitehead *et al.* (2006) used the INCA-N model to show that the droughts might trigger a release of nitrate from the soils and this nitrate would be exported to the river, as illustrated in Fig. 10.3. With climate change predictions downscaled from the HadCM3 model and the A2 emissions scenario, nitrate-N concentrations increased to values close to the EU drinking water limit of 11.3 mg l^{-1}. Falling flow rates and rising nitrate levels could affect water supply and put in doubt plans to improve the water quality and ecology of such a sensitive chalk stream as the Kennet. A series of adaptation strategies was investigated using the model to assess the effectiveness of potential mitigation strategies. For example, reducing agricultural fertilizer use by 50% in the catchment gave the biggest improvement, lowering nitrate concentrations to levels not seen since the 1950s. Reducing atmospheric sources of oxidized and reduced N by 50% reduced the nitrate by about 1 mg l^{-1} compared with the baseline scenario. Constructing water meadows along the river (which is parameterized in the model as an increase in the in-stream denitrification coefficient) would be more beneficial, significantly slowing down the rising levels of nitrate. A mixed strategy of a combination of all three approaches – reducing fertilizer, reducing N deposition by 25% and constructing half the number of wetland areas alongside the river system – also generated significant reductions in nitrate. The realism of these simulations depends on how well the model represents reality, and different approaches to the same system have yielded different results. For example, other GCM simulations predicted an increase in flow rather than the decrease shown in Fig. 10.3. Another study using the same climate and emission scenario but different INCA parameterization predicted a decrease in stream nitrate by 2050 rather than an increase (Skeffington 2008). Nevertheless, the results give some impression of the likely effects of management options, and also show the long response times before any improvement occurs. More explicit modelling of groundwater transport in chalk catchments predicts even longer response times (Jackson *et al.* 2007).

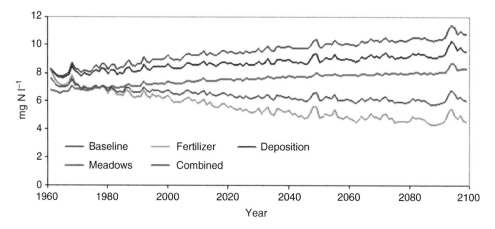

Figure 10.3   The effects of climate change, simulated by the INCA-N model, in the river Kennet for 1960–2100, together with the effects of various adaptation strategies. These were 'fertilizer', reduce N fertilizer input by 50%; 'deposition', reduce atmospheric N deposition by 50%; 'meadows', increase denitrification fourfold by constructing water meadows; 'combined', apply all three strategies but at half the rate.

These reductions also need to be seen in context. Inorganic nitrogen concentrations in rivers draining pristine catchments are often very low (<0.2 mg N l^{-1}; e.g. Perakis & Hedin 2002). The Kennet catchment has been at least partly agricultural land since the Bronze Age, and hence, it is likely to be a long time since conditions were uninfluenced by humans. However, the modelling estimates for the Lambourn (the major tributary of the Kennet) suggest that even in the 1920s, before the widespread use of artificial N fertilizers in the catchment, stream nitrate concentrations were only about a ninth of those currently observed. From Fig. 10.3, this would imply about 1 mg N l^{-1} for a 1920s nitrate concentration in the Kennet. Therefore, even the most effective mitigation option is unlikely to cause a return to pristine conditions.

Dynamic models other than the INCA models have been used and developed in the Euro-limpacs project. Andersen *et al.* (2006) analysed climate change impacts on hydrology and nutrient dynamics in the 113.5 km^2 Gjern River Basin in Denmark with the NAM-MIKE11-TRANS model chain and with regional climate model HIRHAM predictions used for meteorological prediction. HIRHAM predicted a non-significant increase in mean annual precipitation of 47 mm (5%) and a significant ($p < 0.001$) increase in mean annual air temperature of 3.2 °C (43%) between the control (1961–90) and the scenario (2071–2100) periods. The precipitation-run-off model (NAM) used these data to forecast that the mean annual run-off from the river basin would increase by 27 mm (7.5%, $p < 0.05$) between the control and the scenario periods. Changes in the extremes were larger: run-off during the wettest year in the 30-year period increased by 58 mm (12.3%). The most dramatic change in hydrological regime was modelled for the headwater Gelbæk stream draining an 11.1 km^2 loamy catchment. Large (40%–70%) and significant ($p < 0.05$) reductions in run-off during late summer (August to October) were predicted. Conversely, the simulation of run-off in the neighbouring headwater Dalby stream draining a $10.5\,km^2$ sandy, groundwater-dominated catchment suggested almost no change in monthly run-off during the summer. Other results from groundwater-fed catchments (e.g. Jackson *et al.* 2007) also suggest that the effects of climate change will be apparent more quickly in systems with impermeable geology or loamy or clay soils where the hydrology is more responsive than in those catchments underlain by permeable geology or sandy soils where the groundwater will buffer short-term variations. An increase in flood plain inundation from an average of 34 to 51 days per year was predicted for the Gjern catchment. Total N exported from the river basin increased by 7.7% even though N retention in the river system was forecast to increase. This was more than counterbalanced by an increase in the transfer of N from land to surface waters.

Linking models

Many specialized ecosystem models are available which give robust and detailed predictions in a limited scientific field. With climate change, the possible environmental impacts are diverse and interconnected and involve the responses of a myriad of processes in whole catchments. Integrated management strategies will be needed to tackle the problems, and to model responses on a catchment scale individual models also need to be integrated (e.g. Andersen *et al.* 2006; Evans *et al.* 2006). However, linking models poses a number of challenges. The use of the output of one model as the input to the next requires that the models operate on the same spatial and temporal scale and on about the same level of complexity.

For most applications, it is not necessary to programme hard links between the models so that they operate as a single entity. This is difficult as individual models were rarely designed with this in mind, and input and output data formats are unlikely to be compatible. Nevertheless, it is essential for some applications such as those involving multiple model runs, for example, Monte Carlo analysis (see "Uncertainty" Section) or the creation of response surfaces. Manual adjustments of the outputs of one model so that they are suitable as inputs to the next allow independent models to be linked reasonably easily but are time-consuming and can involve considerable skill and scientific judgement. Much more work remains to be done, particularly so that the aspirations of the Water Framework Directive for integrated resource management over whole river basins can be addressed. An example of the approach and potential of linking models to simulate processes at the catchment scale is provided by the study of Kaste *et al.* (2006) on the Bjerkreim catchment in Norway.

The Bjerkreim River Basin ($685\,km^2$) has an average run-off of $2430\,mm\,yr^{-1}$ and discharges into an estuarine area (58°28′N; 5°59′E) near Egersund in south-western Norway. The land cover is dominated by non-forested, mountainous areas (~60%) and is typical of the inner south-western parts of Norway. Water surfaces, peatlands and heathlands make up about 20% of the area, while forests and agricultural land cover 15% and 5%, respectively (Kaste *et al.* 1997). Nitrogen deposition in the Bjerkreim area is the highest in Norway, $15–23\,kg\,N\,ha^{-1}\,yr^{-1}$ (wet + dry) due to both high precipitation rates and relatively high N concentrations (Tørseth & Semb 1997). The main threats to aquatic ecosystems in the catchment are acidification and nitrogen enrichment, with substantial leaching of atmospheric-derived N from soils to surface waters already happening. The question addressed in this study was whether climate change was likely to alter the situation so that more (or less) nitrogen leached into the water, and whether nitrate concentrations in the rivers, lakes and receiving coastal waters were likely to increase or decrease and, if so, by how much.

Data from two GCMs – ECHAM4 from the Max Planck Institute, Germany, and HadAm3H from the Hadley Centre, the United Kingdom, driven with two scenarios of greenhouse gas emissions (IS92a and A2, respectively) – were dynamically downscaled (see above) to a spatial resolution of $55\,km^2 \times 55\,km^2$ and a temporal resolution of 6h using a model related to that used to perform weather forecasting in Norway. The downscaled predictions for control periods were adjusted to match observed data in the catchment before being used to make climate change prognoses. These were for different periods: 2030–49 for the MPI IS92a scenario and 2071–2100 for the Hadley A2 scenario (hereafter called 'MPI' and 'Had', respectively). Predicted changes in nitrogen deposition due to presently agreed legislation were also included in the forecast scenarios. The climate change data were then used to drive four models linked to assess the effects on hydrology and nitrogen concentrations and fluxes in the river and its coastal fjord (Kaste *et al.* 2006; Fig. 10.4). These models were the hydrological model HBV (Sælthun 1996), the water quality models MAGIC (Cosby *et al.* 2001) and INCA-N (Whitehead *et al.* 1998a, b; Wade *et al.* 2002a) and the NIVA Fjord models (Bjerkeng 1994). The flow of data between models is shown in Fig. 10.4. HBV was used to predict flows and other hydrological variables needed to run INCA, MAGIC to calculate N retention in INCA, INCA to calculate stream nitrate concentrations and the Fjord model to assess the biological consequences in coastal waters. The HBV, INCA and MAGIC models were calibrated initially to small sub-catchments within the main catchment area, before extending the calibration to the whole Bjerkreim basin. After calibration, the models were tested against observed data from a control period, in which they performed acceptably, though the means were simulated better than the seasonal patterns, and extreme events and unusual meteorological conditions were more difficult still to simulate (Kaste *et al.* 2006).

The two downscaled climate scenarios projected a temperature increase in the study region of about 1 °C with MPI and about 3 °C with Had, and both predicted increased winter precipitation. Projections of summer and autumn precipitation were quite different between the two models, however: a slight increase with MPI and a significant decrease with Had. Because different models were used for different periods, it was not possible to determine whether this was a model effect or a function of time. Because of higher winter temperatures, the HBV model predicted a dramatic reduction of snow accumulation in the upper parts of the catchment for both the climate change scenarios. This, in turn, led to higher run-off during winter and lower run-off during spring snowmelt. With the Had scenario, run-off in summer and early autumn is substantially reduced as a result of reduced precipitation, increased temperatures and, thereby, increased evapotranspiration. MAGIC and INCA models predicted no major changes in

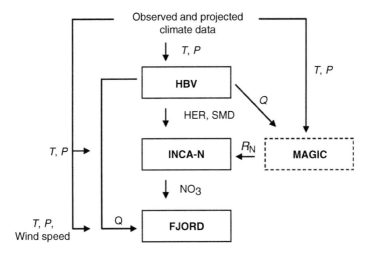

Figure 10.4 Data transfer scheme between models used to model the Bjerkreim River in Norway. The flow of data is indicated by arrows. *T*, temperature; *P*, precipitation; *Q*, water flow; SMD, soil moisture deficit; R_N, nitrogen immobilization as a percentage of input. Other abbreviations in Table 10.1. (From Kaste *et al.* 2006.)

nitrate concentrations and fluxes with the MPI scenario but a significant increase in concentrations and a 40%–50% increase in fluxes with the Had scenario (Fig. 10.5). These results arose from balances between the effects of climate change on nitrogen processes within the catchment. With the MPI scenario, the reduced N deposition was largely compensated by a temperature-driven increase in N mineralization (16%) such that the total available N in the system was nearly constant. With the Had scenario, however, N mineralization increased by nearly 40% compared to the control period. This was partly compensated by reduced N deposition and increased uptake by vegetation but counteracted by a reduction in the basin's ability to retain N.

Among the N retention processes included in the INCA-N model, two opposing factors operate at the same time. First, the long-term accumulation of N in the system leads to a decreased C : N ratio in the organic soil layer, which increases the risk of N leaching. Secondly, the increased temperature promotes vegetation growth and hence uptake of N. The net effect of all these processes was that at the Bjerkreim river outlet, the INCA-N model simulated a 4% decrease in mean NO_3^- concentration but a 4% increase in the mean *flux* with the MPI scenario. This is because the increased temperature accelerated aquatic N retention processes and thus reduced NO_3^- concentrations, whilst increased precipitation and thus stream water flow increased the total NO_3^- export from the basin. With the Had scenario, the stream water NO_3^- concentrations and fluxes were predicted to increase by approximately 50% and 40%, respectively. Here, the predicted decrease in annual flow reduced the NO_3^- export potential relative to the MPI scenario.

A possible consequence of increased nitrate concentrations is that the acidification of the river could increase, thus offsetting ongoing recovery from acidification due

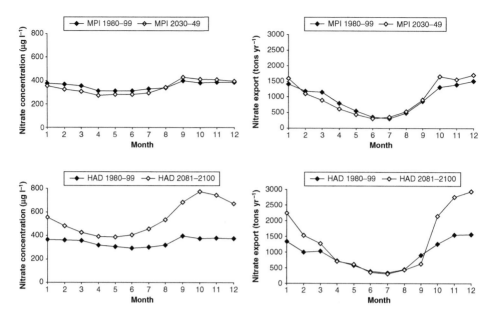

Figure 10.5   INCA-N-simulated NO$_3$ concentrations and fluxes at the Bjerkreim River outlet in 2030–49 and 2081–2100, based on the MPI and Had scenarios, respectively. (From Kaste *et al.* 2006.)

to reductions in acid deposition (see Chapter 7). The increased N loading may stimulate the growth of N-limited benthic algae and macrophytes along the river channels and lead to undesirable eutrophication effects in the estuarine area. Simulations made by the Fjord model indicate that primary production in the summer in the estuary might increase by 15%–20% with the Had scenario. Since this scenario does not entail any increase in NO$_3^-$ fluxes during summer (Fig. 10.5), the production increase may be a result of a longer residence time of the surface layer due to reduced freshwater inputs. It thus seems that the changes in run-off patterns may have a greater effect on algal production than the changes in nutrient conditions *per se*.

The linked-model approach presented here involved several sources of uncertainty. Among these were uncertainties related to (i) scaling of input data; (ii) climate and N deposition scenarios; (iii) model parameterization and calibration; (iv) model structure/ability to simulate key processes and (v) transfer of data with inherent uncertainties between models. All models involved proved to be fairly robust in reproducing current data, but the number of processes which have to be simulated to make predictions mean that the model results should be regarded as possible outcomes and not forecasts, especially beyond 2050.

However, some of the responses predicted by the models are founded in experimental and observational data. For instance, both laboratory- and large-scale experiments have suggested that decomposition and N mineralization show a faster response to a temperature increase than the corresponding N retention processes, at least during an initial phase of the warming process (Kirschbaum

1995; van Breemen *et al.* 1998). This is illustrated in this study by the increase in the N mineralization rate in response to the MPI and Had scenarios for air temperature and precipitation in the Bjerkreim River Basin. Decades with elevated atmospheric N inputs have increased the stores of N in soil, and empirical data have demonstrated a negative correlation between NO_3^- leaching rates and the C : N ratio of soil organic matter (Gundersen *et al.* 1998; MacDonald *et al.* 2002).

Even given current legislation on reducing N deposition, the MAGIC model simulated a slight decrease in the C : N ratio of soil organic matter in the Bjerkreim catchment by 2100. This implies a gradual increase in NO_3^- leaching rates even with no climate change. With the MPI and the Had climate scenarios, the decrease in C : N ratios and thus increase in NO_3^- leaching rates were more pronounced, offsetting the gains from reductions in N deposition. The extent to which this might happen depends on several uncertain factors. Among these are (i) the size of the N pool available for mineralization; (ii) the amount of additional carbon sequestered due to climate change and thereby affecting the soil C : N ratio and (iii) the actual temperature responses of the various N sink and source processes. Additionally, there are still large uncertainties associated with future NO_3^- leaching as a response to decreasing C : N ratios in catchment soils. The empirical model of Gundersen *et al.* (1998), which is included in MAGIC, is based on a large spatial data set, and we currently lack sufficient long-term data on C : N ratios and NO_3^- leaching to confirm the model on a temporal scale.

Overall, this study has shown that the results of linking different models can lead to helpful insights about the effects of climate change on water flows and nitrate leaching in catchments, which are not necessarily clear from running the models alone. Though the results are uncertain and not to be taken as a firm forecast, they highlight possible outcomes and also suggest future research that could be used to improve the model predictions.

Uncertainty

All model predictions are uncertain to some degree, and quantifying and preferably reducing uncertainty is one of the priorities for climate change modelling (e.g. Wilby 2005). The meteorological models used to drive climate change predictions are themselves uncertain, and this can have major impacts on predictions of water quality. For instance, Whitehead *et al.* (2006) and Wilby *et al.* (2006) used three different GCMs to predict changes in flow in the river Kennet for 2050: the results ranged from a 19% decrease to an 80% increase. To add to these, there are the uncertainties associated with water quality modelling structures, parameters and observations.

Though a good model can be calibrated to observed data, its predictions are rendered uncertain because of doubts over whether the model structure and parameters are good representations of the modelled system. It may be possible to calibrate a large number of different structures and parameters to fit the observed data equally well, but these differences lead to very different outcomes when projected into the future. This is known as the 'equifinality problem' (e.g. Beven 2006). There are various methods of estimating and reducing the degree of equifinality so that the number of possible model structures and parameters is reduced. One is to reduce the number of processes in the model, but this conflicts with the requirement for model realism. Water quality modelling is particularly difficult because the large spatial scale required means that it is hard to constrain model parameters. For instance, a water quality model may require the nitrogen concentration in the catchment soil, which is potentially measurable. But given that there will be a wide range of N concentrations in the catchment soils, which should be used? The mean of a large number of observations? Or should soils along flowpaths or close to rivers be given more weight? Or measurements from the upper parts of soil profiles which may have more direct hydrological influence? Experience shows that often models behave best with parameters that cannot be related to measured values in any obvious way.

There are, however, methods for estimating and reducing model uncertainty, and these have been extensively used in the Euro-limpacs modelling programme. One of these is Monte Carlo analysis (e.g. Rubinstein 1981). This is a well-established technique in which instead of running a model once with what is considered the best set of parameters to generate a single result, the model is run many times (typically thousands) with different values for input parameters selected from their potential ranges according to some scheme. The output is then a probability distribution which can be used to generate statistics such as confidence intervals. Moreover, the parameters with most influence on the uncertainty of the final result can be identified (known as sensitivity analysis) and the information used to identify optimum targets for research aimed at reducing uncertainties. The results of Monte Carlo analysis when applied to models can be surprising and counterintuitive. For example, the calculation of critical loads, the deposition thresholds used in pollution control policy, involves models which have 10–20 uncertain parameters. Monte Carlo analysis showed that the uncertainty in the calculated critical loads was typically less than the uncertainty in *any* of the input parameters (Skeffington *et al.* 2007). This behaviour is to be expected where parameters are independent and subject mainly to random errors.

Monte Carlo analysis has been applied to all the INCA models to identify key parameters and place confidence limits on model predictions (Wade *et al.* 2001; Cox & Whitehead 2004; Wilby 2005; Rankinen *et al.* 2006; Jarritt & Lawrence 2007). The results give an idea of the kind of variability that can be expected from water quality climate change simulations. For example, Fig. 10.6 shows the error bounds for predictions of phosphorus concentrations in the river Lambourn in the 2050s, using the INCA-P model (Whitehead *et al.* 2008). The 95% confidence bounds show a relatively narrow band of uncertainty during lower-flow summer months when conditions have been more stable. However, in winter months, the uncertainty increases as flow conditions become more variable and storms generate more run-off of water and nutrients.

One key achievement in the project has been to develop and apply a generalized Sensitivity and Uncertainty Analysis Tool, which uses a variant of Monte Carlo analysis called generalized sensitivity analysis. The specific method applied is that developed by Hornberger and Spear (1980). The Monte Carlo realizations are divided into those that best fit the observed data and the rest. The best results are

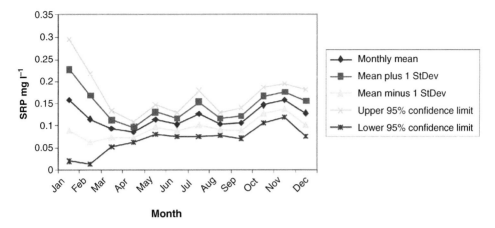

Figure 10.6   Soluble reactive phosphorus concentrations simulated by the INCA-P model for the 2050s in the River Lambourn, together with uncertainty bounds.

used to fit confidence intervals to model predictions. The tool can also be used for sensitivity analysis using Kolmogorov–Smirnoff statistics to produce a list of input parameters ranked in order of their importance in distinguishing good from bad simulations. It is being systematically applied to simulation results produced in the Euro-limpacs project from catchments across Europe in order to assess patterns in overall uncertainty, but this activity was not complete at the time of writing.

Conclusions

The ultimate goal of the catchment modelling in the Euro-limpacs project was to provide a pan-European assessment of the likely impacts of climate change on the flow regimes and water quality of river systems representative of key climate types across Europe. For a comprehensive assessment of water quality, a range of water quality parameters needed to be considered. Those considered were nitrogen, sediment, phosphorus, indicators of acidification (such as pH, alkalinity and acid neutralizing capacity), carbon and mercury. We are still some way from the ultimate goal, but we have made significant progress in three areas.

First, new catchment-scale models have been developed and existing ones improved leading to an improved capability in understanding how different factors and processes are integrated in response to climate or land-use management changes. Each of the INCA family of models has been harmonized in the sense that the versions for each of the different water quality indicators have the same structural representation of the key hydrological stores and pathways. The only differences between the models are the inputs and the biogeochemical cycle incorporated for each water quality measure. Thus, a toolkit of models now exists for a range of key pollutants with a common representation of catchment structure and hydrology.

Secondly, a methodology has been developed to convert outputs of temperature and precipitation from GCMs to values appropriate for use in the catchment-scale models. This standardized method for downscaling can be used with the INCA family of models to assess the likely impacts of climate change on freshwater quality (see "The Euro-limpacs modelling strategy" Section).

And thirdly, a semi-automated sensitivity and uncertainty tool has been developed that can be used to assess the effect of different parameter values on model performance and quantify the spread of modelled outcomes when making projections of future flow and water quality conditions.

The techniques of converting GCM output to meaningful inputs for catchment-scale models and the techniques of sensitivity and uncertainty analysis are not new, and other methods of uncertainty analysis are widespread. The novelty is the creation of a methodology that can now be consistently applied to river systems across Europe to assess changes in the flow and water quality for a broad range of water quality indicators, derive precipitation and temperature inputs to models in a clearly defined way and use a common methodology in sensitivity and uncertainty analysis that can be applied allowing comparison of the results at the pan-European scale.

The questions posed in the introduction can be mainly answered affirmatively. Models are never going to provide exact and unequivocal predictions of the effects of climate change on freshwater ecosystems. But the models developed during the Euro-limpacs project have been able to make useful predictions, qualified as they are by estimates of uncertainty. Models have also been used to explore plausible outcomes, to identify fruitful areas for further research so that future predictions will be less uncertain, to increase understanding of processes and to explore management options. Predictions concerning hydrology seem to be better founded than those concerning water quality, and ecological predictions are the most uncertain of all, reflecting a trend of increasing complexity and reduced understanding. There is still much research to do before we are close to being able to predict the effects of climate change on aquatic ecosystems.

References

Andersen, H.E., Kronvang, B., Larsen, S.E., Hoffmann, C.C., Jensen, T.S. & Rasmussen, E.K. (2006) Climate-change impacts on hydrology and nutrients in a Danish lowland river basin. *Science of the Total Environment*, **365**, 223–237.

Araújo, M.B. & New, M. (2007) Ensemble forecasting of species distributions. *Trends in Ecology and Evolution*, **22**, 42–47.

Arnell, N.W. & Reynard, N.S. (1997) The effects of climate change due to global warming on river flows in Great Britain. *Journal of Hydrology*, **183**, 397–424.

Bernal, S., Butturini, A., Riera, J.L., Vazquez, E. & Sabater, F. (2004) Calibration of the INCA Model in a Mediterranean forested catchment: the effect of hydrological inter-annual variability in an intermittent stream. *Hydrology and Earth System Sciences*, **8**, 729–741.

Berry, P.M., Dawson, T.P., Harrison, P.A. & Pearson, R.G. (2002) Modelling potential impacts of climate change on the bioclimatic envelope of species in Britain and Ireland. *Global Ecology and Biogeography*, **11**, 453–462.

Beven, K. (2006) A manifesto for the equifinality thesis. *Journal of Hydrology*, **320**, 18–36.

Bjerkeng, B. (1994). *Eutrophication Model for the Inner Oslo Fjord*. Report 2: Description of the contents of the model [in Norwegian]. Report no. 3113. Norwegian Institute for Water Research, Oslo.

van Breemen, N., Jenkins, A., Wright, R.F., *et al.* (1998) Impacts of elevated carbon dioxide and temperature on a boreal forest ecosystem (CLIMEX project). *Ecosystems*, **1**, 345–351.

Buisson, L., Grenouillet, G., Gevrey, M. & Lek, S. (2008) *Potential Impact of Climate Change on Fish Assemblages in French Streams. Deliverable No. 266.* Report from EU-FP6 Project Euro-limpacs (Integrated Project to evaluate the Impacts of Global Change on European Freshwater Ecosystems; Project No. GOCE-CT-2003-505540).

Chapra, S.C. (1997) *Surface Water Quality Modelling*. McGraw Hill, New York.

Cosby, B.J., Ferrier, R.C., Jenkins, A. & Wright, R.F. (2001). Modelling the effects of acid deposition: Refinements, adjustments and inclusion of nitrogen dynamics in the MAGIC model. *Hydrology and Earth System Sciences*, **5**, 499–518.

Cox, B.A. & Whitehead, P.G. (2004) Parameter sensitivity and predictive uncertainty in a new water quality model, Q_2. *Journal of Environmental Engineering*, **131**, 147–157.

Durand, P. (2004) Simulating nitrogen budgets in complex farming systems using INCA: Calibration and scenario analyses for the Kervidy Catchment (W. France). *Hydrology and Earth System Sciences*, **8**, 793–802.

Evans, C.D., Cooper, D.M., Juggins, S., Jenkins, A. & Norris, D. (2006) A linked spatial and temporal model of the chemical and biological status of a large, acid-sensitive river network. *Science of the Total Environment*, **365**, 167–185.

Fowler, H.J., Blenkinsop, S. & Tebaldi, C. (2007) Linking climate change modelling to impacts studies: Recent advances in downscaling techniques for hydrological modelling. *International Journal of Climatology*, **27**, 1547–1578.

Froese, R. & Pauly, D. (eds) (2008) *FishBase.* www.fishbase.org (12/2008).

Futter, M.N., Butterfield, D., Cosby, B.J., Dillon, P.J., Wade, A.J. & Whitehead, P.G. (2007a) Modelling the mechanisms that control in-stream dissolved organic carbon dynamics in upland and forested catchments. *Water Resources Research*, **43**, W02424, doi:10.1029/2006WR004960.

Futter, M., Whitehead, P.G., Comber, S., Butterfield, D. & Wade, A.J. (2007b) *Modelling Mercury in European Catchments: Preliminary Process Based Modelling and Applications to Catchments in Sweden and Scotland*, Deliverable No.182. Report from EU-FP6 Project Euro-limpacs (Integrated Project to evaluate the Impacts of Global Change on European Freshwater Ecosystems; Project No GOCE-CT-2003-505540).

Futter, M.N., Starr, M., Forsius, M. & Holmberg, M. (2008) Modelling the effects of climate on long-term patterns of dissolved organic carbon concentrations in the surface waters of a boreal catchment. *Hydrology and Earth System Sciences*, **12**, 437–447.

Gundersen, P., Callesen, I. & de Vries, W. (1998) Nitrate leaching in forest ecosystems is related to forest floor C/N ratios. *Environmental Pollution*, **102** (S1), 403–407.

Hornberger, G.M., & Spear, R.C. (1980) Eutrophication in Peel Inlet—I. The problem-defining behaviour and a mathematical model for the phosphorus scenario. *Water Research*, **14**, 29–42.

Hulme, M., Jenkins, G.J., Lu, X., *et al.* (2002) *Climate Change Scenarios for the United Kingdom: The UKCIP02 Scientific Report*. Tyndall Centre for Climate Change Research, School of Environmental Sciences, University of East Anglia, Norwich.

IPCC (Intergovernmental Panel on Climate Change) (2007) Summary for policymakers l. In: *Climate Change 2007: The Physical Science Basis. Contribution of Working Group I to the Fourth Assessment Report of the Intergovernmental Panel on Climate Change* (eds S. Solomon, M. Manning, Z. Chen, *et al.*). Cambridge University Press, Cambridge and New York.

Jackson, B.M., Wheater, H.S., Wade, A.J., *et al.* (2007) Catchment-scale modelling of flow and nutrient transport in the Chalk unsaturated zone. *Ecological Modelling*, **209**, 41–52.

Jarritt, N.P. & Lawrence, D.S.L. (2007) Fine sediment delivery and transfer in lowland catchments: Modelling suspended sediment concentrations in response to hydrologic forcing. *Hydrological Processes*, **21**, 2729–2744.

Kaste, Ø., Henriksen, A. & Hindar, A. (1997) Retention of atmospherically-derived nitrogen in subcatchments of the Bjerkreim River in Southwestern Norway. *Ambio*, **26**, 296–303.

Kaste, Ø., Wright, R.F., Barkved, L.J., *et al.* (2006) Linked models to assess the impacts of climate change on nitrogen in a Norwegian river basin and fjord system. *Science of the Total Environment*, **365**, 200–222.

Kilsby, C.G., Jones, P.D., Burton, A., *et al.* (2007) A daily weather generator for use in climate change studies. *Environmental Modelling and Software*, **22**, 1705–1719.

Kirschbaum, M.U.F. (1995) The temperature dependence of soil organic matter decomposition, and the effect of global warming on soil organic C storage. *Soil Biology and Biochemistry*, 27, 753–760.

Lawler, J.J., White, D., Neilson, R.P. & Blaustein, A.R. (2006) Predicting climate-induced range shifts: Model differences and model reliability. *Global Change Biology*, 12, 1568–1584.

Limbrick, K.J., Whitehead, P.G., Butterfield, D. & Reynard, N. (2000) Assessing the potential impacts of various climate change scenarios on the hydrological regime of the River Kennet at Theale, Berkshire, South-central England, UK: An application and evaluation of the new semi-distributed model, INCA. *Science of the Total Environment*, 251, 539–555.

MacDonald, J.A., Dise, N.B., Matzner, E., *et al.* (2002) Nitrogen input together with ecosystem nitrogen enrichment predict nitrate leaching from European forests. *Global Change Biology*, 8, 1028–1033.

Mohseni, O., Stefan, H.G. & Eaton, J.G. (2003) Global warming and potential changes in fish habitat in US streams. *Climatic Change*, 59, 389–409.

Nakićenović, N., Alcamo, J., Davis, G., *et al.* (2000) *Emission Scenarios. A Special Report of Working Group III of the Intergovernmental Panel on Climate Change*. Cambridge University Press, Cambridge and New York.

Perakis, S.S. & Hedin, L.O. (2002) Nitrogen loss from unpolluted South American forests mainly via dissolved organic compounds. *Nature*, 415, 416–419.

Rankinen, K., Karvonen, T. & Butterfield, D. (2006) An application of the GLUE methodology for estimating the parameters of the INCA-N model. *The Science of the Total Environment*, 338, 123–140.

Rubinstein, R.Y. (1981) *Simulation and the Monte Carlo Method*. John Wiley & Sons, New York.

Rushton, S.P., Ormerod, S.J. & Kerby, G. (2004) New paradigms for modelling species distributions? *Journal of Applied Ecology*, 41, 193–200.

Sælthun, N.R. (1996) The "Nordic" HBV Model. Description and documentation of the model version developed for the project. *Climate Change and Energy Production*. NVE Publication 7. Norwegian Water Resources and Energy Administration ISBN 82-410-0273-4, Oslo.

Skeffington, R.A. (2008) *The Consequences of Agricultural and Climate Change for Water Quality: Application of INCA-N to the CLUAM Predictions for the Kennet Catchment*. Deliverable No. 113. Report from EU-FP6 Project Euro-limpacs (Integrated Project to evaluate the Impacts of Global Change on European Freshwater Ecosystems; Project No GOCE-CT-2003-505540).

Skeffington, R.A., Whitehead, P.G., Heywood, E., Hall, J.R., Wadsworth, R.A. & Reynolds, B. (2007) Estimating uncertainty in terrestrial critical loads and their exceedances at four sites in the UK. *Science of the Total Environment*, 382, 199–213.

Thomas, C.D., Cameron, A. & Green, R.E. (2004) Extinction risk from climate change. *Nature*, 427, 145–148.

Thomas, C.D., Franco, A.M.A. & Hill, J.K. (2006) Range retractions and extinction in the face of climate warming. *Trends in Ecology and Evolution*, 21, 415–416.

Tisseuil, C., Wade, A.J., Tudesque, L. & Lek, S. (2008) Modelling the stream water nitrate dynamics in 60,000 km^2 European catchment, the Garonne, Southwest France. *Journal of Environmental Quality*, 37, 2155–2169.

Tisseuil, C., Vrac, M., Lek, S. & Wade, A.J. (2009) *Statistical downscaling of river flows*. Deliverable No. 382. Report from EU-FP6 Project Euro-limpacs (Integrated Project to evaluate the Impacts of Global Change on European Freshwater Ecosystems; Project No GOCE-CT-2003-505540).

Tørseth, K. & Semb, A. (1997) Atmospheric deposition of nitrogen, sulfur and chloride in two watersheds located in southern Norway. *Ambio*, 26, 258–265.

Vaughan, I.P. & Ormerod, S.J. (2005) The continuing challenges of testing species distribution models. *Journal of Applied Ecology*, 42, 720–730.

Wade, A.J. (2006). Monitoring and modelling the impacts of global change on European freshwater ecosystems. *Science of the Total Environment*, 365, 3–14.

Wade, A.J., Hornberger, G.M., Whitehead, P.G., Jarvie, H.P. & Flynn, N. (2001) On modelling the mechanisms that control in-stream phosphorus, macrophyte and epiphyte dynamics: An assessment of a new model using general sensitivity analysis. *Water Resources Research*, 37, 2777–2792.

Wade, A.J., Durand, P., Beaujouan, V., *et al.* (2002a) Towards a nitrogen model for European catchments: INCA, new model structure and equations. *Hydrology and Earth System Sciences*, 6, 559–582.

Wade, A.J., Whitehead, P.G. & Butterfield, D. (2002b) The integrated catchments model of phosphorus dynamics (INCA- P), a new approach for multiple source assessment in heterogeneous river systems: model structure and equations. *Hydrology and Earth System Sciences*, 6, 583–606.

Wade, A.J., Whitehead, P.G., Hornberger, G.M. & Snook, D. (2002c) On modelling the flow controls on macrophytes and epiphyte dynamics in a lowland permeable catchment: the River Kennet, southern England. *Science of the Total Environment*, **282–283**, 395–417.

Wade, A.J., Skeffington, R.A., Nickus, U. & Chen, H. (2008) *Methodology for Down-Scaling Climate Data Provided by the PRUDENCE Project for the HADAM3H/RCAO and ECHAM4/RCAO Models.* Deliverable No.433. Report from EU-FP6 Project Euro-limpacs (Integrated Project to Evaluate the Impacts of Global Change on European Freshwater Ecosystems; Project No GOCE-CT-2003-505540).

Whitehead, P.G., Wilson, E.J. & Butterfield, D. (1998a) A semi-distributed integrated nitrogen model for multiple source in Catchments INCA: Part I – Model structure and process equations. *Science of the Total Environment*, **210/211**, 547–558.

Whitehead, P.G., Wilson, E.J., Butterfield, D. & Seed, K. (1998b) A semi-distributed integrated flow and nitrogen model for multiple source assessment in catchments INCA: Part II – Application to large river basins in South Wales and Eastern England. *Science of the Total Environment*, **210/211**, 559–583.

Whitehead, P.G., Wilby, R.L., Butterfield, D. & Wade, A.J. (2006) Impacts of climate change on nitrogen in a lowland Chalk stream: An appraisal of adaptation strategies. *Science of the Total Environment*, **365**, 260–273.

Whitehead, P.G., Butterfield, D. & Wade, A.J. (2008) *Potential Impacts of Climate Change on Water Quality.* Report to the Environment Agency, Science Report – SC070043/SR1, Environment Agency, Bristol, ISBN: 978-1-84432-906-9.

Wilby, R.L. (2005). Uncertainty in water resource model parameters used for climate change impact assessment. *Hydrological Processes*, **19**, 3201–3219.

Wilby, R.L., Dawson, C.W. & Barrow, E.M. (2002) SDSM – A decision support tool for the assessment of regional climate change impacts. *Environmental Modelling and Software*, **17**, 145–157.

Wilby, R.L., Whitehead, P.G., Wade, A.J., Butterfield, D., Davis, R.J. & Watts, G. (2006) Integrated modelling of climate change impacts on water resources and quality in a lowland catchment: River Kennet, UK. *Journal of Hydrology*, **330**, 204–220.

11

Tools for Better Decision Making: Bridges from Science to Policy

Conor Linstead, Edward Maltby, Helle Ørsted Nielsen, Thomas Horlitz, Phoebe Koundouri, Ekin Birol, Kyriaki Remoundou, Ron Janssen and Philip J. Jones

Introduction – Decision making in the context of climate change

There is a priority need in Europe to maintain and improve the quality of our aquatic ecosystems and water resources and the benefits that they provide. Such benefits as clean water, flood risk reduction, fisheries and biodiversity underpin sustainable development, but their links to the natural environment have rarely been core to governmental decision making. The need to reverse this deficiency was recognized in the Plan of Implementation for the World Summit on Sustainable Development at Johannesburg, which has been endorsed by the European Union (EU).

There are three key areas where support is needed: developing strategies for the protection of ecosystems; improving water resource management and the scientific understanding of water; and the promotion of integrated water resource development.

The intricacy of aquatic ecosystems and their connections with land and atmospheric processes make catchment management difficult. Decision making is also complicated because under the EU Water Framework Directive, catchment management must cater for many stakeholders. The need to take account of ecosystem services (e.g. Defra 2007; Royal Society 2009) and implement the 'Ecosystem Approach' (*sensu* Convention on Biological Diversity) makes for further complexity. Uncertainty is inherent in all long-term decision making, but in the case of catchment management and water resources, uncertainty is acutely highlighted by the prospect of climate change.

Uncertainty implies that the outcome of a given policy or management decision cannot be predicted with reliability. Rational decision making handles uncertainty by identifying all possible outcomes and then weighting the value of

Climate Change Impacts on Freshwater Ecosystems. First edition. Edited by M. Kernan, R. Battarbee and B. Moss. © 2010 Blackwell Publishing Ltd.

each with the statistical probability that this outcome will materialize. Thus the choice with the highest expected benefit to one or more groups of people represents the optimal course of action, although this applies only in the immediate future and only takes into account human aspirations. Many or all of the choices included could conceivably be disastrous in a longer term.

This approach, using risk analysis, presumes that we know the probability distribution of different outcomes, which we may not (Kahneman & Tversky 1979). If the mechanisms leading to different outcomes are not sufficiently understood, no one can determine the probability that a particular outcome will follow from a decision (Walker *et al.* 2003). Hence, the poorer the understanding, the more uncertain are the predictions. This presents major challenges to governments and regulatory agencies.

However, uncertainty varies not only by level but also in nature. Uncertainty may reflect inadequate knowledge, or it may reflect inherent variability in human and natural systems (Walker *et al.* 2003). At least three sources of variability may be at play: randomness of natural processes; human behaviour, which often deviates from the rational model of decision making; and interacting social, economic and cultural phenomena. Uncertainty resulting from inadequate knowledge may be remedied or reduced through research. On the other hand, uncertainty owing to inherent variability is beyond management control and cannot be reduced. Uncertainty generally increases as the period of the policy or management decision extends into the future (Brewer 2007). This is because the availability of reliable data diminishes as decisions reach further into the future, and potential variability also increases over longer periods.

For freshwater ecosystems, the understanding of structure and function under current conditions is quite far advanced, and the remaining uncertainty can be reduced through further research. Likewise, the impact of key direct drivers of aquatic ecosystem change is well understood. Such drivers include, for example, temperature, hydrology, nutrients, acid deposition and toxic substances. By comparison, the impact of indirect drivers, such as the effects of climate change on agricultural practices and land use, and other social and economic changes, are less well understood.

Individual and social behaviours vary enormously. For example, the contribution to global food price increases of the recent expansion of biofuels production was not widely foreseen. Land-use patterns and nutrient levels may be affected by policies directly seeking to regulate them, but they are also affected by socio-economic factors that determine the relative costs and benefits of different farming options. Further, human behaviours acting as drivers of ecosystem change are shaped by individual, professional and cultural norms, for instance those concerning good agricultural practices (Nielsen 2009). Thus the impact of human action on aquatic ecosystems represents a significant source of uncertainty.

Climate change is currently the most evident source of uncertainty. Recent assessments of climate changes in Europe conclude that temperatures are likely to increase by 2.1 °C–4.4 °C by 2080 and that precipitation will either increase or decrease depending on the particular region (EEA 2007). Furthermore, it is predicted that extreme weather incidents will become more frequent. However, predictions for both temperature and precipitation changes are characterized as 'highly uncertain' (EEA 2007: 152). This uncertainty interacts with the inherent variability of freshwater ecosystems as well human impacts on the drivers discussed above. For instance, climate change may prompt migration of species and new crop patterns, changing habitats, migration of pest species and establishment of alien species. Likewise, extreme weather events might change the resilience of natural systems, for instance to nutrient influxes. Overall, extreme events could lead to non-linear pressures on existing systems, which increase variability and impede predictability. These uncertainties bear on assessments of mitigation costs and benefits. If the long-term effects of mitigation measures cannot be predicted, it will not be possible to estimate the costs of attaining a certain level of protection either.

Recognizing uncertainty as inherent in environmental policy formulation, the EU applies the precautionary principle to decision making. The principle was instituted in EU law with the Maastricht Treaty of 1992 (Treaty on European Union (92/C 191/01)) and is written in Article 174. The principle is not explicitly defined in the treaty, but at its heart is the notion that policy action against potential threats to the environment may be justified even without deterministic scientific proof of harm, as expressed in the German word *Vorsorgeprinzip*, which refers to acting with foresight (Andersen 2000; EEA 2001: 13). The Commission states that the precautionary principle may be applied as 'a risk management strategy' in some fields, specifically when scientific evidence is insufficient or uncertain and there are reasonable grounds for concern (Commission of the European Communities 2000: 10). Accordingly, measures adopted should be proportionate with a 'desired level of protection' as opposed to zero risk and they should rest on an examination of the benefits and costs of action or lack of it. Finally, the Commission calls for continued scientific evidence to re-examine policy measures. Thus, the EC use of the precautionary principle clearly holds that scientific uncertainty cannot justify a lack of action, even while it holds that scientific knowledge is the *sine qua non* on which to base any decision making.

However, in some circumstances, traditional methods for decision making, i.e. those that aim to identify single optimal choices, may not be appropriate and so alternatives must be found. One alternative is to use scenario-based analyses that use models to explore the consequences of different policy or management decisions under different scenarios. Rather than single-best policies, policy analysts look for robust policies that perform satisfactorily across different scenarios (Walker & Marchau 2003; Popper *et al.* 2005). Furthermore, scenario analysis allows the relative importance of different components to be assessed and makes transparent the trade-offs involved in each strategy. This potentially enables more resilient outcomes.

Scenario building presupposes that, given a particular set of drivers, the future can be predicted well enough to examine the outcomes of different policies across the scenarios. Walker and Marchau (2003) question that robust policies can be identified, particularly for climate change, and therefore question the value of scenario approaches. They advocate instead a stepwise approach: 'Take those actions now that cannot be deferred; prepare to take actions that may later become necessary; monitor changes in the world and take actions when they are needed' (Walker & Marchau 2003: 3).

Given this uncertainty, the maintenance of ecosystem structure and functioning should be an overarching management objective. The rationale for more holistic thinking in the management of water to achieve this objective is embedded within the Ecosystem Approach. The Ecosystem Approach is a strategy for the integrated management of land, water and living resources that promotes conservation and sustainable use in an equitable way and, as such, it forms a methodological framework for implementing projects that is underpinned by 12 principles. The Ecosystem Approach has been endorsed by the Convention on Biological Diversity (CBD) as the primary framework for the implementation of the Convention. It has also been endorsed by the European Commission at the World Summit on Sustainable Development and by the Ramsar Convention, and its principles are highly congruent with elements of the Water Framework Directive (WFD).

The Ecosystem Approach can help to challenge the natural science community to widen its perspective on how detailed science fits into the overarching policy framework (Maltby 1999). Regardless of methodology, however, the common thread is that action is necessary and that action should be based on the best available evidence. This chapter describes tools and decision-making approaches that should help policy makers and catchment managers define more robust strategies, or prudent first steps, in a world where there may be many impacts on freshwaters resulting from climate change.

Tools for decision making and their bases

Modelling

Computer models have a key role to play in supporting policy decision making. They might be used to predict future ecosystem conditions under a changed climate, or the likely outcomes of mitigation measures, and may help in understanding important ecosystem linkages and processes. There are, however, some potential limitations. Difficulties can arise when models are used to predict beyond the range of conditions for which they are calibrated, or if under more extreme external forcings the system passes a threshold whereby the model parameters and the embedded relationships between them no longer correctly represent the system.

In transferring modelling results to decision making, a key challenge lies in incorporating other sources of data, such as economic data, so that robust decisions can be made that consider the full range of implications of any actions for the environment, society and the economy. Models tend to focus on relatively few variables, although model coupling to create a suite of models within a single modelling framework can increase the range of parameters that can be successfully modelled, together with their linkages (e.g. the modelling of the interaction of nitrate and acidification). However, any management decision concerns the system as a whole and is not necessarily restricted to environmental variables. Evaluating the overall range of effects of climate change, or choosing between alternative mitigation options, requires that the relationships of the target system with wider society and the economy are considered and, potentially, trade-offs made between different outcomes (e.g. installing flood defences with the consequent loss of biodiversity, or allowing floodplains to flood and accepting the potential economic losses).

Establishing the system-wide impacts of a policy intervention or some other driver of change can be difficult, requiring an understanding of cause–effect relationships, including those among, and between, ecological, economic and social systems. It also requires these relationships to be quantified. There are varying degrees to which this can be achieved by use of modelling and other techniques, such as collection of monitoring data, space-for-time substitution or expert judgement. While model coupling can capture critical relationships for a limited number of ecosystem variables, modelling of ecosystems and their interactions with social systems and the economy is not yet possible. Thus techniques are required that allow the outputs from models of different system components to be judged in the context of each other. This does not capture the dynamic nature of some of the relationships between system parameters and is probably less accurate than dynamically coupled models, but it can provide valuable insights.

Linking models – a case study

The case study described here illustrates the use of a non-dynamic linkage between the Climate and Land-Use Allocation Model (CLUAM) and the Integrated Catchment Model of Nitrogen (INCA-N) to generate input data for the application of the Decision Support System (DSS) to the Tamar catchment described below. The objective of the Tamar DSS case study is to model the effects of climate change on diffuse pollution of nutrients within the catchment, and to evaluate alternative mitigation measures. While Global Climate Models can provide estimates of changes to temperature and rainfall under different CO_2 emission scenarios, the effect of such climate changes on land-use patterns could be more significant than the direct climate changes themselves on diffuse pollution. The cause–effect relationship between changes in climate and changes in land use is difficult to assess, affected as it is by both the climate requirements of different crops and the global and regional economics of crop prices, which are governed by the effects of climate change and the agricultural response to it elsewhere.

The CLUAM is a linear programming model of agriculture in England and Wales that provides a formal framework within which to examine likely land-use effects of changes in policy, market conditions and changes in climate (see Hossell *et al.* 1995; Parry *et al.* 1996, 1999; Jones & Tranter 2006). CLUAM treats agriculture in England and Wales as a single 'farm' consisting of a range of land types with regional variation. The national 'farm' can produce nine crop and four main livestock commodities, such as the main arable crops and meat and milk production, using a range of inputs and resources (e.g. fertilizer and land) on the different land types. CLUAM partitions the England and Wales land base into 15 Land Classes, each containing a mix of land cover types, i.e. arable, ley (short-term grass), permanent pasture and rough grazing, based on the Centre for Ecology and Hydrology's Land Classification System (Bunce *et al.* 1996a, b). Within each Land Class the land under each of the four land cover types is further subdivided into yield categories, reflecting the range of production potential due to precipitation totals and soil type and, for both arable and grassland, nitrogen input levels. The selection of production activities within each Land Class is constrained by:

1. The availability of land of different qualities within each Land Class (including the possibility of converting one type to another, e.g. ploughing permanent pasture to create arable land) and the ability to switch resources between uses;
2. The total volume of production (reflecting consumer demand) and input use required;
3. Policy constraints that restrict the areas of production activities and input use, or impose specific land-use patterns in certain areas to conform to environmental or other objectives (e.g. quotas, limits to input use in designated areas such as Nitrate Vulnerable Zones).

Within these constraints, land use is determined by the CLUAM according to the maximum profit that can be earned from all the possible activities on all the parcels of land. Both the outputs and inputs, for all the scenarios, were measured in terms of mid-1990s (base year) prices in order to allow direct comparison of results for the future time periods in equivalent value terms.

The CLUAM generates the following outputs, at the Land Class, regional and national levels:

1. Changes in livestock numbers and crop and grassland areas;
2. Areas of land under different land types and the area falling out of agriculture;
3. Areas of land transfers (between land cover types and reflecting land improvement);
4. Change in the use of inputs, including fertilizer and chemical use per hectare and in aggregate.

The CLUAM was adapted for use with river catchments by the inclusion and delineation of the Conwy, Kennet, Tamar and Wye catchments within the model itself. Eight separate model runs have been undertaken. Four of these are 'reference' runs and four are 'scenario' runs. As the object of the modelling exercise is to capture the effect of climate change on land use, the four reference runs represent the future without climate change, but including projections of future social and economic developments derived from the A2 and B2 climate scenarios described in the Intergovernmental Panel on Climate Change (IPCC) Special Report on Emissions Scenarios (SRES; IPCC 2000). The four scenario runs are based on the A2/B2 socio-economic futures, but also include the Hadley Centre Climate Model (HadCM) climate change forecasts for the same periods. Comparison of the scenario and reference runs yields the marginal effect of climate change alone. Provided alongside the results of these eight model runs are the results of a further run, called REF1990s, which represents broadly the current position, before recent (Fischler) Common Agricultural Policy reforms, which were implemented in 2005.

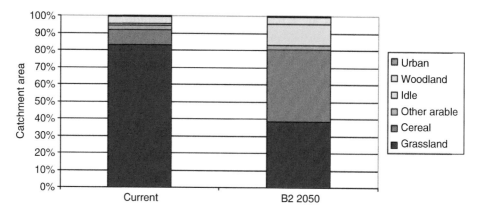

Figure 11.1   Current land-use distribution and projected distribution at 2050 under B2 climate scenario from CLUAM.

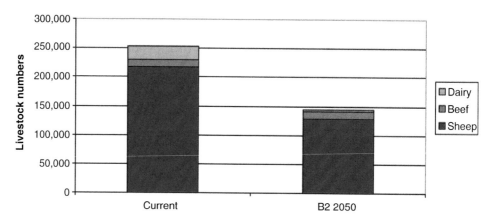

Figure 11.2   Current livestock numbers and projected numbers at 2050 under B2 climate scenario from CLUAM.

In the scenario runs, the CLUAM is driven by the price, demand and technology changes projected by the two global futures (A2 and B2) for both 2020 and 2050, both with and without the HadCM3 climate change projections. However, in the case of the climate change runs, the model is also constrained by the impact of changes to local growing conditions. These local climate changes can have both positive and negative effects, making some areas unsuitable for production of some crops, while making production of 'novel' crops suitable in other areas. For example, lower autumn rainfall levels may make planting of winter cereals difficult in some regions, while a longer and drier summer growing season may make production of crops like grain maize feasible in others.

Figures 11.1 and 11.2 show current land use, as reported in the UK Agricultural census (Defra 2004), and the predicted land use and livestock numbers from the CLUAM in 2050 under the B2 climate scenario. These data provide the inputs to the INCA-N model required to predict the outcome of the direct environmental

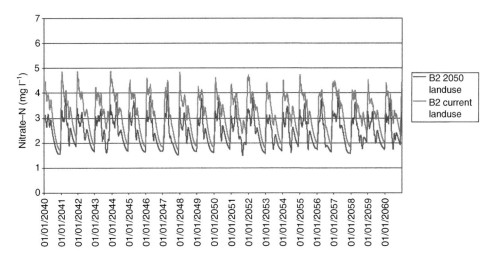

Figure 11.3   Effect of incorporating CLUAM land-use distribution into INCA-N model.

and indirect socio-economic effects of climate change on diffuse pollution. INCA-N, which requires data on organic and inorganic fertilizer application rates, land-use distribution and climate as inputs, predicts *inter alia* daily streamflow concentrations of NO_3–N.

Incorporating the indirect effects of climate change on diffuse pollution in this way can show significant effects on the predicted stream nitrate concentrations from INCA-N. Figure 11.3 shows the stream nitrate concentration at the downstream end of the Tamar catchment (Gunnislake Bridge) for the period 2040–60 under the IPCC B2 climate scenario using the current land use and the land use for 2050 predicted by CLUAM. The CLUAM predicts that the area of grassland and stock numbers in the catchment, particularly of dairy cattle, will be reduced. However, the areas of non-farmed land and cereal land will be increased. Coupled with a predicted reduction in the application of artificial fertilizers of 37% from the SRES projections, the net effect of these changes, as illustrated by Fig. 11.3, is that average stream nitrate concentration will decrease but still remain very high compared with undisturbed conditions.

The use of economic valuation as a tool for decision making

In some regions of the world, water will be an increasingly scarce commodity as precipitation falls and temperature rises. The value of water, however, is not simple to assess, owing to its nature as a 'public good'. Here we explain how the total economic value (TEV) of changes in the quantity and quality of water can be estimated. Such an assessment is integral to the design of economic incentives and institutional arrangements to ensure wise allocation of resources and sustainable management, now and in the future as the impact of climate change increases. The issues involved are illustrated by presenting a case study for the Cheimatitida wetland, located 40 km Southeast of Florina in Northwest Greece, a region where future water resources will be especially threatened by climate change.

Because they are not traded, many of the goods and services generated by lakes, rivers and marginal wetlands are hard to quantify in monetary terms and the consequent risk of their being neglected in policy making has received attention from economists. Water is one of the most important natural resources on which development and indeed survival are based. It is indispensable for domestic use and at the same time of vital importance to agriculture and industry.

Water resources have been degraded and depleted through shifts in climate and by more direct human activities. Integrated water resource management, linking social and economic development with protection of natural ecosystem functioning, is needed if sustainable development is to be achieved. To design and implement efficient and effective policies for water resource management, the TEV of the benefits generated by its several services and functions needs to be determined. Given that many of these benefits (including indirect ones such as amenity, or general ecosystem support) are not reflected in market prices, economists attempt to estimate the true resource value with the use of alternative valuation techniques.

TEV distinguishes between the value that individuals derive from consuming the environmental resource (use values) and the value that they derive even if they do not use it (non-use values). Use values can be classified into direct use, indirect use and option values. Direct use values come from consumption as drinking water, irrigation or industrial raw material. For most private goods, value is almost entirely derived from their direct use. Many environmental resources, however, perform functions that benefit individuals indirectly: indirect use values of aquatic habitats include benefits such as flood control, nutrient retention and storm protection. Finally, option value recognizes that individuals who do not currently use a resource may still value the option of using it in the future. The option value for water resources therefore represents their potential to provide economic benefits to human society in the future.

Non-use values (Krutilla 1967) can be classified into existence value, bequest value and altruistic value. Existence value refers to the value individuals may place upon the conservation of an environmental resource, which will never be directly used by themselves or by future generations. Individuals may value the fact that future generations will have the opportunity to enjoy an environmental resource, in which case they might express a bequest value. Altruistic value means that even if the individuals themselves may not use or intend to use the environmental resource themselves, they may still be concerned that the environmental good in question should be available to others in the current generation.

The challenge for economic valuation is to assign money value to non-market goods and services and thus assist policy makers in determining policy priorities. Over the last several decades economists have developed and refined methods for estimating the non-market values of goods and services. These non-market valuation methods can be categorized as revealed and stated preference methods, depending on whether they are based on existing surrogate markets or constructed hypothetical markets.

Revealed preference methods work by analysing actual markets that are related to the non-market resource under valuation. Information derived from observed behaviour in the surrogate markets is used to estimate willingness to pay (WTP), which represents an individual's valuation of, or the economic benefits derived from, the environmental resource. The essence of the Stated Preference (SP) approach is that the market for the good is 'constructed' through the use of a hypothetical scenario. Consequently, stated preference techniques circumvent the absence of markets for environmental goods and services by presenting consumers with hypothetical markets in which they have the opportunity to pay to use or protect, or accept compensation for the loss of the environmental good or service in question. Two approaches used to derive such values are the Contingent Valuation (CV) method and the Choice Experiment (CE) method.

In a CV study, respondents are asked their maximum WTP (or minimum willingness to accept (WTA) in compensation) for a predetermined increase or decrease in environmental quality. They are offered a change in the quantity or quality of a good at a given cost, and either accept or refuse the payment of the suggested cost. To provide an accurate WTP measure, the survey must meet the accepted standards of survey research (Arrow *et al.* 1993). In a CE study, the environmental resource is defined in terms of its attributes and the levels these attributes would take with and without sustainable management of the resource. Attributes of a water resource could include biodiversity and water quality, whereas levels of biodiversity could, for example, include the number of endangered bird species conserved. A monetary cost/benefit attribute is also included to allow for the estimation of WTP or WTA values. Experimental design methods (Louviere *et al.* 2000) are used to construct different profiles of the environmental good in terms of its attributes and levels of these attributes. Two to three such profiles are assembled in choice sets, which are in turn presented to the respondents, who are asked to state their preferences in each choice option. Statistical analysis is used to estimate the factors that affect choice of an environmental good as a function of its attributes and the levels these attributes take. Respondents' WTP or WTA for a change in any one of the attribute levels is calculated as the trade-off they make between the level of that attribute and the monetary cost/benefit attribute.

The WTP or WTA values estimated from CV and CE studies can then be used in Cost Benefit Analysis (CBA). CBA is an analytical tool based on welfare theory, which is conducted by aggregating the total costs and benefits of a project or policy over both space and time (Hanley & Spash 1995). A project or policy represents a welfare improvement only if the benefits net of costs are positive. Different management options will yield different net benefits and the option with the highest net benefits is the preferred or optimal one.

A CE was implemented to address the issue of efficient management of the Cheimatitida wetland (Birol *et al.* 2006a). Based on expert consultations, literature review and discussions with local people, four wetland attributes that are expected to generate non-use values were selected. These were: (i) biodiversity; (ii) open water surface area; (iii) inherent research and educational values that can be extracted from the wetland; and (iv) values associated with environmentally-friendly employment opportunities. The levels these attributes would take, with and without sustainable management efforts, were determined with wetland experts at the Greek Biotope and Wetland Centre. The fifth attribute was a monetary cost attribute, levels of which were determined through a previous CV study regarding this wetland (Birol *et al.* 2006b). Using these five attributes and their levels, experimental design methods were employed to generate choice sets containing alternative wetland management scenarios and an option to select neither scenario. An example of a choice set is presented below.

Sample choice set

Which of the following wetland management scenarios do you favour? Option A and option B would entail a cost to your household. No payment would be required for 'Neither management scenario' option, but the conditions at the wetland would deteriorate to low levels for biodiversity, open water surface area and research and education attributes, and no locals would be re-trained.

	Wetland management Scenario A	Wetland management Scenario B	Neither management scenario A nor management scenario B:
Biodiversity	Low	High	
Open water surface area	Low	Low	
Research and education	High	Low	I prefer NO wetland management
Re-training of locals	50	50	
One-off payment	€ 3	€ 10	
I would prefer:	Choice A ☐	Choice B ☐	Neither ☐
(Please tick as appropriate)			

The CE survey was administered in February and March of 2005 with face-to-face interviews with 407 members of the Greek public located in eight towns and two cities. These locations were selected to represent a continuum of distances from the Cheimaditida wetland, as well as rural and urban populations. The public's WTP for improvements in each one of the four attributes were estimated and it was found that on average the Greek public is willing, on average, to pay €7–8.4 per person for conservation of high levels of biodiversity; €6.5–10 for provision of higher levels of their open water surface area; €3.2–6.2 for investments in education and research; and €0.07–0.17 for re-training of a local farmer in an environmental-friendly employment. These WTP values represent the economic benefits the Greek public derives from higher levels of these attributes, which when combined represent the cost they are willing to pay for sustainable management of the wetland. When these economic benefits were compared with the costs of providing higher levels of these attributes, it was found that the benefits far exceeded the costs, which means investments in sustainable management of this wetland would bring about welfare improvements in Greece.

The results indicate that there are positive and significant benefits to the sustainable management of the Cheimaditida wetland. The impacts of social, economic and attitudinal characteristics of respondents on their valuation of wetland management attributes were also found to be significant, implying that there is considerable difference of opinion within the Greek public, which should be taken into consideration when assessing the provision of public goods, such as wetlands (Birol *et al.* 2006a).

Transfer of science into policy

Policy makers steer the direction of research by means of research funding policy. For example, in the European Research Programmes FP5 and FP6 several projects have been funded to create knowledge and develop methods for the implementation of the WFD. In FP6 stronger emphasis has been laid on global change. This underlines the possibilities for the EU Commission to channel research activities according to policy needs. The same applies in the member states. However, the link between science and policy is very variable among member states and within different aspects of water management.

Scientists may improve the exchange of knowledge by focusing on research topics that are relevant to the needs of society and policy/decision makers (Quevauviller & Thompson 2005), though in effect this may often be development of pre-existing research, i.e. applied research, rather than new fundamental research. There is no *a priori* way that fundamental research can be predicted to be relevant or not. Appeals to provide 'useful' knowledge, i.e. with direct policy implications, are often aggravating for scientists. The principles of academic freedom may lead to research outputs that do not meet the immediate requirements of policy making in a way contract research would do. Yet, even when not engaging in user-orientated research, scientists may contribute to policy development by entering into public discussions about their research to help define policy problems and to influence future funding (e.g. Day *et al.* 2006).

In the field of water policy, three groups of users can be identified: policy makers, decision makers at the operational level and the public at large. Each of these groups has different information needs. Policy makers need current information on drivers and their impacts; they also need information about the expected effects of possible policies, as well as the costs of implementing them. Further down the line, policy makers will need reviews of the responses of humans and ecosystems to policy instruments, to assess their efficacy.

Decision makers, who are responsible for the practical implementation of policy, require more specific and detailed information on methods, technologies and good practices (Quevauviller & Thompson 2005). Practical tools and models that derive from research activity are useful at this level. Regarding the WFD, methods for involving the public in decision making will likely be much sought after by operational managers. A significant challenge in transferring science into policy is the sectoralism frequently encountered among decision makers and a lack of trans-disciplinarity among scientists. This can lead to conflicting policy objectives (for instance, conflicts between the objectives of agricultural support payments and biodiversity conservation) and insufficient attention paid to ecosystem linkages, and linkages between the environment and society. Attempts are being made to address these challenges by implementing an ecosystem approach in both research and policy formulation.

Finally, the general public will need general information about water quality and water quantity problems and how climate change may affect water quality and water quantity. Such information would enable citizens to follow the debate around water policy making and implementation and make them more responsive to policy instruments. Generally, transparent decision making tends to increase legitimacy of the decisions made. This assessment, therefore, points to the need for a diverse set of tools and methods for the communication of scientific knowledge.

The role of Decision Support Systems

Given the challenge of integrating science into decision making, techniques are needed that can help decision makers balance social, economic and environmental objectives. Such tools should include Decision Support Systems (DSS). Recently a number of reviews of the use of DSS in water management have been carried out by, for example, Horlitz (2006), Evers (2008) and Giupponi et al. (2007). The latter developed 'Guidelines for the development, implementation and application of DSS tools'. A common conclusion of all three works is that users of DSS tools should be involved in their development from the outset. Ideally, they would be involved in financing development projects as a way of increasing their motivation to both develop useful tools and subsequently use them. Many such development projects have encountered problems through lack of practical expertise in support of the research team. This is one of the reasons for the frequent failure of take-up of DSS tools by water managers and policy makers (Giupponi et al. 2007). Steps can be taken (Fig. 11.4) to improve this situation. Ideally, designers and users should meet at the start to discuss the objectives and contact should be maintained throughout. Funding for workshops and other expenses is needed and training of users in use of the DSS should start as early as possible.

In some cases there are unforeseen limits to the levels of cooperation that is possible between designers and intended users. Currently water managers are busy implementing the WFD and are very reluctant to take on the additional work of incorporating other considerations, such as the possible consequences of global change. This is perhaps where scientists have to take the lead and emphasize the extent and implications of the evidence now accumulated.

The Euro-limpacs Decision Support System

As part of the Euro-limpacs project, a DSS has been developed to evaluate catchment management strategies in the context of climate change. The DSS provides a GIS-based framework for integrating social, environmental and economic data through Multi-criteria Analysis (MCA). The DSS is implemented as an extension within the computer program ArcGIS. The resulting framework is intended to address specific and targeted management questions such as:

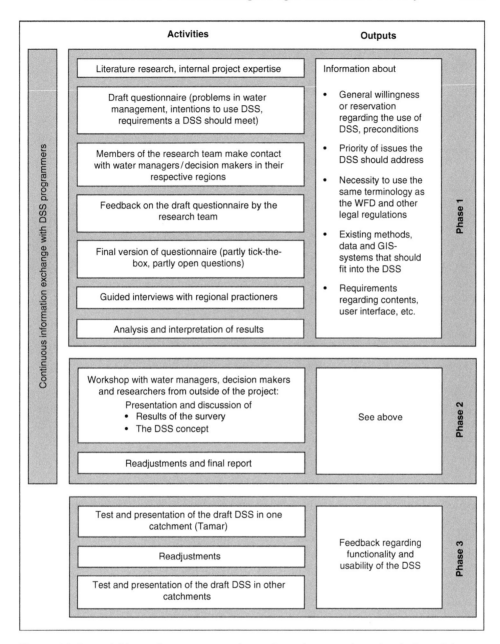

Figure 11.4 Activities to incorporate end-users' requirements into the design of the Euro-limpacs DSS.

- Will climate change affect some parts of a catchment more than others?
- What measures should be taken to mitigate the effects of climate change?
- Which part of a catchment should resources be targeted towards?
- Which measures most effectively tackle the defined problem?

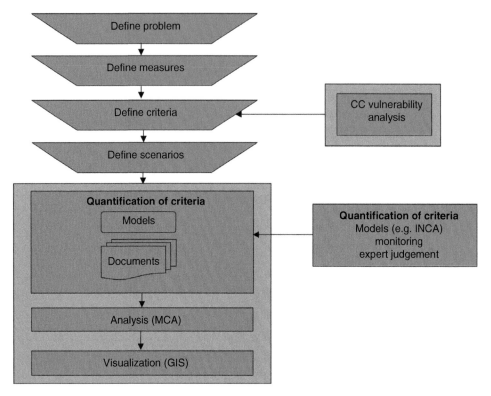

Figure 11.5 Conceptual model of the Euro-limpacs DSS showing the pre-application phases of problem definition and the main phases of the software application. Shaded areas represent stages of the application of the software tool.

There are several steps that the user of the DSS is required to follow, prior to the use of the software tool itself, in order to ensure that the problem to be addressed is structured in a suitable manner (Fig. 11.5). The first step is the definition of the problem in general terms. For instance, it may be defined as diffuse pollution from agriculture, or riparian flooding exacerbated by land-use changes. This step is intended to set the boundaries for the analysis and define the objective of the management strategies to be assessed.

Following this, the user identifies the potential measures that could be put in place to address the problem. Such measures could, for example, be reduction of fertilizer applications, reinstatement of riparian wetlands or reafforestation of upland areas. Summarized reviews of policies relating to different catchment management problems are available within the DSS that illustrate measures previously used.

Criteria are then defined for comparing alternative management options and these include those environmental, social and economic variables that are influenced by the measures that might be put in place. The user selects, from a predefined list, the types of ecosystems that are present in the area in question (e.g. small lakes, large rivers, wetlands) and the DSS tool indicates the ecosystem variables that are likely to be sensitive to climate change. This information has been collated from an extensive review of literature and expert judgement (Hering *et al.* Chapter 5, this volume). The tool also suggests relevant variables that might be used for monitoring these effects and provides a qualitative rationale for the processes involved. In this way the DSS guides the user to include those elements of the catchment that are sensitive to climate change into their analysis.

Using the information from this vulnerability analysis, and current management problems within the catchment, the final preparatory step in the DSS application is the definition of the scenarios, or management measures, that are going to be compared using the DSS. These might be combinations of climate scenarios and management measures applied to different extents. For instance, users may wish to compare the effects of 50%, 100% or 200% increases in the area of riparian wetland under different IPCC climate scenarios.

As the catchment area is divided into subunits (sub-catchments, administrative units, etc.), these preparatory steps in the application of the DSS result in a set of matrices, one for each scenario and management strategy, showing the decision criteria for each spatial unit. The primary purpose of the DSS is to provide a framework to facilitate a structured spatial analysis of these matrices. The user populates the matrices by quantifying the decision criteria for each scenario. This is done outside the DSS and can use a variety of different data sources and approaches, including models, databases or expert judgement. The particular mixture of information used to quantify the decision criteria will vary from application to application, depending on the available data and models.

Once the decision criteria have been quantified, the MCA tools within the DSS can be applied. The value of each decision criterion is converted to a score between 0 and 1 according to a function that is determined by the user of the DSS. The function is intended to convey the relative preference for achieving a certain value for the decision criterion. The form of these functions can be linear or non-linear depending on the decision criteria being considered. Normalizing the scores to the range 0–1 allows decision criteria with different units and relating to environmental, social or economic considerations to be integrated within a single analysis.

The decision criteria are also assigned weights. These weights are set by the user and reflect the importance of the particular decision criteria to the overall comparison of scenarios. For instance, the costs of implementing the measures may be the overriding consideration and this decision criterion would be given a higher weighting than others. Different interest groups may have different priorities and these can be reflected in the choice of weights to determine if the optimal management action is sensitive to the views of different groups. The final MCA score is then calculated as a sum of each of the individual weighted decision criteria scores. The results of the MCA are presented graphically, overlaid on the GIS map showing the spatial units of the assessment.

The DSS has been tested for seven catchments across Europe, of which one is presented here.

Tamar case study

The River Tamar is a small river flowing southwards across the south-western peninsula of England and separating the counties of Devon and Cornwall. Within the catchment, diffuse nitrate pollution from agriculture is a problem and was chosen as the issue to be addressed in this case study application of the DSS.

The management measures selected for representation within the DSS were realistic and reflect previous and ongoing management interventions to address diffuse pollution in the catchment:

• reduction in fertilizer applications and improved fertilizer practices
• reduction in stocking density
• shift from arable to pasture
• restoration of wetlands

Decision criteria were then chosen that could be used to assess the conditions of the sub-catchments under different scenarios of the application of these measures. The decision criteria were:

• nitrate concentration at the outlet of each of the major sub-catchments: yearly average and summer average (as this is an ecologically sensitive period)
• costs of implementing the measures
• biodiversity indicators: area of non-farmed land and area of wetland
• hydrological parameters: mean flow, high flow (Q5) and low flow (Q95)

Three different management options were selected for the application, differing in the extent to which each of the management measures outlined above are applied. These scenarios were:

• Business as Usual (BAU) – Current management and development policies continue unchanged
• Policy Targets (PT) – Policy targets are achieved. This is a 'moderately green' strategy
• Deep Green (DG) – Restoration of the catchment and sustainability are management priority

The climate change scenario years 2050 and 2085 and IPCC climate scenarios A2 and B2 were chosen for use in this exercise. The A2 scenario is characterized by increasing global population, a move towards self-reliance at the national scale and regionally orientated economic development. The B2 scenario is characterized by increasing population, but at a lower level than A2, with a greater emphasis on sustainable development at a local level. When overlaid on the three management options, therefore, there are in total 12 climate/management combinations. For comparison, the current conditions are included as a climate/management combination, making a total of 13.

For each of the climate scenario/management combinations the decision criteria were quantified. These decision criteria were quantified under each scenario/management combination using modelling and expert judgement. Water quality and water quantity criteria were modelled using the INCA-N model. INCA-N predicts stream nitrate concentration and flow using input data for land-use distribution, fertilizer application rates for different land-use types and daily climate variables (Whitehead *et al.* 1998a, b; Wade *et al.* 2002). Climate data for the A2 and B2 SRES scenarios were provided by Sveriges Meteorologiska och Hydrologiska Institut (SMHI) derived from the ECHAM General Circulation Model (Roeckner *et al.* 2003). The ECHAM model outputs were downscaled to the Tamar catchment using the method of Wade *et al.* (2008).

Data on land use, fertilizer application and stocking density for the BAU management options under the A2 and B2 climate scenarios were taken from the CLUAM outputs and modified according to the scheme set out in Table 11.1 for PT and DG options. Table 11.1 summarizes the data sources and assumptions made to quantify the driving forces for diffuse pollution under each climate/management combination. These data then formed the input data to the INCA-N model.

The area of non-farmed land is provided as an output from the CLUAM described above. Data for the potential maximum area of floodplain wetland is taken from a survey of wetland areas (Hogan *et al.* 2001) and this figure was used for the DG option. Current wetland areas are estimated to be approximately 25% of historical extent (Hogan *et al.* 2001). These data were modified according to the scheme set out in Table 11.1 for different management options.

The costs of implementing the measures are estimated using the Gross Margin figures for different agricultural activities calculated by the CLUAM. The cost of implementing the measures for BAU options are considered to be zero, as this is the laissez-faire option. Other management options are costed according to the estimated loss of total gross margin resulting from implementing the management measure. For instance, if there is a reduction in head of cattle between the A2 2050 BAU option and the A2 2050 PT option, then the cost is estimated as the reduction in head of cattle multiplied by the gross margin per head of cattle estimated by the CLUAM in 2050. The assumption is that in order to encourage the reduction in stocking density as set out in the option definition table, the cost, in terms of grants or subsidies to landowners, would have to at least equal the loss of income compared with the BAU option for that particular climate scenario.

Figure 11.6 shows results for the Current, Business as Usual, Policy Targets and Deep Green scenarios under IPCC A2 and B2 climate scenarios for 2050. It illustrates the potential usefulness of the DSS to decision makers at the catchment scale as it shows that, under the A2 climate scenario, putting in place progressively more stringent packages of the management measures defined in the scenarios (Business as Usual, Policy Targets and Deep Green) can improve the overall MCA score, indicating a more positive outcome in some of the sub-catchments. Conversely, under the B2 climate scenarios, the total MCA score decreases with progressively more stringent packages of the management measures in most

Table 11.1   Definition of scenarios and management options

	Current	Business as usual for A2 and B2 scenarios	Policy targets for A2 and B2 scenarios	Deep green for A2 and B2 scenarios
Fertilizer applications	Survey of fertilizer practice (Goodlass & Allin 2004)	Predicted by CLUAM	Lower of: 20% reduction on CLUAM prediction or Nitrate Vulnerable Zone limits	50% reduction on CLUAM prediction
Stocking density	Calculated from Defra (2004)	Predicted by CLUAM	Lower of: 15% reduction on CLUAM prediction or Hill Farm Allowance levels	60% of CLUAM prediction
Change landuse from arable to pasture	Current land-use distribution calculated from Defra (2004)	Predicted by CLUAM	50% reduction in arable area predicted by CLUAM	80% reduction in arable area as predicted by CLUAM
Wetlands	No change on current wetland extent.	50% loss on current wetland extent	50% of floodplains restored	All floodplains restored (Hogan et al. 2001)
Non-farmed land	Current land-use distribution calculated from Defra (2004)	Predicted by CLUAM	50% increase, or 1% of catchment area, whichever is the larger	100% increase, or 2% of catchment area, whichever is the larger

sub-catchments. This is primarily because there is a much higher proportion of cereals predicted under the B2 scenario compared with the A2 scenario by the CLUAM (0.1% and 42% of the farmed area for A2 and B2 Business as Usual scenarios, respectively), so encouraging the necessary transition to grassland to achieve the management targets in the Policy Targets and Deep Green scenarios is more expensive and reduces the preference for these management options. This suggests that under a B2 climate scenario, measures other than those represented in the DSS scenarios should be employed.

Disaggregating the analysis by sub-catchments allows decision makers to target management measures within the catchment. For instance, presenting the outputs as in Fig. 11.6 highlights sub-catchments where the management intervention is of greatest benefit. Under the A2 climate scenario, for instance, the management measures assessed in the DSS are most effectively applied in the middle reaches of the catchment.

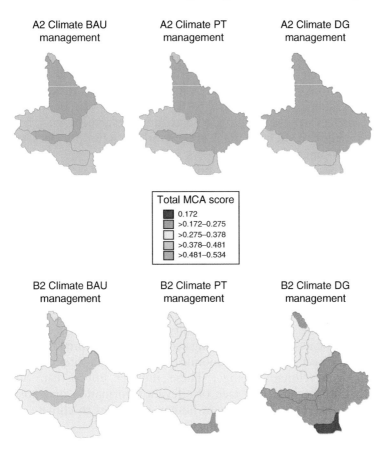

A2 Climate BAU management   **A2 Climate PT** management   **A2 Climate DG** management

Total MCA score
- 0.172
- >0.172–0.275
- >0.275–0.378
- >0.378–0.481
- >0.481–0.534

B2 Climate BAU management   **B2 Climate PT** management   **B2 Climate DG** management

Figure 11.6   DSS outputs for Current, Business as Usual, Policy Targets and Deep Green scenarios under IPCC A2 and B2 climate scenarios at 2050. Maps are coloured according to categorized total MCA scores for each sub-catchment and management/climate combination. A higher score indicates a greater preference of decision makers for that option.

Conclusions

The uncertainty inherent in making policy and management decisions about natural ecosystems presents challenges to both decision makers and to scientists. This uncertainty is set to increase as the consequences of climate change become apparent, making it more imperative than ever that new knowledge gained from research helps to inform policy and decision making. One of the key challenges for scientists is to translate their research into forms that make it useful for decision makers. This will require the development of tools for decision makers, and so these should be high-priority outputs from research projects and programmes. A range of science-based tools that can be applied to managing freshwater ecosystems under climate change already exist, some examples of which are presented here. However, because of their complexity, not all of the available tools are suitable for practical application by decision makers themselves and require application by experts. The chief requirements of decision makers for tools are simplicity, rapidity, transparency and reliability.

A key challenge for decision makers is to act on new scientific evidence and modify current policies and practices accordingly. The potentially rapid shifts in ecosystem conditions that may result from climate change make it imperative that decision makers are responsive to emerging scientific evidence and act quickly. This process would be helped by a greater engagement with the scientific community to guide the development of new tools required for decision making and policy development.

Building bridges between science, policy makers and stakeholders is another important challenge that has to be met if better decision making in the context of climate change is to be achieved. Links between policy makers and stakeholders are becoming more prevalent and embedded within policy development and implementation practice. Such participation has, for instance, been included in the implementation of the WFD and is also a key element of the Ecosystem Approach. New tools to facilitate this are being developed and the Euro-limpacs DSS provides one such example.

Information and approaches from many sources and perspectives should be integrated when addressing current and future freshwater ecosystem management problems. Within the scientific community, integration between the natural sciences and the social sciences, including economics, is essential in order to fully appreciate the indirect social and economic effects on freshwater ecosystems that might result from climate change. These indirect effects can often arise unexpectedly both because of a lack of relevant information on outcomes and because of the uncertainty inherent in human responses to climate change, as embodied in the range of IPCC climate scenarios.

The growing importance of the concept of ecosystem services in national and EU-wide policy frameworks will also make this scientific interdisciplinarity essential: social scientists are needed to understand societal choices and preferences, natural scientists to quantify and understand ecosystem functioning, economists to value the services derived from those functions and the costs of producing and protecting them, and policy makers to develop policy mechanisms to preserve or enhance ecosystem service delivery. All of the tools outlined here, and others, will be needed to achieve this.

References

Andersen, M.S. (2000) Forsigtighedsprincippet – og dets rødder i det tyske Vorsorge-prinzip (The precautionary principle and its roots in the German Vorsorge-Prinzip). *Samfundsøkonomen*, **1**, 33–37.

Arrow, K., Solow, R., Portney, P.R., Leamer, E.E. & Radner, R.H. (1993) Report of the NOAA panel on contingent valuations. Natural Resource Damage Assessment under the Oil Pollution Act of 1990. *Federal Register*, **58** (10), 4601–4614.

Birol, E., Karousakis, K. & Koundouri, P. (2006a) Using a choice experiment to account for preference heterogeneity in wetland attributes: The case of Cheimaditida wetland in Greece. *Ecological Economics*, **60**, 145–156.

Birol, E., Karousakis, K. & Koundouri, P. (2006b) Using economic methods and tools to inform water management policies: A survey and critical appraisal of available methods and an application. *Science of the Total Environment*, **365** (1–3), 105–122.

Brewer, G.D. (2007) Inventing the future: Scenarios, imagination, mastery and control. *Sustainability Science*, **2**, 159–177.

Bunce, R.G.H., Barr, C.J., Clarke, R.T., Howard, D.C. & Lane, A.M.J. (1996a) Land classification for strategic ecological survey. *Journal of Environmental Management*, **47**, 37–60.

Bunce, R.G.H., Barr, C.J., Clarke, R.T., Howard, D.C. & Lane, A.M.J. (1996b) ITE Merlewood land classification of Great Britain. *Journal of Biogeography*, **23**, 625–634.

Commission of the European Communities (2000) Communication from the Commission on the precautionary principle. COM (2000) 1.

Day, J.W., Maltby, E. & Ibáñez, C. (2006) River basin management and delta sustainability: A commentary on the Ebro Delta and the Spanish National Hydrological Plan. *Ecological Engineering*, **26**, 85–99.

Defra (2004) *June Agricultural Survey*. Department for Environment, Food and Rural Affairs, London.

Defra (2007) *Securing a Healthy Natural Environment: An Action Plan for Embedding an Ecosystems Approach*. Department for Environment, Food and Rural Affairs, London.

EEA (2001) Late lessons from early warnings: The precautionary principle 1896–2000. *Environmental Issue Report* No. 22.

EEA (2007) *Europe's Environment*. The fourth assessment. State of the environment report No 1/2007.

Evers, M. (2008) *Decision Support Systems for Integrated River Basin Management – Requirements for Appropriate Tools and Structures for a Comprehensive Planning Approach*. Shaker, Germany. ISBN 3-8322-7515-0 Draft Paper.

Giupponi, C., Mysiak, J., Depietri, Y. & Tamaro, M. (2007) *Decision Support Systems for Water Resources Management: Current State and Guidelines for Tool Development*. Harmoni-CA Report. IWA Publishing, London.

Goodlass, G. & Allin, R. (2004) *British Survey of Fertiliser Practice: Fertiliser Use on Farm Crops for Crop Year 2003*. The BSFP Authority, London. ISBN 1-86190-127-5.

Hanley, N. & Spash, C. (1995) *Cost Benefit Analysis and the Environment*. Edward Elgar Publishing, Aldershot.

Hogan, D., Maltby, E. & Blackwell, M. (2001) *Westcountry Rivers Project Wetlands Report*. Wetland Ecosystems Research Group, Royal Holloway Institute for Environmental Research, London.

Horlitz, T. (2006) *Summarised Report from First Catchment Level Meetings (on Requirements Regarding DSS in Water Management) and Interpretation of Outcomes*. Deliverable No. 132 for Euro-limpacs, EU-FP6-Project no. GOCE-CT-2003-505540.

Hossell, J.E., Jones, P.J., Rehman, T., *et al.* (1995) *Potential Effects of Climate Change on Agricultural Land Use in England and Wales*. Research Report No.8. Environmental Change Unit, University of Oxford.

IPCC (2000) *Special Report on Emissions Scenarios*. Intergovernmental Panel on Climate Change.

Jones, P.J. & Tranter, R.B. (2006) *Impacts of economic drivers on agricultural practices under climate change*. Centre for Agricultural Strategy, University of Reading, Reading. Euro-limpacs Project Deliverable 133, http://www.eurolimpacs.ucl.ac.uk/content/view/158/33/

Kahneman, D. & Tversky, A. (1979) Prospect theory: An analysis of decision under risk. *Econometrica*, **47** (2), 263–292.

Krutilla, J.V. (1967) Conservation reconsidered. *American Economics Review*, **57** (3), 777–786.

Louviere, J.J., Hensher, D.A., Swait, J.D. & Adamowicz, W.L. (eds) (2000) *Stated Choice Methods: Analysis and Applications*. Cambridge University Press, Cambridge.

Maltby, E. (1999) Ecosystem approach: From principles to practice. In: *The Norway/UN Conference on the Ecosystem Approach for Sustainable Use of Biological Diversity* (eds P.J. Schei, O.T. Sandland & R. Starnd), pp. 30–40. Norwegian Directorate for Nature Management/Norwegian Institute for Nature Research, Trondheim.

Nielsen, H.O. (ed.) (2009) *Bounded Rationality in Decision-Making. How Cognitive Shortcuts and Professional Values May Interfere with Market-Based Regulation (Issues in Environmental Politics)*. Manchester University Press, Manchester.

Parry, M.L., Hossell, J.E., Jones, P.J., *et al.* (1996) Integrating global and regional analyses of the effects of climate change: A case study of land use in England and Wales. *Climatic Change*, **32**, 185–198.

Parry, M.L., Carson, I., Rehman, T., *et al.* (1999) *Economic Implications of Global Climate Change on Agriculture in England and Wales*. Research Report No. 1., Jackson Environment Institute, University College London.

Popper, S.W., Lempert, R.J. & Bankes, S.C. (2005) Shaping the future. *Scientific American*, **292** (April 2005), 66–71.

Quevauviller, P. & Thompson, K.C. (eds) (2005) *Analytical Methods for Drinking Water – Advances in Sampling and Analysis – Water Quality Measurements*. John Wiley & Sons, Chichester.

Roeckner, E., Bäuml, G., Bonaventura, L., *et al.* (2003) *The Atmospheric General Circulation Model ECHAM5 Part I: Model Description*. Report No. 349: 1–127, Max–Planck Institute for Meteorology, Hamburg.

Royal Society (2009) *Ecosystem Services and Biodiversities in Europe*. EASAC Policy Report 09. www.easac.eu.

Wade, A.J., Durand, P., Beaujouan, V., *et al.* (2002) A nitrogen model for European catchments: INCA, new model structure and equations. *Hydrological Earth Systems Science*, 6, 559–582.

Wade, A.J., Skeffington, R.A., Nicus, U. & Chen, H. (2008) *Methodology for Down-Scaling Climate Data Provided by the PRUDENCE Project for the HADAM3H/RCAO and ECHAM4/RCAO Models*. Eurolimpacs Project Deliverable No. 433.

Walker, W.E. & Marchau, V.A.W.J. (2003) Dealing with uncertainty in policy analysis and policy making. *Integrated Assessment*, 4 (1), 1–4.

Walker, W.E., Harremöes, P., Rotmans, J., *et al.* (2003) Defining uncertainty: A conceptual basis for uncertainty management in model-based decision support. *Integrated Assessment*, 4 (1), 5–17.

Whitehead, P.G., Wilson, E.J. & Butterfield, D. (1998a) A semi distributed nitrogen model for multiple source assessments in catchments (INCA): Part I. Mode structure and process equations. *Science of the Total Environment*, 210–211, 547–558.

Whitehead, P.G., Wilson, E.J., Butterfield, D. & Seed, K. (1998b) A semi distributed nitrogen model for multiple source assessments in catchments (INCA): Part II. Application to large river basins in South Wales and eastern England. *Science of the Total Environment*, 210–211, 559–584.

12

What of the Future?

Brian Moss

Introduction

Cassandra, the soothsayer of Greek mythology, was given the power to predict the future accurately by the god Apollo. There was a condition, however, but she refused to bestow upon him the sexual favours he demanded in return. As punishment he decreed that she should still be able to predict correctly, but that no one would believe her. Thus it was that the wooden horse was wheeled into Troy and the city fell to the Greeks despite her warning. Intellectual endeavour is about using information from the past and the present to foretell the future as closely as possible. It inevitably falls short of Cassandra's divine powers in terms of accuracy, for there are huge elements of complexity, chaos and uncertainty, especially where matters of environment and human activity meet. But to be even a little forewarned is to be forearmed, though, like Cassandra's, the predictions may be scoffed at where they deviate from society's short-term aspirations for its comfort and convenience.

Predictions of climate change, based on physical models of the Earth's atmosphere and oceans are now generally accepted and, wisely, degrees of confidence are stated for them. Confidence is very high that major changes will occur in the current century, but lower when the details of exact changes in temperature and precipitation are concerned. Recent data showing that the extent of melting of the Arctic sea ice in 2007 was greater than anticipated by the most pessimistic Intergovernmental Panel on Climate Change (IPCC 2007) scenarios by about 40 years are salutary. The fact that the predictive models are purely physico-chemical and do not incorporate the effects of positive biological feedback makes the proposition of major change stronger but the details even more speculative. It is possible that the effects of warming in increasing respiration rates in the Earth's ecosystems more than photosynthetic rates will increase the flux of

Climate Change Impacts on Freshwater Ecosystems. First edition. Edited by M. Kernan, R. Battarbee and B. Moss. © 2010 Blackwell Publishing Ltd.

carbon dioxide to the atmosphere from respiration of carbon stores in soils and sediments, thus reinforcing the warming trend. It is not impossible that the converse, major negative feedbacks, may eventually occur in what is a colossal and complex earth chemical system to annihilate the roots of change in a grand manifestation of Le Chatelier's principle. And if the rates of processes, on which experimentation is possible to help the predictions, have a considerable uncertainty, the collective future behaviour of human societies, each with different cultures, historical conditioning, wealth and aspirations, is virtually unforeseeable, except in the very general terms of archaeological patterns.

Is it therefore of any value at all to scan the future? Not to do so is to abandon future generations to fate, to rest in a self-satisfaction and irresponsibility that might make things much worse. There are lessons of history that natural climate change, in concert with intrinsic factors, has often destroyed whole cultures and civilizations. Archaeologists might argue that civilizations repeatedly rose again, that we are merely in part of a usual cycle, but ecologists will point out that what was then local was reparable elsewhere, but that there is no 'elsewhere' for what is global. It would be a betrayal not to try to foresee consequences when even very imperfect predictions give some chance of avoidance.

Consequently we have attempted a foresight of how freshwater systems, and particularly their consequences for human societies, might change given 2 °C and 4 °C temperature increases within the 21st century. We contemplated the effects of an even greater increase now that 4 °C seems very likely whilst at least 2 °C seems inevitable. The potential devastation that would be created by a greater than 4 °C rise (widespread melting of the polar glaciers, rises in sea level of the order of metres, release of enormous quantities of methane from the northern peatlands and tundras), however, makes any prediction too uncertain. Complete chaos is, by its very nature, completely unpredictable. We decided to hold the optimistic view that human societies will do what is necessary to halt the rise at 4 °C and the over-optimistic view that temperatures might be held down to 2 °C. The chances of the latter, failing extensive popular insurrection and complete collapse of current economies, seem low at present. We also bore in mind that climate change will not be the only major problem confronting human societies in the next several decades. Population increase and the increasing depletion of crude oil are two further horsemen of the Apocalypse.

Our approach was to use discussions among experts from within Euro-limpacs to create scenarios for the present, and for these two temperature markers likely to be reached in the mid and late 21st century (and possibly rather earlier), respectively, in five broad geographical areas: Arctic–Boreal, e.g. northern Sweden or Finland; mid-latitude continental e.g. Poland, France, Germany; maritime peninsulas and islands, e.g. Iceland, the United Kingdom, Ireland, Denmark; Mediterranean lands, e.g. S. France, Iberia, Italy, Greece; and finally the high mountains of the Alps, Pyrenees and Carpathians. To standardize this thought experiment, we envisaged for all but the mountains a model catchment in each zone that is about 10^4 km^2 and about half-covered with natural or semi-natural vegetation and half with appropriate agriculture/ pasturage. It had some villages and a small town with 100,000 people and a maximum elevation of 1,000 m falling to a flat plain. What follows is, of course, a work of collective creative imagination. But it emphasizes the consequences of links between freshwaters and the human societies that universally depend on them. It gives a flavour rather than a precise menu, but this is what we think might happen.

The Arctic/Boreal zone

In the very early 21st century, people in the catchment of the River Erehwomos enjoyed a varied landscape, for the conifer forests of the lowlands thinned out to an arctic–alpine tundra on the mountains, where reindeer still grazed in a National Park of stark beauty, covered with lakes and peatlands. It was cold, of course, with a mean annual temperature of 5 °C, falling in winter to −10 °C and rising to 15 °C in summer, but the inhabitants were hardy outdoor people for the most part. The upland soils were permafrosted but the ice left the rivers and the many lakes in May. The rivers gushed, as the snowpack melted and soon there was a surge in activity. The salmon fishery, the fish farms and the tourism, logging and paper industries emerged from the winter quiet. Some of the rivers in the lowlands had been engineered and cleared of woody debris to smooth the passage of logs floated down in the past and there were several large hydroelectric dams generating power for more southerly latitudes. Farming was not prosperous but in the warmer lowland river valleys, mixed farms, with small herds of dairy cows, a few sheep and fields of oats and rye, separated the villages, with their ice-broken road surfaces, from the forests. The winter was lengthy with the days shortening to only an hour or two by December, but long and bright in summer. Most of the precipitation of around 1,000 mm fell as snow in winter leaving little to fall as summer rain. The snag, however, was that the low evaporation rate of only 200 mm left water so plentiful on the landscape that mosquitoes thrived in the many pools and wetlands, and the streams harboured so many blackfly larvae that a major summer irritation came with their hatch into aggressively biting adults.

As temperatures rose steadily towards the middle of the 21st century, the guides who led parties to shoot deer and moose in the forests had noticed that the tree line had crept upwards by several hundred metres. Their season had started much earlier, and spring flowers carpeted the soils in late April, rather than the early June of the 20th century. Foresters had noted increased tree growth and deer had become more plentiful. To some extent this was creating more damage around the tree line and restricting the otherwise inexorable advance of the forest on the upland tundra. Some of the birds, the golden plover for example, had disappeared northwards but others had come in from the south.

The winter snow depth had increased but the melt came over a month earlier and the rivers rose much higher than previously, cascading down to threaten the stability of some of the older hydroelectric dams and to flood the farmlands. There had been severe erosion of the lower-lying fields, and winter housing for the cattle had had to be reconstructed in newly bought land (for the few farmers who could afford it) in cleared forest at higher levels. From their viewpoint in the valleys, people noted grey patches of rock in summer where, 50 years before, the mountain glaciers had glistened white and the peaks had reflected the orange light as the sun set in the late summer evenings. The loss from the glaciers was somehow depressing and the mood was worsened by the rivers, which had been running with greyer water as the glacier-eroded deposits had steadily washed away in early summer. Later the water was turbid brown as the upland peats had unfrozen and eroded, and a tea-like stain had appeared even when the water was filtered. Sometimes, in the rapid melt of spring, large dams had formed from blocks of river ice, backing up huge quantities of melt water so rapidly that villages had had to be rapidly evacuated. In one case, 20 people refusing to leave their homes had been drowned when the dam gave way and a whole village was obliterated.

There had been some problems with the drinking water supply also. Where it came from upland lakes, the levels of some heavy metals had increased to above tolerable levels as the debris below the glaciers, previously frozen solid, had melted and weathered. The analysts had picked up also an increase in the conductivity and nitrogen concentrations in the river water, but these had not yet caused any perceptible problem.

There had been ecological changes in the rivers. Arctic charr, once a favoured sport fish, had declined, and even disappeared, in parts of the lowlands, though they could still be caught in the mountains, and some new fish were starting to move up from the main stem river further south, of which the Erehwomos was a tributary. Salmon were much scarcer and smaller and the spring runs were not so prominent, though their decline had been anticipated for many other reasons than warming. Rising human populations the world over and their need for food had meant that sea fish stocks had dwindled to almost nothing. The river bottoms had become more vegetated with willow shrubs, and moose and deer were found there more frequently. Wolves were heard a little more frequently as they too responded to the increased production.

The mosquito and blackfly problems had become worse, and the increase in flooding risks had led to some people moving out. Small farmers had been unable to cope but the wealthier ones were doing better as they bought up small farms and amalgamated them into bigger units and in some years risked planting new crops like barley and wheat that they could not have grown before. The power generators and foresters were happy too. Hydroelectric power was favoured by a government now realizing that action to curb carbon emissions should have been taken more vigorously and much earlier and there was investment to replace the ageing machinery. But the brown summer water, with the chocolate-coloured banks of sediment left in August and September when flows had fallen to levels rarely before experienced, worried the ecologists who were studying the catchment. Just what was happening to the huge carbon store of the upland plateaux and mountains? And though the forest growth had increased, so had the rate at which the debris on the forest floor had rotted. Where did the carbon balance now lie?

By the late 21st century, that question had been partly answered. Warming had been experienced rather faster than anticipated because carbon had been respired from the huge peatlands of the north as they warmed and dried in summer. An inhabitant of the Erehwomos catchment in the 20th century would barely have recognized it in the late 21st century. Much of the upland tundra had gone to scrubby forest, clinging to soils that had eroded rapidly; the reindeer had moved further north; the lowland forests, now with noticeably more deciduous trees, had been increasingly logged and were patched with new farms where slopes permitted it. The loss of agricultural potential in the much hotter world far to the south had meant that almost any farmable land in the northern latitudes had been cleared, even if the crops were poor. The season was still very short for although warming favoured growth, daylengths were unaffected and growth could still not begin until the early summer. Prices, however, had risen enough to make even a meagre harvest worth having. And many more people had moved northwards from Southern and Central Europe, resulting in a polyglot community that was struggling to reconcile its many different aspirations and customs. The area, indeed much of the north, had become a new frontier.

It rained much more and the winter snowpack was thin and lasted only a month or two in December and January. River flows were more even throughout the year, and water levels higher in spring, though the summer was dry. The Arctic charr had disappeared and the salmon were very scarce. Most fish were lowland species and attempts to replace species that had gone had misfired as a handful of these now predominated, all of them coarse fish (cyprinids). They were edible if not delectable. The sport fishing industry had declined and the increased population had struggled to feed itself, supplementing its diet with poached deer and moose, both of them now nearly extinct locally as a result. When ecological surveys had been carried out, it was found that the river communities had changed with loss of specialist cold-water species, such as stoneflies and the native crayfish, though a now rampant introduced species of crayfish had replaced it. There was much more algal growth on the rocks and bigger patches of sediment, for the intensified farming released more nutrients and the more even water flows no longer scoured out the bed as the spring freshets had done previously, when a billion tonnes of snow had melted over just a few weeks. Submerged plants colonized the sediments in summer and blackfly were scarcer, though still, like the lake mosquitoes, a nuisance. The local people had taken the law into their own hands on the latter, however, and insecticide was used rather liberally. One of the consequences had been the loss of cladoceran grazers from the lakes in the lowlands and an increase in algal growth. Some of the fish were now health hazards too because of the pesticide residues they contained.

This was a pity, for fish were generally more abundant in the lakes; the winterkills that had been frequent in shallow waters under the previous long winter ice no longer occurred. The period of ice cover was now quite short, but the fish were of small species, feeding on the bottom, rather than the tasty large salmonids that had previously been prized. Not that people cared very much now; food was food. And things were still changing, more rapidly year by year, it seemed. All over the northern latitudes, permafrost had been lost and methane in the peaty soils was being released whilst even more carbon dioxide was respired away from the peat itself. Each year, almost, was the warmest year yet. But life was tolerable. News from the tropics, that millions had died for lack of food, water and dry season temperatures that could barely be contemplated, and from the far north, where traditional hunting peoples had been unable to adjust and had died out, made even this now tattered landscape something of a haven.

The mid continental latitudes

High summer, in the catchment of the River Gutfluve, at the turn of the 21st century was close to idyllic. Its place, in the centre of mainland Europe, had brought the blessings of deep fertile soils, deposited 15,000 years earlier by the glaciers, still large patches of deciduous forest, a rich local culture, with notable festivals celebrating beer and music, wine and food. It was the culmination of 5000 years of human settlement that had, of course, removed the native wild forest ecosystem with its aurochs, bison, bear and wolf, but replaced it with a cultural landscape that was still interesting, pleasant to look at and equable to live in.

On the hills, a cold winter at $-4\,°C$ still led to an average yearly temperature of $9\,°C$ and summer maxima that could reach $20\,°C$. There were still mixed forests of conifers and hardwoods, and a forestry industry that produced timber from plantations for building and burning. In the lowlands it was generally above freezing in winter and rose to a sometimes uncomfortable $35\,°C$ in mid summer, but generally it was around $25\,°C$. Snow lay for 2 months on the hills but for a much shorter period in the lowlands, so that crops of wheat, maize, potatoes, and even some vines, graced the fields and the cattle were fat. Total precipitation of 700 mm gave a well-watered spring, but with 400 mm of evaporation, the late summer was dry and the crops needed to be irrigated from the river or the groundwater. Reservoirs had been constructed along the river for storage of water for drinking. Indeed, the river was mostly engineered in the lowlands and the flood plain had been lost to agriculture and riverside towns, but at the foot of the hills was a stretch where a system of traditional flood plain fishponds, managed with the experience of hundreds of years, produced succulent fish for the table. There was industry too, and a power station, close to the sea where the river discharge was highest, took river water for cooling its plant and put it back a few degrees warmer.

A few decades on, a $2\,°C$ rise in temperature and a shift in rainfall towards the winter had brought noticeable change in the landscape, but only small adjustments in the culture. Central Europe had coped with centuries of political turmoil; hotter summers, even the occasional whirlwind, were taken in their stride. Of course everything had changed but always to a degree; as yet no cliff had been fallen over. The snow lay only for a short time on the mountains; mostly it rained and the downpours were often torrential. The flood defences had been strengthened where the river abutted towns and villages, but not otherwise, so that quite often the agricultural land was turned back to a flood plain in winter. Not that this mattered much; the floods brought down a load of fertile silt from the eroding mountainsides and the more damaging summer floods were rare, for summers were now drier and hotter. There was much more maize growing and even a couple of attempts at sugar cane, but really it was yet too cool for that. The demands by government that more fuel come from biomass and ethanol had encouraged experimentation, however.

The upstream reaches of the rivers dwindled in summer and occasionally dried out. There was no longer a distinctive spring flood, just a modest rise when it had snowed and the invertebrate and fish communities of the river had changed to resemble much more those of the downstream reaches. New dragonflies and other insects had appeared and the surrounding hill forests were changing. There were more clear-felled areas as demand for biofuels increased, and more soil erosion. Outbreaks of insect pests had devastated some of the plantations and the European beech was spreading in the remaining forests along with more exotic trees, whose seeds had escaped from garden plantings. Ecologists found some adaptation through natural selection in algae and invertebrates but this had had very little measurable effect on the overall functioning of the rivers, ponds and reservoirs. They regretted the loss of highly specialist species of mountain springs, which now dried up in summer, but the populace in general was more concerned with the increasingly frequent summer heat waves. Sometimes summer temperatures became unbearable outdoors between mid morning and late afternoon.

The government had encouraged more food growing, for imported supplies had dwindled as other countries looked to their own needs and the previous practice of flying in fresh food from other continents had been made uneconomic by the hugely increased cost of fuel, following the passing of peak oil production 20 years before. The consequence was an increased leakage of nutrients to the rivers and their reservoirs, and to the groundwater. Algal blooms in the reservoirs had become more frequent, leading to greater costs of purifying drinking water. Rotting scums at the edges, as the water levels fell, even resulted in a resurgence of botulism in birds and poisoning of unfortunately thirsty sheep and dogs. Ponds in the catchment were increasingly drying up in summer, with a loss of habitat for amphibians and some large invertebrates that were unable to coexist with fish in the larger ponds and reservoirs. Some of these were now very rare. The attempts of the European Community to improve ecological quality through the Water Framework Directive of 2000 had been frustrated, at first by the general preference to maintain economic growth over environmental quality but then simply by the climate-induced changes in biological communities that had exposed the somewhat dated approach to ecology with which the Directive had been saddled. The scramble to grow more food, to protect buildings from flooding and to cling on to a car culture had all been nails in its coffin. There was administrative confusion. Warming was a global problem but international attempts to mitigate it had always been frustrated by the special pleading of one powerful nation or another. Successive government policies thus grasped at straws and followed fashions as the temperature continued to rise. Such confusion was nothing new. The people were used to it. But there was an air of foreboding.

Towards the end of the 21st century confusion still reigned. Temperatures were rising inexorably because sufficient measures to curb carbon emissions had been taken too modestly and too late, and biological feedback from the increased respiration of organic soils and loss of methane from the northern permafrost had greatly accelerated the net accumulation of greenhouse gases in the atmosphere. Despite mounting evidence, the lesson that conditions for human comfort on the planet were ultimately determined by huge biological forces and the colossal mobilization of the common chemical elements rather than by the puny mechanics of man-made economics had still not been finally accepted.

The Gutfluve catchment suffered torrential rainstorms and saw no snow in winter. In summer it sweltered and the streets of the towns and villages were deserted after early morning, coming alive again only in the evening. Electricity was rationed, so air conditioning was too expensive to use. Desperation had urged a conversion of every scrap of cultivable land to maize growing. Other crops did less well in the dry summer heat and fertilizer was used abundantly so that the downstream river and reservoirs were permanently green except immediately after torrential storms when the flood had temporarily washed the algae downstream and turned the water deep brown. The power stations demanded as much fuel from crops as they could get, for natural gas and oil supplies were now so low that they were reserved for public transport and manufacturing. In upstream sections, aquatic plants and filamentous blanket weed nearly blocked the flow as they grew on the accumulated banks of sediment, and were it not for the storms washing them out from time to time, it seemed that the rivers might become just a series of eutrophicated wetlands dominated by a few genera like *Typha*, which were such avid competitors that they overcame everything else. The fish fauna was almost entirely of carps and catfish, some of them deliberately introduced, and in the hottest years, tropical floating species like water cabbage and water hyacinth took hold in the downstream reaches and the edges of the reservoirs. Those ponds that were left, especially those surrounded by cropland, were festering masses of floating duckweed, foetid and near anaerobic below the surfaces and harbouring only the animals previously associated with gross organic pollution.

There had even been outbreaks of water-borne diseases not recorded in the area for centuries or ever before. For one thing, new species of mosquito had arrived and become infested with malaria from immigrants and travellers from Asia and Africa. For another, new viruses had invaded or mutated in the warmer conditions. And for a third, despite desperate attempts to maintain energy supplies through solar panels, wind and biomass burning, there were frequent power cuts, when the processes at the water purification works were interrupted. Cholera and typhoid were still not common but a variety of gut bacteria made life even more difficult for a population whose former culture had been deeply undermined. Life went on, though the frequent interchange that had typified the street and café culture previously had been replaced in summer by long withdrawn siestas in the heat of the day that led to a more isolated society preoccupied with family affairs. There was less interest in communality; the heat led to a sapping of energy and the landscape was so intensively cultivated that much of its charm had been lost. Summer walks through the fields and remaining woods were no longer so appealing and the confidence that had formerly been placed in the central government's ability to solve problems had been lost. It seemed there was a danger of lapse into the near anarchy of former city states, each looking to its own affairs.

Peninsulas and islands

On the western fringes of Europe, the River Seventine was the main waterway of the island of Hibscotia. The island had had a wet and windy climate for many centuries with cool winters and slightly warmer summers. Sometimes the summer temperatures reached 20 °C but typically they were around 15 °C and in winter they were generally above freezing. Though snow dusted the hilltops for a few days at a time, it usually soon melted and only about 400 mm of the 1,400 mm of precipitation evaporated. It was a wet place of grey skies, sodden pastures, dense alder and willow woodlands in those river valleys that had not been drained. Hibscotia was an odd, independent sort of island, for despite the humidity of its air and the quality of its mains drinking waters, the salesmen of moisturizing creams and bottled waters had, against all logic, reaped rich rewards. But then, it had traditionally been governed by the graduates of schools of economics, philosophy and law and not by those of science, so faced with the compelling evidence of climate change in the late 20th century, its inhabitants murmured regrets and good intentions and then continued to look after the economy and buy the bottled water and moisturizing creams.

The landscape was green, turning a yellowish green sometimes in the driest summers but its pastures yielded wholesome cattle and sheep, large crops of grass for the dairy industry and an abundance of wheat, other grains, roots and oilseed rape. The hills had long been grazed down to open grasslands, where naturally oak woodlands would formerly have stood, and any sense of landscape and ecological continuity had been obscured by the innumerable hedges and fences that divided the land into a chessboard of fields, with an occasional square covered by some still managed remnant of what the mediaeval landscape had been. These remnants were relatively few and mostly conserved as nature reserves. All semblances of intact ecosystems had long gone and there was a deep distrust of any move to reinstate birds or mammals that had once been native, but shot out long ago in the interests of one landowner or another. The catchment of the Seventine was thus largely bare of forest, apart from plantations of exotic conifers in the uplands, and almost entirely farmed. The island location, however, added a strong maritime dimension and the surrounding seas had long been overfished, whilst the coastline, perhaps the island's wildest feature, was indented with bays and rocky headlands, sand dune systems and estuaries pressing deep into the hinterland.

Flooding was a major preoccupation and most of the funds of its national agency for the environment went on sea defences, the leveeing of rivers and the drainage of flood plains.

Its once abundant ponds had been decimated several times over, as agriculture had become intensive in the last half of the 20th century and ponds had become inconvenient for the passage of large machines. A traveller crossing the coast by air would thereafter never be out of sight of a town and tracts of arable or pasture. It was such a heavily populated island that many rivers had been dammed for water supply, and in the slightly drier south-east the ground waters had been heavily mined through an exuberant policy of irrigation.

The inhabitants first noticed the effects, in the early 21st century, of warmer, wetter winters as mosses grew prolifically on the drives up to their houses and the rocks of their rockery gardens and the weeds in the flower beds persisted in the rarity of the frosts that had previously and conveniently killed them. Gardening was a national preoccupation, much in line philosophically with the controlled approach to the landscape in general. People also noticed, as the years went on, a tendency for water to become scarce in late summer as the levels in the many reservoirs went down, leaving unsightly edges. But all of this was merely inconvenient. Occasionally there were shocks, when violent rainstorms closed major roads and flooded railways, and, as sea levels rose, chunks of soft coastline were lost to the sea, taking with them isolated houses and hamlets.

But the first real awakening came when, not without precedent in the past two centuries, even the strengthened sea defences of some 5 km of coast were undermined one night and the drained flood plain of the River Seventine was reconverted to a large shallow lake for some 20 km inland. The flood plain had been drained for so long that its soils had shrunk and oxidized and much of it was 2 or 3 m below the high tide level. This resulted in a vigorous debate as to whether the defences should be reinstated at very high cost or whether the situation should simply be accepted. The latter was the outcome simply owing to the time it took to carry out inquiries and consultations, at the end of which the local communities had adjusted to the situation, new coastal activities had been established and further turmoil was unwelcome.

There were worries then that the system of ocean currents that brought warm sub-tropical water northwards might change direction as the oceans warmed and very cold water no longer moved down from the Arctic Ocean to force the warmer water towards Europe. That might mean that the island became much colder, rather than much warmer. Optimists thought the two trends might usefully cancel each other out and this gave heart to the non-scientists in the government that all might work out very nicely without too much diversion of attention from the economy.

In the event, the island did see some substantial changes by the middle of the 21st century. The wetter winters meant that many rivers overtopped their floodbanks every few years. More and more funds were invested in flood protection, with the result that building, and other development, continued on flood plains, so that the risks of damage remained much the same. Drier summers resulted in shortages of drinking water and pressure to repair old leaking water mains, hitherto losing nearly a third of the water they carried, on the one hand, and in previously almost unheard of rationing in some areas on the other. The sales of bottled water boomed. The heavy rains eroded much peat from the uplands, staining the water and giving rise to much hand-wringing in government conservation agencies about loss of carbon storage. Part of the problem, however, was creation of bare erodible areas by burning to encourage heather (*Calluna vulgaris*) to grow to feed grouse, which could, very profitably, be shot every August and September. So the hands wrung but did not much turn to action.

Nitrate concentrations in the rivers had long been a problem and the rains washed even more downstream boosted by the same pressures for increased agriculture that were affecting other regions too. It was a little like the Second World War when, in the interests of growing as much food as possible, every possible patch of land was ploughed. This even included gardens that had previously been the traditional lawn and flower beds. Individuals were much more concerned about their situation than the government.

Other substances were increasing in concentration too. The westerly winds had always brought in salt from spray drawn up into the atmosphere in ocean storms. Indeed the higher mountain lakes on the harder rocks had long been a dilute solution of sea water. In turn, the slightly salty water had provided sulphates and chlorides that acidified the soil when the bases in the water were taken up by plants. Other forms of acidification had been a problem in the late 20th century as a result of coal burning in power stations (producing sulphur oxides) and from vehicle engines (producing nitrogen oxides and eventually nitric acid in the atmosphere) and through release of ammonia from the decomposing excreta of intensive stock keeping. Scrubbing the sulphur oxides from the chimney gases had solved the first, and the second was helped by the high cost of fuel in the 21st century, leading to many cars resting idle on driveways.

The third problem was still extant, and the increased rainfall and the sea spray it brought were amplifying the acidification in the uplands. It led to losses of fish and invertebrates, even of some stream birds like dippers, and expectations a few decades earlier of its complete solution now seemed over-optimistic. Eutrophication symptoms had increased everywhere with warming, and acidification was not solved; river engineering continued apace to try to avoid flooding; and traces of agricultural and industrial chemicals still contaminated the waters, though the line of discharge of gross pollutants had been held. There were subtle changes in fisheries as conditions in the south became too hot for some species and, with no possibilities for invasion of warm-adapted species from the mainland, communities became impoverished. There were more accidentally or deliberately introduced exotic species and since most came from warmer regions, many were growing well.

Sometimes only professional ecologists noticed such species, if they were small, but the damage wrought by carp, as it converted clear plant-dominated lakes to turbid ones, and the overweening growth of some floating plants were very obvious. In terms of problems in freshwaters, it was as if no real progress had been made in 50 years, despite much legislation. Only at the coast, where the innocently rising sea level masked the exertion of huge forces, was there much reinstatement of more natural conditions. The residents of coastal villages were very concerned.

That concern was universally shared as temperatures rose by 4 °C over those in the 20th century by the late 21st century. It was not as if life was impossible, but it was very difficult. There was immense pressure on water supplies, for millions of refugees from the now intensely dry and hot southern parts of Europe had poured in to take advantage of the cooler offshore climate. Their entry was legal until eventually the Government withdrew from the European Union so as to stem the flow. Indeed the Union was in great danger of breaking up entirely as northern states looked to their borders and southern states fell into chaos. Travel had become very expensive, so the mutual understanding brought about by frequent visits and meetings among people of different countries had been denied to the new generations, which were lapsing into a historically precedented isolationism. The land was much more built over, food was rationed; the patchwork of small fields in the landscape had entirely gone, except in the hills, where every piece of land had some chickens and sheep. Maize was grown on an industrial scale in huge fields and people's diets were much less varied than they had previously been. Most ponds had been filled in; amphibia had virtually disappeared. The common toad was now a very rare toad. In the warmer parts, every lake and river was overwhelmingly dominated by common carp skulking under a thick cover of floating weed. Even the tropical scourges of water cabbage and water hyacinth gave problems.

Paradoxically, the difficulties and shortages had led to more local community cooperation. To be sure there was a great deal of black market skulduggery, but the much warmer weather had brought about more contact between people and a great deal of ingenuity was emerging in solving the pervasive small problems of existence. Bottled water was no longer affordable for taxes had had to be raised greatly to pay for the services needed by a hugely increased population, many of them unemployed, though demands for workers on the land were increasing. The dry summers nonetheless held the market for moisturizing creams!

The Mediterranean zone

In the late 20th century, white-walled houses with red-tiled roofs, in small villages and hamlets, graced the hillsides of the upper and middle sections of the valley of the River Graecerina, whilst its lower reaches held large flood plain lagoons before it discharged into the blue waters of the Mediterranean Sea, on a coast studded with holiday resorts. The uplands, where the headwater streams flowed, at least in the winter, bore a scrubby evergreen forest of oak, juniper and pine, where the goats and sheep had not prevented its regeneration. Wolves and bears were still living in the National Park that occupied a large tract of the upland. In the middle sections were terraces of orderly vines and olive trees. The lagoons were shallow and some still retained a prolific bird life whilst others, isolated from the river, dried down to salty mudflats, pecked over by waders in summer. The fauna was still rich with reptiles, amphibia and fish, many of them endemic to the region and, especially among the amphibia, dependent on the temporary seasonal nature of the ponds in the flood plain.

The rural population was small because farming had been under pressure from lack of water for some time, and most of the population had moved to exploit the opportunities of the tourist industry at the coast. Pilgrims visited holy springs that permanently flowed from the limestone rocks, but this was a dry landscape and river flows were very low in summer, though they had not yet ceased entirely. Winter brought an equable mean temperature of about $8\,°C$ and most of the rainfall of $550\,mm$. Summer temperatures averaged $28\,°C$, sometimes rising well into the thirties or even more than $40\,°C$, but the real problem was the lack of water because evaporation accounted for nearly 90% of all the rain that fell. Former inhabitants had stored the precious water in cisterns underground and understood the dynamics of the wells. Economic pressures in the 20th century, however, had undermined this ancient wisdom and water had been mined from the ground to irrigate wheat, and market gardening for high-value vegetables and fruit on the coastal plain, and some industry along the lower, now heavily engineered reaches of the river, but even more to serve the demands of residents and visiting tourists for swimming pools and power showers.

By the mid 21st century, a rise to $480\,ppm$ in carbon dioxide, with a concomitant increase in methane from the melting permafrost far to the north, and the accumulation of nitrous oxide from denitrification of farm fertilizers and car use throughout Europe, had created an increasingly desperate situation. The upland forests had dwindled to scrub; the charismatic mammals of the National Park had moved northwards and been shot by farmers, for with a greater rise in temperature than had been realized on average in Europe, evaporation rates had soared and their surface water sources had all but disappeared. The river only reached the sea in 1 year out of 5. The olive groves and vineyards had been abandoned. Crops failed in most years and the farmers moved away leaving a low maquis scrub to reclaim the land. There were even some patches of desert with mostly bare rock and a seasonal cover of annual weeds. What little water there was now commanded a high price and was affordable only by the market gardeners and the coastal towns. Many ponds, rendered dry one summer, never recovered in the following winter and the unique fish and amphibian fauna had suffered loss of at least half of its former species. Moreover, in the upland scrubs and remaining poor forest, fires raged most years, accelerating the change to semi-desert and encroaching further towards the rich coastal resorts with every year. With denudation of the land, the winter rain, falling now in a few torrential storms, washed the soil away, making any future recovery very difficult, and clogged the river with mud banks littered with the debris of human industry and abandonment. There were frequent landslides, and deep gulleys were appearing in the hillsides. The abandoned villages had an even deeper sadness about them. The lesson that water is the most precious of commodities for human societies had been forgotten but was now being painfully taught yet again. The church bells no more tolled over a warm and picturesque land.

In the coastal lowlands, activity was maintained by using desalinated sea water, but that had the side effect of creating hypersaline lagoons where the huge quantities of evaporated salt were discarded. They were sometimes colourful with the bright orange of tolerant algae, but they somehow jarred in the landscape and birds rarely could use them, let alone any other animals beyond the microscopic protozoa. Where freshwater, or rather, now mildly saline lagoons still persisted, there had been encroachment at the edges by aggressive reedy plants like *Typha*, *Impatiens* and *Arundo*, and sometimes near-monospecific tracts completely covered the former open water, pushing out the submerged species and the invertebrates and fish that depended on them. Alien crayfish had escaped from fish farms and were now also wreaking havoc on the remaining plants.

By the end of the 21st century, with temperatures locally up by 4–5 °C and rainfall very sporadic, there was the complete collapse of the local society and economy that had affected much of the tropics and subtropics some decades earlier. The landscape was mostly semi-desert; the river flowed only for a few days following a storm; the coastal resorts had failed for they were too hot now for comfort and the former cheap air transport that had sustained them was no more as fuel costs had risen. Refugees from North Africa and beyond eked out a hand-to-mouth existence in camps around the crumbling concrete of the apartment blocks and hotels. The market gardens had been converted to small plots by individual refugee families who managed to grow just about enough cassava to survive and could take a few fish from the sea. There was little organized infrastructure. The Mediterranean coast had been reduced to the status of 20th-century Saharan Africa.

High mountain zone

The high mountains had always been iconic in Europe for the photographic panoramas of snow-capped peaks rising to a blue sky, the ski slopes in winter and the alpine meadows studded with flowers in summer, let alone the images from the world's most viewed film *The Sound of Music*. Set deep in the mountains, the River Ppartnov had a small catchment of $300\,km^2$ that epitomized the region. Winter temperatures in the early 21st century averaged around freezing point and in summer about 25 °C, but changed a great deal from the tips of the highest peaks at 3,000 m to the valley floor 2,400 m below. Precipitation likewise was much higher, 900 mm, on the mountains but only 400 mm in the valley. Much of the water ran off in streams because much of the snow melted in spring and for some years the mountain glaciers had been retreating from their much greater extents a century or more before, when they had challenged the early visiting mountaineers. Evergreen forests of spruce and fir, with the deciduous larch, gave out at an altitude of 2,000 m to grasslands and tundras, where marmots and, on the rocky heights, the rarer ibex could be seen. An occasional brown bear still wandered through the forest and fed on berries at the tree line. Small farms, usually only a few hectares in area, kept some cows and grew grass and potatoes or a small plot of grain, and even some maize up to 1,200 m and on the flat valley floor. Tourists came for winter sports and in summer to walk up to the small tarns in the corries and along the lively rivers. They spent their euros on woodcarvings and embroidery and schnapps in the picturesque town in the valley, where the river had been long embanked and canalized to preserve the scarce land from any flooding. The flood plain had prehistorically occupied the entire valley floor. Just above the town the river had also been dammed to provide some local hydropower, not least to drive machinery at the local sawmill.

By the mid-20th century, the precipitation had not changed much but had shifted towards a greater winter snowfall and a much drier summer as temperatures had increased. The snow line had moved some 200 m upwards, leaving the lower ski slopes abandoned and the glaciers below the peaks had noticeably retreated. Much more water was cascading down the streams in spring, and the forests sometimes resounded to the crash of large boulders moving as the force of the water recarved the channels. The forest too had changed with dwarf pine, beech and other deciduous trees starting to take over from the previous evergreens. The water was still relatively cool and the arctic charr and brook trout still persisted but there had been some subtle chemical changes as higher temperatures had increased the weathering of the rock debris ground previously by the glacier movement. Levels of some heavy metals had risen and the local authorities had condemned the water for direct human supply and for stock watering, so that farming patterns had changed and the sound of cow bells was no longer heard as it had been when they came in for milking in the evenings. Farming was under pressure in any case, as overheads had increased in a global market that did not favour small operators, despite the generally increasing world shortage of food. Tourism still thrived, for the higher slopes were still covered with snow for several months and a refrigeration plant had been built, based on a small new reservoir, to create artificial snow to feed some of the lower slopes when winters were particularly warm. Tourism was still the main source of income for the area; the community had adjusted, even if it was worrying that the iconic snow-capped peaks looked less impressive when an increasingly wide bare black band of rock was exposed above the grasslands each year.

Some decades later, with average temperatures around 4 °C higher than those that had started the influx of foreign mountaineers to scale the peaks in the Victorian and Edwardian periods, things were very different. The mountain glaciers were virtually gone. Only small patches of ice persisted on the very highest peaks and the winter snow lay barely long enough for the skiing season to start before it had to close down. A panorama of summer snow-clad peaks was no more. Trees were encroaching more and more and a sparse vegetation had established on the exposed rocky debris. The forest trees grew better but there was a reduced market for their timber because people had drifted away. The tourist income had disappeared, both in winter for lack of snow and in summer because it had still been the snowy peaks that had epitomized the area. The bare rock was less able to attract people now that air travel was so much more expensive than it had been at the turn of the century. Many wooden chalets had been abandoned and the farmland had been amalgamated into bigger units that were profitable now that food was expensive everywhere.

The local population could get by, indeed it had started to increase again as newcomers moved in from lowland areas that were becoming too hot for comfort, but there were few luxuries. The forests were at first undisturbed by loggers or human visitors and the bears and even some wolves began to move back from areas that had been protected for nature conservation during the period of intense tourism. But this was to be short-lived. The temperature was still rising and word spread among refugees that the alpine valleys and lower slopes were equable. Soon the area was busy again with a relatively self-contained community, learning rapidly to manage with a spartan lifestyle. The houses were repaired and forestry revived to supply the materials locally. Food was still a problem, partly alleviated by chickens and goats and vegetable patches and a system of bartering for goods and services.

For the moment the future looked more hopeful. There had been major catastrophes elsewhere: coastal cities, even huge tracts of land flooded by the rising sea levels, millions of people dead from a combination of water shortage, water-borne disease and inadequate nutrition coupled with heat stroke. Former centres of civilization had been mostly abandoned and millions, who had not died where they had lived, had tried to move and only a few had had the funds to be able to settle in more equable places. But now new centres were emerging and the archaeologists had new material for their models of cultural loss and resurgence – if only the temperature would soon stop rising.

Index

a priori classification systems 207, 209
A2 scenario 252, 278, 279
abbreviations list 238
acid episodes 154, 162–166, 174
acid neutralizing capacity (ANC) 154, 155, 162–166
acidification 152–175
  acid episodes 154, 162–166, 174
  computer modelling 155, 158, 159, 161, 165, 167–168, 170–173
  confounding variables for recovery detection 216–220
  dissolved organic carbon 166–169
  drivers 163, 171
  future predictions 294–295
  lakes
    Elison Lake, Cape Herschel 30
    monitoring 88
    sulphur deposition 31
  nitrate leaching 156–166
  rivers, monitoring 97
  Saharan dust 169
  SWAP 215
acronyms list 238
actual evapotranspiration (AET) 240
Additive Modelling (AM) 30
Afon Gwy, Wales 163, 171
Afon Hafren, Wales 163, 164
Africa 20–21, 22–24
Agapetus ochripes 75
agriculture, Northern Europe 120
air pollution, acidification 170

Akkadian Empire, Mesopotamia 23
alien species, rivers, monitoring 100
Alnion glutinosae 227
Alpine lakes 27, 59–60
aluminium ions 153, 154, 162, 170
AM *see* Additive Modelling approach
Anabaena 141
ANC *see* acid neutralizing capacity
Anisops 21
annual discharge of rivers 65–66
Antarctic, ice shelves collapse 15
Anthropocene period 15
anthropogenic effects *see*
    human activities
AO *see* Arctic Oscillation
Arctic Oscillation (AO) 18
Arctic/boreal zone 287–290
artificial plant bed studies 123–125
assessment *see* monitoring
Asterionella formosa 28, 142, 143
Atlantic salmon (*Salmo salar*) 103
atmospheric analysis, Earth/planets 3
atmospheric sulphur deposition 55
Aulacoseira spp.
  A. ambigua 31, 142, 143
  A. angustissima 31
  A. granulata 31, 142, 143
  A. subarctica 142, 143
Austria 59–60, 74–76

B2 scenario 278
Baikal, Lake 46, 52

bankfull discharge 73
Barbus barbus 246, 247
baseline changes 216–220
Becva River, Czech Republic 74–75
behavioural adaptations 72
benthic macroinvertebrates
    acidification 173–174, 217–218
    Becva River, Czech Republic 74–75
    Lambourn River, southern
        England 71–72
    restored streams 77
    rivers, monitoring 101, 135–137
berillium-10 22
Berula 71
Betula pubescens 137
bicarbonate ions 153
biodiversity
    artificial plant bed studies 125
    fish 92–93
    future predictions 295
    Loch Leven, United Kingdom 142
    monitoring 84–113
    rivers/streams
        current velocity patterns 71–72
        Erehwomos River 289
        monitoring 98, 104–106
        paired stream studies 137
        restored streams 77
        threats to 65, 66
biological feedback 285–286
biomes 4–5, 8
biosphere 4, 6
biota
    acid deposition effects 173–174
    monitoring 84–113
        lakes 85–93
        rivers 94–106
        wetlands 106–113
    water significance 6–8
bird migration 110, 112
Birkenes catchment, Norway 156,
    164–165
Bjerkreim River Basin,
    Norway 251–255
blackfly, Erehwomos River 288
Bohemian Forest, Czech
    Republic 161–162, 170
boundary conditions 17
Brachycentrus montanus 75
break-up *see* ice cover
Breisjon, Lake (Norway) 50

brine flies (*Ephydra*) 21
brown trout (*Salmo trutta*) 50–51,
    93, 246, 247

caddis fly (Trichoptera) 71, 74–76,
    101–103, 174
Caenidae 71
Cairngorm Mountains 26
calcium ions, acidification 153
Callitriche 71, 212
Calthion 227
Canada 29, 46, 166, 172, 174
Canonical Correspondence Analysis
    (CCA) 30
carbon
    see also dissolved organic carbon
    balance, Erehwomos River 288
    carbon-14 variability 21, 22, 23
    pool, acidification, soil, nitrate
        leaching 158, 159
carbon dioxide 4, 296–297
Caricion spp. 227
Cassandra (Greek mythology) 285
catchment areas
    acidification
        Czech Republic 165
        forest whole tree harvest 172–173
        Norway 156–158, 164–165
        Welsh streams 163, 164
    modelling
        see also CATCHMOD; INCA;
            MAGIC
        downscaling 239–242
        linking models 251–255
        responses to climate change
            236–258
        statistical models 242–248
        uncertainty 255–257
    rivers/streams 67–70
    stream restoration 78
CATCHMOD hydrological model 242
CCA *see* Canonical Correspondence
    Analysis
centennial timescale 21–26
Ceratophyllum demersum 126, 128
change drivers
    acid episodes 163
    acidification 171
    effects on major biomes 7–8
    experiment design 10
    uncertainty 263

channel morphology 74–76
Chaoborus flavicans 219
Chara spp. 25, 211
chemical impacts 53–60
chemical oxygen demand (COD) 54
chemical studies, Earth/planetary
    atmospheres 3
China, Neolithic cultures 23
Chironomidae 137, 174
chlorophyll *a* 128–129, 130, 140–141
Cladium mariscus 19
climate gradients 10
climate region differences 144
CLIMEX (Climate Change
    Experiment) 154
CLIMOOR (Climate Driven Changes in
    the Functioning of Heath and
    Moorland Ecosystems) 57
Clocaenog (North Wales) 57–59
CLUAM outputs 279, 280
COD *see* chemical oxygen demand
cold-stenothermic species 101, 102, 103
common reed (*Phragmites australis*)
    131, 132
confounding variables 216–220
connectivity of a site 224, 225
Constance, Lake 46
continuity of habitat 7
Coregonus albula 92
cost assessment 6, 11–12
crenal zone insect species 102, 103
crop production, Northern Europe 120
CSs (polychlorostyrenes) 181
current velocity 73, 74
Cyclotella spp. 28, 29, 142, 143
Cymbella spp. 30
Cyp1A (cytochrome P450 1A
    enzyme) 192–195
Cyprinidae 93, 100
cytochrome P450 1A enzyme 192–195
Czech Republic
    acidification
        aquatic biota 173–174
        Bohemian Forest 170
        DOC in streamwater 167–168
        Litavka 171
        Lysina catchment 165
        nitrate leaching 160, 161–162
    Becva River 74–75
    reference conditions
        establishment 208

Daphnia spp.
    D. *cucullata* 92, 93
    D. *galeata* 92, 93
    D. *magna* 134
    D. *pulicania* 196
    lakes 122–123
    Loch Leven, United Kingdom 141, 142
DDT insecticide 181, 183, 186,
    189, 190
decadal change, Holocene period 26–27
decision making 262–282
    Decision Support Systems 274–281
    precautionary principle 264
    scenario-based analysis 264, 278–281
    science transfer into policy 273–274
    tools 265–273
    uncertainty 262–264
Decision Support Systems (DSS)
    274–281
denitrification 138–140, 145
Denmark
    artificial plant bed studies 123–125
    Gjern River basin 250
    ice cover and fish abundance 122
    mesocosm eutrophication
        experiments 125–126, 128–131,
        133–134
deoxygenation-tolerant fish 128
Diatoma elongatum 142, 143
diatoms
    alpine lakes 28
    laminated sediments 26–27
    Loch Leven, United Kingdom
        142, 143
    Norwegian lakes 30
dibenzofurans 181
Dinkel catchment, Netherlands 227–229
dioxins 181
direct impacts of climate change 38–60
    chemical 53–60
Directive EC/2000/60 *see* Water
    Framework Directive
discharge of rivers/streams
    65–66, 67–68
dispersal, species 223–225
disturbances, freshwater organisms 7
DOC (dissolved organic carbon)
    acidification 166–169, 172
    acidification recovery
        detection 219–220
    lakes, monitoring 88

reference conditions establishment 215
surface waters 53–59
downscaling, modelling 239–242
drivers *see* change drivers
droughts
river species coping strategies
71, 72
Saharan region, Holocene period
20, 23, 24
summer droughts, Clocaenog
experiment 57–59
winter, Lambourn River, southern
England 68
DSS (Decision Support Systems)
274–281
dynamic catchment scale models
248–251

EARWIG (Environment Agency Rainfall
and Weather Impacts
Generator) 240, 241
Ebre (Ebro) River 181, 197, 199
EC/2000/60 *see* Water Framework
Directive
ECHAM General Circulation Model
157, 158, 252, 279
Ecosystem Approach 262, 265
ecosystems 4, 204, 223–225
Eemian period 17
Egypt, Old Kingdom collapse 23
Elatine hexandra 210
Eleocharis acicularis 212
Elison Lake, Cape Herschel 30
Ellesmere Island, Canada 29
Elodea spp. 126, 129, 130, 131
Emys orbicularis 19
endemic species monitoring 101,
102, 103
endosulfans 182, 193
ENSO *see* Southern Ocean Oscillation
Environmental Protection Agency
(EPA) 185
Ephemeroptera 71, 101–103, 174
Ephydra 21
equifinality problem 255–256
Erehwomos River 287
Esox lucius 246, 247
Estonia, Lake Vortsjarv 69
Eukiefferiella minor 137
Euro-limpacs research
programme 8–11

catchment scale modelling 236,
237–248
abbreviations list 238
climate–nutrient interactions
120–121, 140, 142
Decision Support System 274–281
dynamic models 248
future climate scenarios 44
paired site eutrophication
studies 134–140
scenario-based models 286
Swiss rivers/streams 50–51
Europe
see also individual countries; Water
Framework Directive
acidification 152–156, 159–160,
170–174
alpine lakes 27–29, 33, 59–60
atmospheric sulphur deposition 55
crop production 120
fish monitoring 103–106
freshwater systems 8–13
future predictions 295, 296–299
Holocene period 22–24
inland waters investigation 8–11
insects monitoring 101–103
land use changes 66–67
Maastricht Treaty of 1992 264
persistent organic pollutants 181, 184,
186–191, 193–195, 197–199
predictions 40–44
reference conditions establishment
methods 206
rivers monitoring 94, 101–106
rivers/streams discharge 65–66
stream restoration 76
European pond turtle (*Emys
orbicularis*) 19
eutrophication 119–147
artificial plant bed studies 123–125
climate change effects 145
future predictions 295
lakes 121–125
food webs monitoring 93
key questions 121
sediment analysis 26–27, 28, 31
mesocosm experiments 125–134
mitigation methods 146
paired site studies 134–140
palaeolimnology 140–142, 143
river monitoring 96

evolution 3, 7, 133–134
experimental design 10
Experimental Lakes Area, Canada 46
expert judgement 205–206, 207
external forcing 18
extreme events 41–43

FAME (Fish-based Assessment Method for
    the Ecological status of European
    rivers) 104
Fennoscandia, ice cover records 52
fertilizers 249, 250, 278–281
Finland 49–50, 52, 167, 172–173
fish
    among-stream variance 221
    Cyp1A expression 192–195
    lakes
        European, food web changes
            121–122, 124
        monitoring 90–91, 92, 93
        PCBs/PBDEs 188, 189, 190
    mesocosm eutrophication
        experiments 126, 127, 128
    persistent organic pollutants 188, 189,
        190, 191
        mercury levels 185, 198
        organochlorines 186–187,
            188, 190
    rivers
        Erehwomos River 288, 289
        French streams statistical modelling
            tests 243–246
        monitoring 98, 103–106
        Rhone River, France 104–105
        trait modalities monitoring 104–106
        variability change effects 73
Flix, Catalonia, Spain 181, 197, 199
flooding risk, Erehwomos River 288
flow regimes, rivers 101
food webs
    lakes
        eutrophication 121–122, 124
        monitoring 91, 93
        persistent organic pollutants 187–188
    wetlands 110
forcing mechanisms
    see also change drivers
    external forcing 18
    greenhouse gases, IPCC Fourth
        Report 40–41
    intra-Holocene climate change 18

Greenland lakes 25–26
    palaeo-evidence 32
forests, acidification 172–173
fossil record see palaeoecology
Fragilaria spp. 28, 142, 143
fragmentation of landscape 226
France 104–105, 134, 243–246,
    248–249
freeze coring, laminated sediments 26
freeze-up 51
    see also ice cover
future climate see predictions for the
    future

GAMs (Generalized Additive
        Models) 105, 106
Gårdsjön, Sweden 167–168, 198
Garonne River, France 248–249
Gasterosteus aculeatus 126,
    127, 128
GCMs see General Circulation Models;
        Global Circulation Models
General Circulation Models (GCMs)
    catchment scale modelling 236,
        237, 239–240, 241, 252, 258
    ECHAM 157, 158, 252, 279
    fish species distribution 243, 246
    Tamar River, United Kingdom 279
    warming predictions 40–41
Generalized Additive Models
        (GAMs) 105, 106
generalized sensitivity analysis 256
geothermal areas, Iceland 135–137
Germany 77, 160
Gjern River basin, Denmark 250
glaciers, Erehwomos River 288
Gleissberg 87-year solar cycle 21
Global Circulation Models
        (GCMs) 44
global distillation effect 186
Glossosomatidae 71, 75
Goeridae 71
Gossenköllesee, Alps
        191–192, 197
Graecerina River 296
Greek mythology, Cassandra 285
greenhouse gases
    forced warming, difficulty
        distinguishing 32–33
    forcing, IPCC Fourth Report 40–41
    highest now 15

Special Report on Emissions
  Scenarios 45, 239, 240
Greenland lakes 24–25
Groby Pool, England 209, 210, 211,
  212–213

habitat scale 71–73, 74–76
Hadley Centre models 242,
  246, 247
  Had A2 scenario 252, 255
  HadAM3 global climate
    model 157, 158
  HadAm3H 252, 254
Hale 22-year solar cycle 21
Hallstatt 2200-year solar cycle 21
Hälsjärvi, Lake, Finland 49–50
HBV hydrological model 252–253
health & safety 180–200
heat waves, projected change 41
Hengill, Iceland 135–140
hexachlorocyclohexanes (HCHs) 181,
  182, 189, 190
Hibscotia island 292–296
high mountain zone 59–60, 192,
  298–299
  lake pollution 186–189, 190,
    191–192
HIRHAM predictions 250
historical information 209–214
  see also palaeo-reconstruction
Hollandsegraven sub-catchment,
  Netherlands 227–229
Holocene period
  climate during 16–18
  low latitudes 19–21, 23
  multi-millenial scale climate
    change 18–21
  natural climate variability 16–33
Hubbard Brook, New Hampshire 160
human activities/effects
  change drivers 263
  civilizations 286
  climate reference standards 11
  Euro-limpacs research programme 8–11
  evidence for climate
    impact 15–16
  Holocene period 17–18
  IPCC Fourth Report 1
  newsworthiness 12–13
  persistent organic
    pollutants 199–200

population increases/crop
  production 120
reference conditions
  establishment 204–216
remote regions, recent change
  27–29, 33
toxic substances exposure 183
Hutchinson model 3, 4
hydraulics, rivers/streams 71–73,
  74–76
hydrogen ions, acidification
  153, 154
hydrologically effective rainfall
  (HER) 240
hydrology monitoring
  Kennet River 241
  lakes 86, 87–88
  wetlands 106–107, 111
hydromorphology of freshwater
  systems 65–78
  catchment scale 67–70
  habitat scale 74–76
  influencing factors 65–66
  lakes 69–70
  Lambourn River, southern
    England 67–68, 71–72
  land use changes 66–67
  reach scale 71–73
  restoration success 76–78
  Vecht River, Netherlands 66–67
hypolimnetic lake temperatures
  46, 47, 48

ice cover 16–17, 51–53, 87, 122
ice shelves, Antarctic 15
ice-rafted debris (IRD) 22, 23
Iceland, Hengill 135–140
IIASA (International Institute for Applied
  Systems Analysis) 170
IMAGE (Integrated Model to Assess the
  Global Environment) 170
INCA (INtegrated CAtchment) models
  catchment scale 248
  harmonization 257–258
  INCA-C model 172
  INCA-N model 240, 241, 242,
    252–253
    Kennet River 248–250
    Tamar River, 279
  INCA-P model 157
  Monte Carlo analysis 256

indicators
  acidification recovery 216–217
  monitoring
    lakes 85, 86, 87, 88
    rivers 94–96
    fish 104
    wetlands 106–112
Indus Valley Civilization, India 23
industrial organochlorines 186–187
insect species
  Ephemeroptera 71, 101–103, 174
  monitoring 101–103
  mosquitoes 288
  Trichoptera (caddis fly) 71, 74–76,
    101–103, 174
insecticides 181
inter-annual change 26–27
interglacial warm periods 16–33
Intergovernmental Panel on Climate
    Change (IPCC)
  A2/B2 scenarios 252, 278, 279
  Arctic ice melting predictions 285
  climate change predictions 3–4
  computer simulations 10
  Fourth Assessment Report 2007 1–2,
    39, 40
    extreme events 41–43
    heat waves 41
  freeze-up definition 51
  persistent toxins,
      precipitation 197–198
  physical models creating
      underestimates 147
internal variability modes 18
Intertropical Convergence Zone
    (ITCZ) 19–20
invertebrates
  see also benthic macro-invertebrates;
      insect species
  river monitoring 101–103, 135–137
investigation approaches
  artificial summer droughts 57–59
  dissolved organic carbon study 56–57
  European freshwater systems 8, 9–11
  whole-lake manipulation
      experiments 49–50
IPCC see Intergovernmental Panel on
    Climate Change
IRD see ice-rafted debris
IS92a scenario 252
islands, predictions 292–296

Isoetes lacustris 210, 211
Italy 31, 159–160, 170
ITCZ (Intertropical Convergence
    Zone) 19–20
Izaña, Tenerife 191–192

Kennet River, United Kingdom 240–242,
    248–250

lacustrine sediments 189
Lagarosiphon major 126, 128
Lago Magiorre, Italy 31
Lahn River, Germany 77
lakes
  acidification 167, 172–173
    Bohemian Forests 161–162
    effects on biota 174
    Scandinavia 152–153
  Europe 16–17, 19, 20–21, 22–24,
    152–153, 161–162
    alpine, recent change in sediments 27
  eutrophication 121–125
    climate region differences 144
    nitrogen 145
  high mountain lakes 59–60, 186–189,
    190, 191–192
  hydromorphology 69–70
  ice cover records 51–53
  monitoring 85–93
  North African 20–21, 22–24
  pollution 29–31
    nutrient pollution 28, 31
    persistent organic 181–183,
    186–189, 190, 191–195, 197, 198
  reference conditions establishment 205,
    209–210, 211
  seasonal change 25–26
  temperature effects 24–26
  thermal regimes 45–48
  THERMOS whole-lake mixing
      experiment 49–50
  Yoa 21
Lambourn River, southern
    England 67–68, 71–72, 250, 257
laminated sediments 26–27
land ice/sea surface temperature 17
land use 66–67, 120
Lange Bramke, Germany 160
Langtjern, Norway 156
Lasiocephala basalis 75–76
Le Chatelier's principle 186

lead-210 30
leaf decomposition 98, 101, 131–132, 137
lemnids 126, 127, 128
Lepidostoma basale 75
life strategies 99
linking models 251–255
Litavka, Czech Republic 171
Litorella uniflora 210, 212
litter *see* leaf decomposition
littoral zone
  assemblages 217, 218
  food webs 124
  reed stand mesocosm 131–133
Lobelia dortmanna 210, 211
Loch Leven, Scotland 31, 140–142,
    209, 210
Long-Range Transboundary Air Pollution
    (LRTAP) convention 154
loss on ignition (LOI) records 25, 26
Lovelock model 3, 4, 6
low latitudes, Holocene 19–21, 23
LRTAP *see* Convention on Long-Range
    Transboundary Air Pollution
Lymnaea peregra 137
Lype phaeopa 75
Lysina, Czech Republic 167–168

Maastricht Treaty of 1992 264
macro-invertebrates 135–137
  see also benthic macro-invertebrates
MAGIC (Model of Acidification of
    Groundwater in Catchments) 155,
    252–253
  Afon Gwy, Wales 171
  DOC in stream water 167–168
  Litavka, Czech Republic 171
  nitrate leaching 158, 159, 161, 165
  Plastic Lake, Ontario 172
  Scottish loch pH
    reconstruction 214–215
magnesium ions 153
major ecosystems, MEA (2005) 4–5
Max Planck Institute for Meteorology
    model *see* ECHAM General
    Circulation Model
mayflies (Ephemeroptera) 71,
    101–103, 174
MCA (Multi-criteria Analysis) 274
mean similarity analyses 77
Mediterranean zone 296–298
Mendota Lake 52

Menyanthes trifoliata 212
The Merchant of Venice (Shakespeare) 6
mercury pollution 180, 185,
    196–197, 198
mesocosm experiments 10,
    125–134
mesohabitats 71–72
methylmercury 185
microevolution experiments 133–134
mid continental latitudes 290–292
mid-European lakes 22–24
Millennium Ecosystem Assessment
    (2005) 4–5, 7–8
mixing of lake waters 46–48
modelling/simulation
  see also ECHAM; INCA; MAGIC
  acidification 154, 157, 158,
    170–173
  catchment scale 236–258
    dynamic models 248–251
    Kennet River case study 240–242
    linking models 251–255
    model creation 237
    statistical models 242–248
    uncertainty 255–257
  EARWIG 240, 241
  experimental design 10–11
  IMAGE 170
  Lambourn River 67–68
  MyLake (Multi-year Lake simulated
    model) 49–50
  PROTECH phytoplankton community
    model 141
  RAINS 170
  RCAO 42
  reference conditions establishment
    205, 207–209
  SMART 155
moisture change-driven climate effects
  see also droughts; precipitation
  Holocene period 16
  North African lakes 20–21
  palaeo-evidence 32
  Sahara–Sahelian region 24
monitoring 84–113
  biota
    lakes 85–93
    rivers 94–106
    wetlands 106–113
monsoon system, Holocene period 23
Monte Carlo analysis 256

morphology *see* channel morphology; hydromorphology of freshwater systems
mosquitoes, 'Erehwomos' River 288
mountain lake pollution 186–189, 190, 191–192
MPI IS92a scenario 252–253, 254, 255
Multi-criteria Analysis (MCA) 274
multi-millennial scale climate change 18–21
multiple palaeo-indicators 210–214
MyLake (Multi-year Lake simulated model) 49–50
Myriophyllum alterniflorum 210, 211, 212

NAM-MIKE11-TRANS model 250
NAO *see* North Atlantic Oscillation
natural climate variability 16–33
Netherlands 66–67, 76, 227–229
Nitella flexilis 211
nitrate
    see also INCA-N model
    leaching
        climate change effects 156–166
        Czech Republic 161–162
        Northern Italy 159–160
        Norway 156–159
        United States, New Hampshire 160
    pollution, Tamar River, United Kingdom 278–281
nitrogen
    see also denitrification; total nitrogen
    acid deposition 152–175
    Bjerkreim River Basin, Norway 251–255
    Bohemian Forest Lakes 161–162
    Europe 1880–2030 153
    Kennet River, United Kingdom 249, 250
    lakes eutrophication 145
    North American lakes 28
    saturation 160
    soil leaching 158, 159
    stream eutrophication 137
    wetlands 138–140
North African lakes 20–21, 22–24
North America
    acidification 152–154, 159–160, 166, 172, 174

    atmospheric sulphur deposition 55
    Canada 29, 46, 166, 172, 174
    ice cover 51
    lakes 24, 28
    United States 28, 46, 160
North Atlantic Oscillation (NAO)
    acidification recovery detection 220
    characteristics 18, 45
    hypolimnetic temperatures in lakes 46
    IPCC Fourth Report 39–40
    sea salts and DOC solubility 55
    Vortsjarv Lake 69
Northern Europe
    see also Norway
    crop production 120
    Iceland 135–140
    lakes, Holocene period 16–17
    predictions 40–44
    Scandinavia 54, 152–153, 185, 198
Northern hemisphere snow cover 2
Norway
    acidification
        Birkenes catchment, acid episodes 164–165
        CLIMEX experiment 154
        nitrate leaching 156–159
    Bjerkreim River Basin, Norway 251–255
    brown trout, lakes monitoring 93
    dissolved organic carbon seasonal variations 55
    Lake Breisjon, whole-lake manipulation experiment 50
    lake diatom records 30
null models 208
nutrients
    see also eutrophication
    additions
        paired stream studies 137
        paired wetland studies 138–139
    concentration rises 120
    pollution
        lakes 31, 33
        North American lakes, recent change 28
Nymphaea alba 212

ocean currents 294
OCs *see* organochlorines

OLU (operational landscape units)
approach 224–229
operational landscape units (OLU)
approach 224–229
orbital change
intra-Holocene climate change 18
long-term 21, 23
organic matter (OM) 54
organobromines 186–187
organochlorines (OCs) 181, 186–188,
190, 197–198
organohalogen compounds 180–184,
186–195
Ottershagen floodplain 227–229
Øvre Neadålsvatn, Norway 191–192, 197
oxygen
Danish mesocosm eutrophication
experiments 128, 129
depletion, lakes, monitoring 88
oxygen-18, temperature effects on
lakes 25
temperature rises 119
Øygard catchment, Norway 158, 159

paired site experiments 134–140
palaeo-reconstruction
eutrophication studies 140–142, 143
natural climate variability 16–33
recent history 9–10
reference conditions establishment 205,
207, 210, 211, 212, 213
usefulness 146
PBDEs (polybromodiphenylethers) 181
pBDM (one point Boreal Dilution
Method) 165
peninsulas and islands 292–296
Percidae 93
perihelion 18
Persicaria amphibia 212
persistent organic pollutants
(POPs) 180–200
cytochrome P450 1A enzyme
192–195
human activities 199–200
human exposure 183
lakes 181–183, 186–195, 197, 198
mercury 180, 185, 196–197
organohalogen compounds 180–184
polycyclic aromatic hydrocarbons 184,
195–196
precipitation change effects 197–199

temperature change effects 185–197
types 181
water management strategies 199–200
pesticides, organohalogens 181
pH reconstruction 214
phenol oxidase 58
phosphorus 137, 140, 157, 257
Phragmites australis 131, 132
Phragmition 227
physical characteristics
of water 6–7
physical impacts on freshwater
systems 44–53
physicochemistry monitoring
lakes 86, 88
rivers 96–97
wetlands 107–109, 111
phytoplankton
acidification recovery detection 218,
219, 220
diatoms 26–27, 28, 30, 142, 143
lakes
food web changes 121–122
monitoring 86, 88–93
PROTECH phytoplankton community
model 141
rivers/streams 69
pike (*Esox lucius*) 246, 247
Pinnularia balfouriana 29, 30
planetary function 3–6
plants
see also phytoplankton
artificial plant bed studies 123–125
forests acidification 172–173
pollen records 210, 211, 212
reedbeds 131–133
Plastic Lake, Ontario 166, 172
Plecoptera 101–103, 174
Pluhuv Bor, Czech Republic 167–168
Poland, ice cover records 53
Polar Circulation Index (PCI) 22
pollen records 210, 211, 212
pollution
see also nitrate; persistent organic
pollutants
climate change effects separation 29–32
freshwaters 7
North American lakes 28
RAINS 170
polybromodiphenylethers (PBDEs) 181,
186–187

polychlorobiphenyls (PCBs)
  fish 188, 189, 190
  remote site waters 183
  stability 181
  temperature effects 186–187, 193
polychlorostyrenes (CSs) 181
polycyclic aromatic hydrocarbons
    (PAHs) 184, 195–196
POPs *see* persistent organic pollutants
potamalization 97, 99
Potamogeton spp. 210, 211, 212, 213, 214
  P. crispus 129, 130, 131
  P. natans 126
potential evapotranspiration (PET)
    240, 242
Ppartnov River 298
precautionary principle 264
precipitation
  see also droughts; moisture-change-
    driven climate effects
  Holocene period 23
  persistent organic pollutants 197–199
  projected change 41–43
  recent change 39
  river fish traits 104–106
  seasonality changes 66
predictions for the future 40–44,
    285–299
  high mountain zone 298–299
  Mediterranean zone 296–298
  mid continental latitudes 290–292
  peninsulas and islands 292–296
  predictors, reference conditions 220
primary production
  Danish mesocosm eutrophication
    experiments 129–130
  monitoring
    lakes 86, 89–90, 92
    rivers 94, 97
    wetlands 109, 111–112
  Swiss mesocosm experiments 133
profundal assemblages 217, 218
PROTECH phytoplankton community
    model 141
PRUDENCE project 41, 157, 158, 239
Pyrenean lakes 29, 186–189, 191–195

r-strategies, river monitoring 99
RAINS (Regional Air Pollution
    Information and Simulation
    Model) 170

Ranunculus spp. 71, 212
Rasass See, Alpine lake 59–60
RCAO (Rossby Centre coupled regional
    climate model) 42
RCMs (regional climate models) 42,
    239–240
reach scale 71–73
recent changes 9–10, 27–29, 30, 31,
    33, 39
Redon Lake, Pyrenees 191–192, 193, 197
  Cyclotella 29
reedbeds 131–133
REFCOND 206
reference conditions
  approaches 205–209
  detecting recovery with changing
    baseline conditions 216–220
  determination/restoration 203–230
  ecological relationships
    variance 221–222
  establishment 203–216
  historical information 209–214
  multiple palaeo-indicators
    210–214
  predictors 220
  REFCOND 206
  restoration 220–229
  success determinants 203
  uncertainty 214–216
  variance 221
reference standards 11
regional climate models (RCMs) 42,
    239–240
reindeer 288, 289
remote sites
  high mountain lakes 59–60, 186–189,
    190, 191–192
  persistent organic pollutants 182, 183,
    184, 186–189, 190–192
  recent change 27–29, 33
residence times 69
resource predictability 69
respiration rates 97, 285–286
restoration
  ecosystem connectivity 223–225
  freshwater ecosystems 203–230
    OLU approach 224–229
  large woody debris use 222–223
  reference conditions 220–229
  rivers/streams 76–78, 222–223
  species dispersal 223–225

Rhone River, France 104–105
rivers
    acidification, nitrate leaching 160
    Bjerkreim River Basin,
        Norway 251–255
    Erehwomos River 287–288
    eutrophication 120
    Garonne River, France 248–249
    hydromorphology 65–78
        modification effects 66
        restoration success 76–78
    ice cover records 51–53
    Kennet River, United
        kingdom 240–242
    Lambourn River, United
        Kingdom 67–68, 71–72, 250, 257
    monitoring, biota 94–106
    riparian vegetation 69
    Swiss, water temperature 50–51
    Tamar River, United Kingdom 278–281
RIVPACS-type model reference
    sites 208
Rocky Mountains 28
Rossby Centre coupled regional climate
    model (RCAO) 42

Sagittaria sagitifolia 212
Saharan dust, acidification 169
Sahara–Sahelian region 20, 24
saline lagoons 297
salinization
    sea salts
        acid episodes 163, 164
        acidification 168–169, 171–172
        future predictions 294–295
        organic carbon seasonality 55
    Southern Europe 120
Salmonidae
    Erehwomos River 288
    lakes monitoring 93
    Rhone River, France 104
    rivers monitoring 100
    Salmo salar monitoring 103
    Salmo trutta (brown trout) 93, 196,
        246, 247
SAR *see* sediment accumulation rate
scale of experiments 10, 221, 236
Scandinavia 54, 152–153, 185, 198
scenario-based models 40, 44, 264,
    278–281, 286
Schmidt stability 48

Schwabe 11-year solar cycle 21
Schwarzsee ob Sölden 59–60
Scotland 31, 140–142, 209, 210, 214
SDSM *see* Statistical Downscaling
    Method
sea level changes 1–2
sea salts
    acid episodes 163, 164
    acidification 168–169, 171–172
    future predictions 294–295
    organic carbon seasonality 55
sea surface temperature (SST) 17, 18–19
seasonality
    Becva River, Czech Republic 74–75
    Danish mesocosm eutrophication
        experiments 130
    Holocene period 26–27
    lake hydromorphology changes 70
    rivers/streams hydromorphology 66, 68
    summer droughts 57–59
secondary production monitoring 90–94,
    97–99, 109–110, 112
sediment
    high mountain lakes, persistent organic
        pollutants 192
    records
        eutrophication 28, 30, 31
        Greenland lakes 25
        recent change 9, 27–29, 30, 31
        reference conditions
            establishment 205, 210–214, 215
        Sahara–Sahelian region 24
        seasonal time-scale 26–27
    Scotland 26, 141–142
    transport, river species coping
        strategies 72–73
sediment accumulation rate (SAR) 70
Sensitivity and Uncertainty Analysis
    Tool 256
Seventine River 292–296
Shakespeare, William 6
short-scale change 21–26
Simocephalus 133–134
Simulum vittatum 137
Slovakia, ice cover records 53
SMART (Simulation Model for
    Acidification Regional
    Trends) 155
snow cover 2, 56–57
snowline 298–299
snowmelt 165

snowpack 156–157, 289
soil
　acidification
　　carbon/nitrogen pool, nitrate
　　　leaching 158, 159
　　chemical processes 169–170
　dissolved organic carbon 56–59
　erosion, lake catchments 70
　frost 56–57
solar variability
　Cairngorm Mountains sedimentary
　　records 26
　carbon-14 variability 21
　climate change mechanisms
　　18, 21, 22
　European/North African
　　lake-levels 23
solutes, high mountain lakes 59–60
source areas, OLU approach 226
Southern Europe
　Italy 31, 159–160, 170
　Mediterranean zone 296–298
　nutrient concentration 120
　Spain 181, 199
Southern Ocean Oscillation (ENSO) 18
space-for-time analysis 10, 121–123
Spain, Flix reservoir 181, 199
spates 68–69, 71–73, 74
spatial approaches 205, 206, 207
Special Report on Emissions Scenarios
　(SRES) 45, 239, 240, 279
species distribution 223–225, 243–246
species diversity *see* biodiversity
Sphaeriidae 71
SRES (Special Report on Emissions
　Scenarios) 45, 239, 240, 279
SST (sea surface temperature) 17, 18–19
stakeholder involvement 262
statistical models 242–248
Statististical Downscaling Method
　(SDSM) 239, 240
Stephanodiscus spp.
　S. hantzschii 142, 143
　S. parvus 142, 143
stepwise change 23
The Stern Report (2006) 11–12
sticklebacks (*Gasterosteus aculeatus*)
　126, 127, 128
Stockholm Convention 181
stoneflies (Plecoptera) 101–103, 174
Storgama, Norway 156

stratification, lakes 87
streams
　see also rivers
　acidification
　　Czech Republic 173–174
　　Norway, nitrate leaching 156–159
　　Wales 163, 164, 171
　fish species distribution 243–246
　Hollandsegraven sub-catchment,
　　Netherlands 227–229
　hydromorphology 65–78
　　reach scale 71–73
　paired site eutrophication
　　studies 134–137
　restoration
　　large woody debris use 222–223
　　success 76–78
　Swiss, water temperature 50–51
　thermal regimes 45–48
sublittoral assemblages 217, 218
subsystems of real communities 10
subtropical climate 123–125
Suess 210-year solar cycle 21, 26
sulphate 88, 153, 166
sulphur
　acidification recovery detection 217,
　　218
　atmospheric, dissolved organic
　　carbon 55
　Czech Republic, nitrate
　　leaching 161–162
　deposition 31, 152–175
　Europe 1880–2030 153
　UNECE protocol 154
summer droughts 57–59
SWAP (Surface Water Acidification
　Programme) 215
Sweden
　Gårdsjön experimental lake 198
　　DOC in stream water 167–168
　lakes
　　acidification recovery detection 217,
　　　218, 220
　　ice cover records 52
　　Kassjön, laminated sediments 27
　reference conditions establishment 208
　soil frost/snow cover effects on
　　DOC 56–57
Switzerland
　mesocosm eutrophication
　　experiments 125–126, 131–133

predictions 41
rivers/streams water temperature 50–51
Zurich Lake thermal regime 46–48
system-A variables 207, 208, 209

Tahoe, Lake 46
Tamar River, United Kingdom 278–281
Tansley, Arthur 4
Tatra Mountains
   ice cover records 53
   lake pollutants 186, 187–189, 190,
      191, 193–195
Teide, Tenerife 197–198
temperate climate 123–125
temperature
   annual global mean surface temperature
      since 1850 39, 40
   centennial-scale effects, Holocene
      period 24–26
   effects on biological systems 119
   eutrophication 119–147
   future predictions 295, 297
   Holocene period 16
   increase underestimation 4
   IPCC Fourth Report 1–2
   lakes 45–50, 89–93
   latitudes, palaeoclimate
      reconstructions 18–19
   mercury 196–197
   mesocosm eutrophication
      experiments 125–134
   organohalogen compounds 186–195
   palaeo-evidence 32
   persistent organic pollutants 185–197
   polycyclic aromatic
      hydrocarbons 195–196
   predictions 299
      heat waves 41
   RCAO simulations 42
   rivers
      fish trait composition 103, 104–106
      invertebrates 101
      monitoring 96–98
      Swiss 50–51
   The Stern Report (2006) 11–12
   thermal regimes 45–48
temporal approaches 205, 207
terminology 204–205
thermal regimes 45–48
THERMOS whole-lake mixing
   experiment 49–50

Tolypella glomerata 211
total nitrogen (TN) 121–122
total organic carbon (TOC) 89, 167
total phosphorus (TP) 121–122
toxic substances *see* nitrate; persistent
   organic pollutants; pollution
TP *see* total phosphorus
traits, river fish 104–106
Trapa natans 19
Trichoptera (caddis fly) 71, 74–76,
   101–103, 174
trophic structures *see* food webs
typology-based approaches 207–209

uncertainty
   catchment scale modelling 255–257
   decision making 262–264
   reference conditions 214–216
UNECE *see* United Nations Economic
   commission for Europe
United Kingdom
   Groby Pool 209, 210, 211, 212–213
   Kennet River 240–242, 248–250
   Lambourn River 67–68, 71–72,
      250, 257
   mesocosm eutrophication
      experiments 125–126,
      133–134
   reference conditions establishment 208,
      209, 210, 211, 212–213
   Scotland 31, 140–142, 209,
      210, 214
   Tamar River 278–281
   Wales 57–58, 163, 164, 171
United Nations Economic commission for
   Europe (UNECE) 154
United States 28, 46, 160
Uruguay 123–125
Utricularia vulgaris agg. 210, 212

Valkea-Kotinen Lake 49
variability
   climate, *see also* natural climate
      variability
   fish species distribution models 246
   inherent in systems 263
variance, reference conditions 221
Vecht River, Netherlands 66–67
volatility, organochlorines 186
volcanic activity 18
Vortsjarv Lake, Estonia 69

VULCAN (Vulnerability assessment of shrubland ecosystems in Europe under climatic changes) 57

Waldaist River, Upper Austria 74–76
Wales 57–58, 163, 164, 171
water chestnut (*Trapa natans*) 19
Water Framework Directive
    (EC/2000/60)
  monitoring 84, 85
  reference conditions 210, 223
    definition 204
    spatial approaches 207
  rivers monitoring 94
  stakeholder involvement 262
  water monitoring requirements 11
water level, lakes 22–24, 69, 87
water management strategies
    199–200
wetlands
  eutrophication 120

Swiss mesocosm
    experiments 131–133
  monitoring, biota 106–113
  paired site eutrophication studies 134,
    138–140
  restoration, OLU approach 224–229
WFD system-A models 207, 208, 209
whole tree harvest (WTH) 172–173
whole-lake manipulation
    experiments 49–50
winter sports 298–299
woody debris use in restoration 222–223
World Health Organization (WHO) 185
World Summit on Sustainable
    Development (Johannesburg) 262

Yoa, Lake, NE Chad 21

Zanichella palustris 211, 212
zooplankton 90–93, 121–122
Zurich, Lake, Switzerland 46–48

Printed and bound by CPI Group (UK) Ltd, Croydon, CR0 4YY

16/04/2025

14658829-0002